OPTIMIZATION
IN CONTROL THEORY
AND PRACTICE

OPTIMIZATION IN CONTROL THEORY AND PRACTICE

BY

I. GUMOWSKI

Fellow of Clare Hall 1965–66,
University of Cambridge

AND

C. MIRA

Electrical Engineering Research Laboratory,
University of Toulouse

CAMBRIDGE
AT THE UNIVERSITY PRESS
1968

Published by the Syndics of the Cambridge University Press
Bentley House, P.O. Box 92, 200 Euston Road, London, N.W. 1.
American Branch: 32 East 57th Street, New York, N.Y. 10022

Library of Congress Catalogue Card Number: 68–12059

Standard Book Number: 521 05158 4

Printed in Great Britain
at the University Printing House, Cambridge
(Brooke Crutchley, University Printer)

CONTENTS

ACKNOWLEDGEMENTS

The authors wish to express their thanks to Clare Hall and the Engineering Department of the University of Cambridge as well as to the National Research Council of Canada for their generous help which rendered this publication possible.

They are also greatly indebted to the Electrical Engineering Research Laboratory of the University of Toulouse.

The idea to discuss the optimization methods used in control engineering could never have been put into practice without the encouragement of Professor J. F. Coales of the University of Cambridge.

INTRODUCTION

The study of a practical optimization problem requires a realistic representation of the physical system by means of a suitable mathematical model and the explicit or implicit formulation of an appropriate performance criterion. The mathematical model must describe correctly at least the qualitative features of the practical system in the complete range of the probable operating conditions, and the optimality criterion must be a valid representation of the practical meaning of optimality.

By examining any non-trivial practical case, for example the operation of a petroleum refining plant with the object of maximizing the octane number of the final product, it becomes quite obvious that a successful determination of an appropriate model and of a satisfactory analytical performance criterion requires both a profound knowledge of the technology involved and of the mathematical methods presently available. Because generally it appears quite unlikely that the enormous amount of knowledge required to optimize a practical system could be assembled by any one individual, a collaboration between two groups of specialists, the 'technologists' and the 'theoreticians' becomes indispensable. The former are usually physicists, chemists and practising engineers and the latter essentially applied mathematicians. In actual teams there might be some overlapping in the above classification. For simplicity we will lump the physicists, chemists and practising engineers under the name of designers.

Unfortunately, in the last two decades an ever-increasing gap has developed between the designers and the theoreticians. This gap can be attributed to the general fact that the designers and the theoreticians do not encounter, study and try to solve the same problems in the same order. In optimal theory there appear certain difficulties, of an essentially mathematical nature, which have no immediate counterpart in practical designs, and it is quite natural that much effort should be devoted to them by the theoreticians. And quite similarly, in design practice there appear some difficulties which have been insufficiently analysed experimentally to be the subject of a theoretically fruitful study. For various reasons the designer may frequently decide to change the design originally contemplated, and thus bypass a certain difficulty which he has no time to study in detail. Because of this divergence of interests the designers and the theoreticians are no longer able to communicate efficiently, and in many instances they have even started to develop their own and separate professional jargon.

This book, although written from a mainly theoretical point of view, has as its prime objective the study of such problems which are likely to stimulate an exchange of ideas between the designers and the theoreticians. It is hoped that such an exchange of ideas will help to decrease the aforementioned communication gap. No attempt will be made to cover all possible problems giving rise to communication difficulties, but it is hoped that no really important problem has been omitted. Whenever possible, illustrative examples will be chosen from the field of control engineering. It is supposed that the reader is acquainted with elementary control theory and practice and that he has some knowledge of non-linear mechanics.

In spite of its length the first chapter serves only to formulate the problem, i.e. to specify the conditions, which it is hoped, will render a theoretical optimization problem meaningful to a designer. It is mainly concerned with some aspects of the theory of models, the properties of admissible solutions, and the sensitivity of models to various types of perturbations. If a mathematical model possesses low sensitivity to perturbations, likely to be encountered in practice, then it can be used for immediate design purposes. Such a model will be said to possess the a priori property. The need for this rather strange term will become obvious from the argument leading to low sensitivity with respect to structure-perturbations.

The second chapter deals with the basic properties of functionals, which appear in elementary problems. A parallel is kept between the problem of finding extrema of ordinary functions and that of functionals. The main objective of this chapter consists in showing why functionals do not admit in general a Weierstrass-type theorem on the existence of extrema, and what consequences this entails for control theory and practice.

In the third chapter it is shown that optimization problems lead naturally to a boundary-value problem associated with a partial differential equation, first suggested by Carathéodory. By introducing the notion of 'geodesic equidistants' Carathéodory has shown that the Euler and Hamilton–Jacobi formulations of an extremal problem can often be considered as particular cases of the generalized Huyghens principle. This line of argument is continued and extended to control theory.

In chapter 4 it is shown that all presently available optimization methods, and in particular Euler's equations, Pontryagin's maximum principle and Bellman's dynamic programming, are merely particular cases of the boundary-value problem considered in chapter 3.

The fifth chapter deals with approximate methods of calculating extremal solutions, either by means of the direct methods of the calculus of variations or by numerical algorithms associated with ordinary and partial boundary-value problems.

Note. References to works listed in the Bibliography are given in the text within square brackets, thus: [ ].

CHAPTER 1

THE GAP BETWEEN CONTROL THEORY AND CONTROL PRACTICE

§ 1. PROPERTIES OF PLANTS AND TYPES OF MATHEMATICAL MODELS

Many authors have called attention to the fact that there exists a gap between contemporary control theory and the art of realizing practical systems (see for example pp. 99–101[L 10],[L 12],[F 10]). Practical systems are as a rule very complex, whereas a theory has to be at least conceptually simple if it is to have any adepts. But if the simplicity is carried to a point where a theory is no longer realistic in numerous problems of current interest, then the designers, having little time to spare, will lose interest in it. Such a state has been reached in contemporary theory and the principal reasons responsible for this state can be deduced from the following comparison[F 10]:

> It is well known experimentally that most plants of control systems are time-variable, non-linear, with essentially distributed parameters and that they contain sources of noise.
>
> In contemporary control theory such plants are analysed as a rule by means of models composed of ideal, lumped and noiseless elements. The mathematical tools consist of ordinary differential equations and of the linear theory of stability. The non-linearity and noise are then taken into account in a rather rudimentary way. (1.1)

Quoting Fuller (p. 291[F 10]): 'In conclusion it can be said that at present there exists no theory of control which allows simultaneously for the noisiness, variability, partial non-linearity and jelly-like behaviour of plants.' Because of the discrepancy illustrated by (1.1), contemporary control theory has methods to deal separately with some real plant properties, but the results obtained are not always realistic and appropriate for design purposes. Especially the more refined methods of linear stability theory, for example the root-locus analysis, lead to controllers which saturate with noise and which tend to operate very closely to their theoretical stability boundary. Unless the degree of plant variability is accurately known, the designers rely, quite rightly, more on their experience than on theoretical results.

In many cases this course of action leads to a decrease of dialogue between designers and theoreticians, and in extreme cases it may even reach a non-objective state, characterized by the argument that theoreticians do not appreciate fully some physical or technological aspects of the practical design, or that the designers, relying mainly on very rudimentary theoretical concepts, like for instance the Nyquist plot, do not appreciate fully some important abstract theorem.

Because the gap between control theory and practice appears to be merely a particular version of the classical gap between theoretical and experimental physics, it should not be allowed to become an obstacle towards further progress. The authors of this book believe that to reach a better understanding between the designers and the theoreticians the first step should come from the latter. Before discussing technical details of how this objective can possibly be attained it is useful to examine briefly the historical origin of the communication gap in control engineering.

It is a very striking fact that as recently as the First World War theoreticians and designers were not working separately. In fact, in most cases they were the same persons. At that time the design and realization of engineering projects was not sufficiently specialized to allow an extreme division of labour. The same statement could be even made about mathematics; pure mathematicians did not then refrain from carrying out their own numerical calculations.

As far as these authors could ascertain, the starting signal for the separation of engineers into theoretical and practical factions was given in electrical engineering by the publication in 1926 of the work of Cauer[C8]. Cauer obtained some very remarkable results[C9], at first unfavourably received by both the engineers and the mathematicians, by using systematically the notion of ideal circuit elements, considered *independently* of any concrete physical system. The notions of ideal elements and ideal circuits permitted to translate into the language of electrical engineering many well-known mathematical results of the theory of linear differential equations with constant coefficients and of the theory of meromorphic functions. This innovation in obtaining new engineering results, which were not necessarily useful in engineering practice, was followed enthusiastically in practically all countries and it led in particular to the extreme popularity of Laplace transforms, poles-and-zeros techniques and root-locus diagrams. Probably the most enthusiastic proponents of these techniques worked in the United States, and their collected results became known under the name of modern linear system theory.

Perhaps, unfortunately, this modern linear system theory was introduced gradually into control engineering, and with it the study of ideal control systems for their own sake, without regard to any concrete practical need. The study of ideal systems is of course intrinsically very important, but if carried to extremes, it cannot help but be detrimental to a dialogue between theoreticians and designers. A routine exchange of results is discouraged and as a consequence both groups progress at a lesser rate.

Fortunately the trend to study excessively abstract systems was not universally followed, the exception being research groups in non-linear mechanics. The better known gathered around Van der Pol in Holland, Lefschetz, Minorsky and Friedrichs in the United States, Miss Cartwright in the United Kingdom, and Mandelshtam, Andronov, Krylov, Bogoliubov and Malkin in the Soviet Union. Whereas the study of non-linear mechanical and electrical systems was carried out in the western countries more or less sporadically and by personal motivation, in the Soviet Union this study was carried out systematically in response to governmental decrees. The research in question was mainly concentrated in two research institutes, nominally concerned with vibrations, one in Leningrad under Mandelshtam and later under Andronov, and the other in Kiev, under Krylov and Bogoliubov. Some particular research, inspired by the work in the Leningrad institute, which now would be considered as a part of control engineering, was carried out in Moscow by Bulgakov, Lure, Letov and their disciples. It is significant that in the books of Lure[L 16], and Letov[L 11], specifically devoted to control systems no use whatever is made of block diagrams or of any other ideal elements of modern linear system theory. In these books control systems are classified by means of the structure of equations describing them, i.e. indirectly by the nature of phenomena which can occur in these control systems.

After this brief historical review let us return to a discussion of the technical reasons contained in (1.1), which are more specifically responsible for the gap between control theory and control practice. The practical designer is highly interested in the fact that the complete control system, and in particular the plant, must be protected against any mode of operation which is likely to damage it or to cause a premature failure. Knowing from experience at least the rough extent of variability of concrete plants, the practical designer cannot rely on a stability boundary calculated with the assumption that variability is absent. Furthermore, a stability boundary calculated by means of linear stability theory is not always technologically significant. For example, some very efficient class-B and class-C radio-frequency power amplifiers are designed in such a manner that they are not completely free of parasitic oscillations. Complete neutralization of these parasitic oscillations is rare and it is difficult to maintain during the life cycle of the thermionic valves. Theoretically such radio-frequency amplifiers operate beyond their stability boundary, but since the power dissipated by the parasitic oscillations is small compared to the useful amplifier power output, this small power loss is practically unimportant. Consequently, a theoretically calculated stability boundary may sometimes be crossed without danger to the practical system, and it would be useful to know when such a favourable situation is likely to occur.

A partial solution of the problem of dangerous and not dangerous stability boundaries was given by Bautin[B3], who has related the nature of the stability boundary to the nature of plant variability. At least some of the plant variability is systematic, i.e. it can be expressed analytically in the form of a function describing the degradation of performance with time, temperature, power input, etc.

It was shown by Malkin (§ 70 and § 74[M2]) that the presence of noise has no qualitative effect on the practical operation of a system, if the latter is strongly asymptotically stable in the sense of Liapunov[L15]. With some mathematical complications these conclusions can be extended to systems with distributed parameters, provided not more than a few of the possible modes of operation are utilized at a time (see for example § 37 of [H4]).

Consequently, plant variability, noisiness, etc., are not *fundamentally* responsible for the gap between control theory and control practice. In the opinion of these authors the fundamental reason for this gap resides in the relationship of the above-mentioned factors to the theory of excessively abstract systems, which evolved from modern linear system theory. In fact, if a mathematical model of a practical system is based on highly idealized elements, this model will only apply if the design of the practical system has been carried to a successful completion. In other words, it will apply only *after* the designer has eliminated all possible causes of trouble. If the practical system does not yet operate properly, its mathematical model composed of idealized elements will not offer any clue permitting to locate the cause of trouble. Consequently, following the terminology proposed by Tomovic[T5], such mathematical models are said to be of the a posteriori type, i.e. they are only appropriate to describe the finished product. A theory based on a posteriori models will be called an a posteriori theory.

Even a superficial analysis will show that a very substantial part of contemporary control theory is of the a posteriori type, and thus it has relatively little to offer to the practical designer. In particular, it is not of much help in the selection of practical components permitting to attain a given design goal. It is therefore quite natural that the practical designers are not very enthusiastic about contemporary control theory. What they would like to have is a theory which is capable of deducing from a reasonable knowledge of component properties a reasonable knowledge of properties of component combinations. Such a theory, making possible the *prediction* of realistic system properties from the knowledge of realistic properties of its components may be called, using a terminology inspired by Tomovic[T5], an *a priori theory*, and the corresponding mathematical model *an a priori model*. An a posteriori model is not intrinsically useless because, as we will show later, it is frequently the first step toward an a priori model.

There does not seem to exist a fundamental obstacle preventing the development of an extensive a priori control theory, but such a theory is impossible without a detailed knowledge of properties of real components. Consequently, an a priori control theory is impossible without a dialogue between the theoreticians and the designers. These authors hope that the material which follows will encourage this dialogue.

§ 2. THE PROBLEM OF ESTABLISHING A PRIORI MATHEMATICAL MODELS

Two distinct but essentially equivalent methods are available to establish an a priori model of a concrete physical system. These methods constitute merely a particular adaptation of the classical microscopic and macroscopic viewpoint of physics, leading to what might be called inductive and deductive models, respectively.

Inductive model. Starting from a detailed knowledge of component properties, elementary interactions between them are determined, and from these interactions properties of the whole system are deduced. The variability of the system is thus expressed explicitly in terms of the variability of individual components.

Deductive model. Starting from the observed properties of the complete system, variability included, a global (phenomenological) model of its behaviour is established. The variability of individual components is expressed implicitly by the variability of the complete system.

The determination of an inductive model is quite straightforward, but it leads generally to a very unwieldy mathematical formulation. This unwieldiness is due to the fact that all properties of each individual component do not have an equally important effect on the properties of the complete system. If a certain set of properties of the complete system is to be represented by a model of a specified precision, then many properties of the individual components may be safely neglected. The difficulty of this approach, used generally by the designers with a rather rudimentary mathematical apparatus, consists in the determination of the *minimal* set of component properties necessary to characterize adequately the properties of the complete system. It is of course advantageous to neglect as many component properties as possible, because this reduces the complexity of the final mathematical model, but if the simplification process is carried too far, the final mathematical model becomes physically unrealistic. Even in moderately complex situations, such as that of even the simplest practical control process, the boundary between an unwieldy and an unrealistic inductive model is very hard to perceive.

At first sight the determination of a deductive model appears to be quite artificial, but if properly applied it will generally lead to a simpler mathematical formulation than the inductive method. Within the specified precision, the agreement between the mathematical model and the chosen set of observed system properties is established by a process of successive refinement. The refinements, deduced from a phenomenological interpretation of the system properties, are stopped when the error of the mathematical representation has become sufficiently small. The advantage of this approach consists in the fact that it is possible to start from a very coarse mathematical representation, and then add features as required to match the set of system properties chosen. Such a procedure avoids dealing with component properties, known or unknown explicitly, which have a negligible effect on the behaviour of the complete system.

In many cases an a posteriori model, based on the ideal elements of modern linear system theory, can be used as a starting-point of the process of successive refinement, and the solution of the corresponding a posteriori equations can be interpreted as a first term, i.e. as a Poincaré 'generating solution', of a sequence of functions which converge, at least asymptotically, to a physically realistic solution. The difficulty of this approach, used generally by the control engineering theoreticians with a rather rudimentary process of successive refinement, consists in the determination of the *minimal* set of system properties necessary to account for the inherent component variability. As in the case of inductive models, there exist only a narrow path dividing the unrealistic from the unwieldy, when the refining process is carried too far.

These authors believe that the distinction between deductive and inductive, a posteriori and a priori models is essential in order to diminish the communication gap between designers and theoreticians. The designers are mainly interested in inductive a priori models, whereas theoreticians work mostly with deductive a posteriori models. More common interest will develop when the theoreticians will try to improve inductive a priori models by the use of more sophisticated tools of the theory of approximation. Designer interest in deductive a priori models can be stimulated by establishing a comprehensive set of correlations between elementary physical phenomena, i.e. system properties such as resonance, limit-cycle oscillation, synchronization, amplitude or frequency jumps, etc., and the basic forms of the corresponding mathematical equations. The beginning of such a set of correlations can be found in all classical works on non-linear mechanics and recently also in some control engineering papers (see for example[B 18],[G 8]). The practical usefulness of correlating specific physical phenomena with specific equation types consists mainly in the possibility of reducing the amount of otherwise indispensable experimental data. Experimental tests performed on complex systems may involve higher expenses than time spent on a more refined theoretical analysis.

§ 3. MODERN LINEAR SYSTEM THEORY AND A PRIORI MODELS

It is a well-established fact that modern linear system theory has been highly successful when applied to numerous practical cases, in spite of the fact that it is based on idealized elements. But it is also a well-established fact that these practical cases are all of a very particular type: the physical systems in question possess an asymptotically stable constant steady state (equilibrium state) surrounded by a relatively large zone of influence. In the preceding sentence stability is understood in the Liapunov sense and the zone of influence is the domain of admissible instantaneous perturbations, which will not prevent the system from returning to its equilibrium state[L 15],[C 13],[M 2].

Consider, for example, a physical system which under some specified conditions can be described realistically by two ordinary first-order differential equations of the form

$$\frac{dy}{dt} = f(x,y), \quad \frac{dx}{dt} = y, \tag{1.2}$$

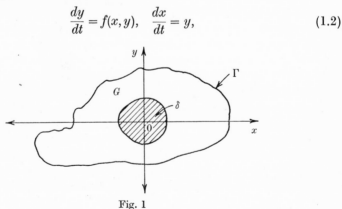

Fig. 1

where $f(x,y)$ is a real-valued continuous and differentiable function of its arguments. If (1.2) admits an asymptotically stable equilibrium state $x(t) \equiv 0$, $y(t) \equiv 0$, then the zone of influence of this equilibrium state can be represented in the phase-plane by some area G (Fig. 1). The boundary

Γ of G, which in certain cases may degenerate into the 'point' at infinity of the phase-plane, is called the stability boundary of the equilibrium state (see for example §10, Ch. 5, [A 6]).

Consider (1.2) with the initial conditions $y(t_0) = y_0$, $x(t_0) = x_0$ and let $x(t) = \phi(t, t_0, x_0, y_0)$, $y(t) = \psi(t, t_0, x_0, y_0)$ be the corresponding solution of (1.2). Asymptotic stability of $x(t) \equiv 0$, $y(t) \equiv 0$, implies that $|x(t)|$ and $|y(t)|$ will be as small as desired for all $t > T \geqslant t_0$, where T is a sufficiently large positive constant, provided the point (x_0, y_0) is located inside the area G. Liapunov has also shown that, excluding certain critical cases[L 15], the non-linear equations (1.2) may be replaced by their linear approximations

$$\frac{dy}{dt} = ax + by, \quad \frac{dx}{dt} = y, \tag{1.3}$$

where a and b are real constants, provided the initial conditions $y(t_0) = y_0$, $x(t_0) = x_0$ are confined to a sufficiently small neighbourhood δ (shaded in Fig. 1) of the point $(0, 0)$.

If the operation of the physical system described by (1.2) is such that the probable instantaneous perturbations will not displace the system state outside of the region δ, then the linear model (1.3) is realistic, even though it may be known that the physical system contains some essentially non-linear components. In view of what has been said in § 2, the linear model (1.2) will be of the a priori type. Unfortunately, this linear model will become progressively less realistic as larger deviations from the equilibrium state are considered, and (1.3) will become inherently unsatisfactory for system states located near the stability boundary Γ of (1.2). This result was proved by Liapunov in 1892 and it has been frequently verified by observing large amplitude stages in power amplifiers. The scepticism of designers when confronted by results obtained by means of linear stability theory is thus well justified, even if the plant variability argument is completely disregarded. The designer does not only wish to know whether there exists a non-vanishing zone of influence of a stable equilibrium state, but also how large this zone is, compared to the domain of the probable operating conditions. The first problem can be solved by linear stability theory, but unfortunately not the second, and this conclusion was already known by Liapunov (see [L 15] and the introduction of [C 13]). In fact Liapunov has shown [L 15] that for any initial conditions located in G, the solution $x(t) = \phi(t, t_0, x_0, y_0)$ of (1.2) can be represented by the convergent series

$$x(t) = \sum_{j=1}^{2} A_j \exp[\alpha_j(t - t_0)] + \sum_{i=2}^{\infty} B_i \exp[\mu_i(t - t_0)], \tag{1.4}$$

where A_j, B_i are constants depending on x_0 and y_0,

$$\mu_i = \sum_{j=1}^{2} k_{ij}\alpha_j, \quad \sum_{j=1}^{2} k_{ij} = i,$$

k_{ij} are either zero or positive integers, and α_j are roots of the characteristic polynomial of (1.3), defined by

$$\begin{vmatrix} -\alpha & 1 \\ a & b-\alpha \end{vmatrix} = 0. \tag{1.5}$$

Liapunov has shown that similar series expansions exist for general autonomous or non-autonomous systems[L 15].

The speed of convergence of the series (1.4) decreases as the point (x_0, y_0) is displaced gradually from the immediate neighbourhood of $(0,0)$ toward the stability boundary Γ. The first term of (1.4), which represents the solution of (1.3), becomes gradually less important and the 'principle of superposition' of modern linear system theory loses gradually its validity. Near the stability boundary Γ the first term of (1.4) becomes negligible compared to the second, and the usefulness of modern linear system theory becomes nil.

Liapunov has shown that entirely different approximations of (1.2) are required near Γ. Such approximations can be obtained by means of Poincaré's theory of bifurcation (see for example[M 23],[A 7]) and by the Liapunov theory of critical cases[L 15],[C 13],[A 6],[M 2]. Many authors have called attention to this fact, and in particular Letov[L 10],[L 11].

§ 4. GENERAL PROPERTIES OF A PRIORI MODELS NOT CONTAINING FUNCTIONALS EXPLICITLY

Cumulative experience has led to the generally accepted conclusion that a successful practical design must be relatively insensitive to small variations of its components. A realistic mathematical model of this design must have an analogous behaviour: at least the qualitative properties of solutions must be relatively insensitive to small changes in the equations constituting the model. This insensitivity to small changes can be considered as another characterization of an a priori model, and it is obviously equivalent to that given in § 2. Low sensitivity to structure perturbations implies the existence of a non-vanishing region of operation, where at least the qualitative behaviour of the system can be predicted from its model.

The structure of a mathematical model can be changed in a variety of ways, some significant and others not. A significant change of structure manifests itself by the appearance of bifurcation solutions, which in turn can be produced, for example, by the presence of small time constants, by the presence of pure delay and by the effect of noise. The mathematical tools appropriate for the study of significant structure perturbations of models can be found in the classical works of Poincaré, Hadamard, Hilbert, Malkin, Andronov, Pontryagin and others (see for example[M 23], [H 3], vol. II[C 15],[M 2],[A 6],[A 4] [P 9],[T 1], [V 3]).

Before discussing the mathematical aspects of model structure perturbation we will cite the well-known fact that the majority of physicists, engineers or even mathematicians were not led to their discoveries by a process of deduction from general postulates or general principles, but rather by a thorough examination of properly chosen particular cases. The generalizations have come later, because it is far easier to generalize an established result than to discover a new line of argument. The famous mathematician Halphen is known to have often complained that non-essential generalizations are overcrowding the publication media. Not wishing to be included in this complaint, these authors will investigate the problem of sensitivity of mathematical models by pursuing only the key ideas, illustrated whenever possible by particularly transparent examples.

4.1   Sensitivity to small parameter variations

The problem of sensitivity to parameter variations has arisen initially in mathematical physics, and in particular in the study of boundary-value problems. Since many physical phenomena can be described by boundary-value problems associated with either ordinary or partial differential equations, the first problem was an essentially mathematical problem of the existence and uniqueness of solutions. In the process of establishing the required existence and uniqueness theorems it became soon apparent that it was necessary to take into account the fact that at least some numerical constants, and especially those occurring in the boundary conditions, were obtained by means of measurements. But a number representing the result of a measurement, for example, s, the distance between two material points, is not a number in the mathematical sense, but only a sort of representative member of the dense set of numbers located in the interval $s - \alpha < s < s + \alpha$, where $\alpha > 0$ is the so-called absolute precision of the measurement of s. Consequently, if a boundary-value problem is to represent a physically significant situation, its solution must exist and be unique not only for isolated values of parameters appearing in the boundary conditions, but also for any other slightly different parameter values, located in the respective precision intervals. Furthermore, all corresponding solutions must be qualitatively of the same type, and quantitatively their difference must be small if the parameter variation is small.

In order that a boundary-value problem be physically significant Hadamard has formulated the following conditions[H 3]:

(a) a solution should exist,
(b) this solution should be unique,
(c) the unique solution should be continuous with respect to the data contained in the boundary conditions.
$\left.\begin{array}{c}\\\\\\\\\end{array}\right\}(1.6)$

If a boundary-value problem satisfies the three conditions (1.6), it is now said to be correctly set (or correctly posed) in the sense of Hadamard.

The first two conditions of (1.6) are mathematically self-evident. Physically they mean that neither 'too much', i.e. mutually incompatible properties, nor 'too little', i.e. excessive latitude, should be asked from the solution. The third condition of (1.6) states that the boundary-value problem should not only admit a solution, say y, for an isolated boundary function, g_0, specified on an isolated boundary Γ_0, but also for any other 'close' boundary function g_ϵ, specified on a 'close' boundary Γ_ϵ. In other words, the solution y should be at least a locally continuous function of its arguments g and Γ, or using a more precise terminology, y should be at least a locally continuous *functional* of the function g and of the boundary Γ.

Since the continuity of functionals is a rather complex property, to be discussed later, we will suppose at present that the set of admissible functions g and of the admissible boundaries Γ can be represented by means of a parametric family, and we will examine the continuity properties with respect to the parameter or parameters thus introduced. Condition (c) of (1.6) states then that the boundary-value problem should not only admit a solution $y = y_0$ for an isolated set of parameters, say $\lambda = \lambda_0$, but also in a sufficiently small neighbourhood of λ_0. In other words, a boundary-value problem can be said to be correctly set in the sense of Hadamard if it admits at least one parametric family of solutions y_λ in which the nominal solution y_0 is imbedded. The family y_λ, in the choice of which there is usually considerable latitude, must be such that any neighbouring solution y_{λ_1}, approaches y_0 as λ_1 approaches λ_0. Consequently, from a physical point of view, condition (c) of (1.6) is a key condition, allowing to reject from all possible mathematical problems those which are not physically significant.

Since by tradition the objective of the theory of boundary-value problems was limited to the determination of nominal solutions y_0, the study of correctly set problems remained essentially qualitative. When examining the practical sensitivity of a given mathematical model a quantitative aspect must be added by asking how fast the nominal solution y_0 varies when one or more parameters of the set λ_0 are given slightly different values. Depending on the nature of the parametric imbedding process used, the resulting 'sensitivity coefficients', for example the partial derivatives $\partial y_0/\partial \lambda_i$[T 4],[G 14], will be valid 'in the large' or only 'in the small'. Because the nature of the boundary-value problem conditions the choice of the parametric imbedding, the limits of validity of the corresponding sensitivity coefficients shed light on the extent of the domain, in parameter space, where a solution y_0 of a specific qualitative type can exist. The limits of validity of a specific parametric imbedding are thus seen to be related to the bifurcation solutions of the original problem. A value λ_0 of a parameter set λ is said to be a bifurcation value if y_λ undergoes a qualitative change for $\lambda = \lambda_0$ [A 6],[M 23]. The nature of the detectable qualitative change depends of course on the nature of the parametric imbedding process symbolized by y_λ.

As an illustration of the fact that all correctly formulated problems, i.e. problems pertaining nominally to a physical situation and not containing any mathematical error, are not correctly set in the sense of Hadamard, consider the following example, taken from [H 3]:

Determine a function $u(x, y)$ satisfying

$$\Delta u = \frac{\partial^2 u}{\partial x^2} + \frac{\partial^2 u}{\partial y^2} = 0 \tag{1.7}$$

in the half-plane $y > 0$, and

$$u(x, 0) = 0, \quad u_y(x, 0) = g(x) = \frac{\sin nx}{n} \tag{1.8}$$

on the boundary $y = 0$, where n is a parameter and u_y designates the partial derivative $\partial u/\partial y$. It can be easily verified by substitution that for a fixed value of n the solution of (1.7), (1.8) is given by

$$u(x, y) = \frac{1}{n^2} \operatorname{sh} ny \sin nx. \tag{1.9}$$

This solution satisfies the conditions (a) and (b) of (1.6), but it does not satisfy (c). In fact, for $n \to \infty$ the boundary conditions (1.8) become $u(x, 0) = 0$, $u_y(x, 0) = 0$, to which corresponds the unique solution $u(x, y) \equiv 0$, whereas letting $n \to \infty$ in (1.9) yields a non-vanishing result

$$\lim_{n \to \infty} u(x, y) = \tfrac{1}{2} \lim_{n \to \infty} \frac{1}{n^2} e^{ny} \sin nx.$$

Consequently $u(x, y)$ is not a continuous functional of $g(x)$, in spite of the fact that the difference between $g(x) = (1/n) \sin nx$ and $g(x) \equiv 0$ can be made as small as desired. Physically this result can be interpreted as follows: If (1.7), (1.8) are considered as a model of a system, where $u(x, y)$ is the output and $g(x)$ the input, then an observer of the system would conclude that the input–output relationship is entirely random, at least if the input $g(x)$ is confined to the neighbourhood of the system equilibrium state $g(x) \equiv 0$, $u(x, y) \equiv 0$. The above interpretation was suggested by Poincaré [H 3].

Let us now show that the conditions (1.6) are necessary, but generally not sufficient, if a boundary-value problem is to represent an a priori model of a physical system. Consider in fact the very particular boundary-value problem

$$\ddot{y}(t) + y(t) = 0, \quad y(0) = \alpha, \quad \dot{y}(0) = \beta \quad (t \geqslant 0), \tag{1.10}$$

admitting the obvious solution

$$y(t) = \alpha \cos t + \beta \sin t, \tag{1.11}$$

α and β being two real constants. The solution (1.11) of (1.10) satisfies all three conditions of (1.6), because $y(t)$ is a continuous function of the variables α and β. But (1.10) cannot be considered as an a priori model of any macroscopic physical system during an indefinite interval of time t, because such systems dissipate energy, and the slightest amount of energy dissipation will render the solution (1.11) physically non-realistic. To take into account the dissipation of energy it is necessary to introduce into (1.10) a term of the form $f(t, y, \dot{y})$, and it is well known that (1.10) is not insensitive to such a perturbation of its structure. For example, if $f(t, y, \dot{y}) = \epsilon(\dot{y})^k$, $0 < \epsilon \ll 1$, $k > 0$ an odd integer, then the differential equation $\ddot{y} + \epsilon \dot{y}^k + y = 0$, replacing $\ddot{y} + y = 0$ in (1.10), no longer admits a non-trivial periodic solution. Consequently, an a priori model of a system must not only be correctly set in the sense of Hadamard, but it must also be insensitive to certain types of structure perturbations. For example, if the boundary-value problem

$$\ddot{y} + f(t, y, \dot{y}) + y = 0, \quad y(0) = \alpha, \quad \dot{y}(0) = \beta \quad (t \geqslant 0), \tag{1.12}$$

analogous to (1.10), is to be an a priori model of a macroscopic physical system, then the solution y of (1.12) must be a continuous function of α, β and t and a continuous *functional* of the function f, at least in a small neighbourhood of some fixed function $f = f_0$.

4.2 Sensitivity to small variations of structure not affecting the order of a system of differential equations

Before discussing any method permitting to determine the sensitivity of a model with respect to a small change of its structure, let us recall the obvious fact that all physical laws are necessarily approximate and all mathematical models of physical systems necessarily incomplete. When a physicist or an engineer states that a given component, or a concrete system, is described by say a differential equation $Ly = 0$, then this statement is not meant in a mathematically strict sense. Quite to the contrary, similarly to the case of an isolated measurement, the equation $Ly = 0$ must be considered as a nominal member of a dense set of equations $\bar{L}y = 0$, which are close in some sense to the nominal equation $Ly = 0$. The 'closeness' of $\bar{L}y = 0$ and $Ly = 0$ is specified implicitly by both the physical and mathematical aspects of the argument which determines the precision of $Ly = 0$. An essential difficulty is encountered at this stage. Whereas it was quite straightforward to define the dense set of numbers $s - \alpha < s < s + \alpha$, by introducing the notion of a distance between two points, the definition of a physically significant distance between two equations is not so obvious. An equation being a far more complex entity than a number, the distance between two equations can be defined in many different ways. For example, in principle it is possible to metricize either the operator space implicitly defined by $\overline{L}y = 0$, or the functional space defined by the corresponding solutions. It is unlikely that the choice between these two alternatives can be made without recourse to physical evidence, but the opinion of practically all classical authors seems to favour the metrization of the solutions of $\overline{L}y = 0$. Perhaps this tendency is merely a matter of convenience, motivated by the fact that operators are more complex than functions, but as a result of many observations there is strong evidence that a metrization of the solution-space is usually quite adequate. In fact many nominal equations, for example the Laplace equation with Dirichlet-type boundary conditions, have a remarkable 'structural stability', i.e. the nominal solutions of these equations turn out to be physically satisfactory, in spite of the lack of a completely convincing mathematical proof that this should indeed be so.

A very ingenious approach to the problem of structural stability is given by the Andronov–Pontryagin theory of inert systems [A 4],[A 6] (in Russian *Grubye systemy*). To illustrate this theory consider an autonomous second-order system

$$\dot{x}(t) = P(x, y), \quad \dot{y}(t) = Q(x, y), \tag{1.13}$$

defined in a fixed domain G of the x, y-plane. The system of equations (1.13) is called inert if in the phase-plane x, y the topological structure of the trajectories of

$$\frac{dy}{dx} = \frac{Q(x, y)}{P(x, y)} \tag{1.14}$$

does not change for small changes of P and Q. Quantitatively this can be expressed, for example, in the following way: (1.13) is an inert system in the domain G if for any $\epsilon > 0$, no matter how small, it is possible to find a $\delta > 0$ such that for any regular pair of functions $p(x, y)$, $q(x, y)$ which satisfy in G the inequalities

$$|p| < \delta, \quad |q| < \delta, \quad \left|\frac{\partial p}{\partial x}\right| < \delta, \quad \left|\frac{\partial p}{\partial y}\right| < \delta, \quad \left|\frac{\partial q}{\partial x}\right| < \delta, \quad \left|\frac{\partial q}{\partial y}\right| < \delta, \tag{1.15}$$

there exists a topological transformation of G into itself, which transforms every trajectory of (1.14) into a trajectory of

$$\frac{dy}{dx} = \frac{Q(x, y) + q(x, y)}{P(x, y) + p(x, y)}, \tag{1.16}$$

and vice versa, in such a manner that any pair of corresponding points are at a distance less than ϵ.

An a priori model of a physical system, existing in nature or successfully designed, must of course be described by an inert system of equations. The equations (1.14) and (1.16) are analogous to the nominal equation $Ly = 0$ and the set of close equations $\bar{L}y = 0$ discussed earlier. There does not exist yet a systematic theory of inert systems of an arbitrary order. The original work of Andronov and Pontryagin was limited to second-order systems and they have obtained a certain number of particular results[A 4],[A 6].

Consider the trace σ and the Jacobian Δ of (1.13),

$$\sigma(x,y) = \frac{\partial P}{\partial x} + \frac{\partial Q}{\partial y}, \quad \Delta(x,y) = \begin{vmatrix} (\partial P/\partial x)(\partial P/\partial y) \\ (\partial Q/\partial x)(\partial Q/\partial y) \end{vmatrix}, \quad (1.17)$$

and let $x = \phi(t)$, $y = \psi(t)$ be a parametric representation of a closed curve $\mathscr{L}$ with the associated parametric period T, then, using the terminology of non-linear mechanics, the following theorems can be formulated:

Theorem I. In an inert system there is no equilibrium point for which $\Delta = 0$, i.e. the variational equations of (1.13) do not admit a singular equilibrium curve or a complex singular point.

Theorem II. In an inert system there is no equilibrium point for which $\Delta > 0$ and $\sigma = 0$, i.e. constant solutions of (1.13) do not constitute a Liapunov critical case.

Theorem III. In an inert system there are no closed trajectories $\mathscr{L}$ for which

$$h = \frac{1}{T} \int_0^T \sigma(\phi, \psi) \, dt = 0,$$

i.e. periodic solutions of (1.13) do not constitute a degenerate Liapunov critical case.

Theorem IV. In an inert system there is no separatrix which leads from a saddle point to another saddle point, i.e. an inert system cannot be conservative.

Theorem V. The system (1.13) is inert when the trajectories of (1.14) satisfy theorems I–IV, i.e. when (1.13) is non-conservative and when it admits only non-degenerate constant and periodic steady states.

Theorem VI. If the system of equations (1.13) is inert and the functions $P(x,y)$ and $Q(x,y)$ are regular in a finite subdomain of G, then in the phase-plane region corresponding to this subdomain there exists only a finite number of cells, each filled with qualitatively identical trajectories.

Theorem VI is extremely important, because it suggests that a second-order system, no matter how complex, can have only a finite number of *distinct* modes of operation, each mode corresponding to one cell in the phase-plane. The characterization of a system by its modes of operation is essential for the theory of a priori deductive models. It will be shown that such a characterization plays also a key role in the theory of optimal control, because it provides a physical meaning to the abstract notion of attainable phase states (see for example [M5]). The boundaries of the phase-plane cells, i.e. in the case of (1.14) the separation curves between the various possible modes of operation of the system, are determined by the bifurcation values of the initial conditions $x(t_0)$ and $y(t_0)$ of (1.13).

Suppose now that the equations (1.13) contain also a parameter, i.e. let $P = P(x, y, \lambda)$, $Q = Q(x, y, \lambda)$, and suppose that for $\lambda_1 < \lambda < \lambda_2$ P and Q remain regular in G. If there exists an $\epsilon > 0$, no matter how small, such that the topological structure of the trajectories of (1.13) is different for $0 < |\lambda - \lambda_0| < \epsilon$ and for $\lambda = \lambda_0$, then λ_0 is a bifurcation value, and (1.13) is not inert for $\lambda = \lambda_0$. As was already pointed out by Andronov and Pontryagin[A 4], non-inert systems are rather exceptional, in the sense that they occur mainly as limiting cases of inert systems. It is noteworthy, however, that in theoretical control engineering papers they are more a rule than an exception. As a result, these theoretical papers are not as well received by designers as they would otherwise deserve to be. In fact, non-inert systems are extremely useful in studying the transition between two different modes of operations. Suppose, for example, that the bifurcation value $\lambda = \lambda_0$ describes the critical loop gain of a control system, such that for $\lambda < \lambda_0$ the control system is stable, i.e. operates as an 'amplifier', and for $\lambda > \lambda_0$ unstable, i.e. breaks into self-excited oscillations. Then if the corresponding bifurcation solution is suitably imbedded, it will be valid in a non-vanishing interval

$$\lambda_0 - \alpha < \lambda_0 < \lambda_0 + \alpha \quad (\alpha > 0),$$

and will thus describe the control system in either the amplifying or the oscillating state.

From theorems I to IV it is obvious that non-inertness is associated with Liapunov critical cases. Consequently, the study of Liapunov stability is a convenient starting-point for the sensitivity analysis of a priori models in the neighbourhood of parameter bifurcation values.

To illustrate the use of bifurcation solutions consider the Van der Pol equation

$$\ddot{y} - \lambda(1 - y^2)\dot{y} + y = 0, \quad y(0) = 2, \quad \dot{y}(0) = 0 \quad (t \geqslant 0), \quad (1.18)$$

admitting $\lambda = 0$ as a bifurcation value, because (1.18) becomes then a conservative system. The Liapunov-stable bifurcation solution

$$y(t)_{\lambda=0} = 2 \cos t \tag{1.19}$$

is of course useless in itself, but some of its value becomes immediately apparent when (1.19) is imbedded in the parametric family of periodic solutions

$$\left.\begin{array}{l} y(t) = 2\cos\tau + \lambda(\tfrac{3}{4}\sin\tau - \tfrac{1}{4}\sin 3\tau + \ldots) + \ldots, \\ \tau = t(1 - \tfrac{1}{16}\lambda^2 + \ldots). \end{array}\right\} \tag{1.20}$$

The bifurcation solution (1.19) appears thus as a limiting case where Liapunov stability of (1.20) undergoes a qualitative change. The periodic solutions (1.20) are asymptotically stable for $\lambda > 0$ and unstable for $\lambda < 0$.

More properties of (1.18) can be deduced by replacing the initial condition $y(0) = 2$ by $y(0) = A$ and imbedding the resulting bifurcation solution $y(t)_{\lambda=0} = A \cos t$ in the Bogoliubov–Mitropolsky asymptotic expansion[B 15]

$$\left.\begin{aligned} y(t) &= a \cos \psi - \tfrac{1}{32}\lambda a^3 \sin 3\psi + \dots, \\[2mm] \psi(t) &= t - \tfrac{1}{8}\lambda^2 \int_0^t [1 - a^2(\tau) + \tfrac{7}{32}a^4(\tau)] \, d\tau, \\[2mm] a(t) &= \begin{cases} 2 & \text{for} \quad A = 2 \\ A \, e^{\frac{1}{2}\lambda t}[1 + \tfrac{1}{4}A^2(e^{\lambda t} - 1)]^{-\frac{1}{2}} & \text{for} \quad A \neq 2. \end{cases} \end{aligned}\right\} \tag{1.21}$$

The expansion (1.21) constitutes a solution of the Liapunov critical case in the neighbourhood of the bifurcation value $\lambda = 0$, and it solves completely the corresponding structural sensitivity problem of (1.18).

When the independent variable is limited to a finite, and generally a rather small interval, there exists an alternate method for investigating the closeness of two differential equations, or two systems of differential equations. This method is based on a refinement of the Picard existence theorem (see for example [K 1]). To illustrate its use consider the two first-order equations

$$\dot{y}(x) = f(x, y), \quad y(x_0) = y_0, \tag{1.22}$$

and

$$\dot{y}(x) = g(x, y), \quad y(x_0) = y_0, \tag{1.23}$$

defined in a fixed domain G for the x, y-plane and satisfying there the following conditions:

(a) $f(x, y)$ and $g(x, y)$ are continuous with respect to x and y;

(b) $f(x, y)$ satisfies a Lipschitz condition with the constant $M > 0$, i.e. $|f(x, y_1) - f(x, y_2)| \leqslant M|y_2 - y_1|$ for any two points (x, y_1) and (x, y_2);

(c) a constant $\delta > 0$ exists such that $|f(x, y) - g(x, y)| \leqslant \delta$.

If $y = \phi(x)$ and $y = \psi(x)$ designate the solutions of (1.22) and (1.23), respectively, then the inequality

$$|\phi(x) - \psi(x)| \leqslant M(e^{M|x-x_0|} - 1) \tag{1.24}$$

holds in a certain subdomain $\bar{G}$ of G. $\bar{G}$ may coincide with G, provided neither $\phi(x)$ nor $\psi(x)$ reach the boundary of G.

As a concrete example consider

$$\left.\begin{aligned} f(x, y) &= xy, \quad g(x, y) = \sin xy, \quad x_0 = 0, \quad y_0 = 0\cdot 1, \\ G &: |x| < \tfrac{1}{2}, \quad |y| < \tfrac{1}{2}. \end{aligned}\right\} \tag{1.25}$$

Then $\phi(x) = 0 \cdot 1 \, e^{\frac{1}{2}x^2}$ and $M = \frac{1}{2}$, because in G

$$\left|xy_2 - xy_1\right| = x \left|y_2 - y_1\right| \leqslant \tfrac{1}{2} \left|y_2 - y_1\right|.$$

Letting $xy = u$ and noting that in G $|u| < \frac{1}{4}$,

$$|f - g| = |u - \sin u| \leqslant \left|\tfrac{1}{6}u^3 - \tfrac{1}{120}u^5\right| < \tfrac{1}{384} = \delta,$$

and hence

$$\left|\psi(x) - 0 \cdot 1 \, e^{\frac{1}{2}x^2}\right| \leqslant \tfrac{1}{192}(e^{\frac{1}{2}|x|} - 1) < \tfrac{1}{320}\,|x| < \tfrac{1}{640}. \tag{1.26}$$

The estimate (1.26) holds in the whole of G, because $|\phi(x)| < \frac{1}{2}$ and $|\psi(x)| < \frac{1}{2}$, i.e. neither solution reaches $|y| = \frac{1}{2}$. The practical disadvantage of this method is that both M and δ depend *strongly* on the size of G.   -

4.3   Sensitivity to small terms containing high-order derivatives

The notion of an inert system, developed in detail for equations of form (1.13), remains valid for the more general case

$$\begin{aligned}\dot{x}(t) &= P(x,y) + p(x,y,\dot{x},\dot{y}), \\ \dot{y}(t) &= Q(x,y) + q(x,y,\dot{x},\dot{y}), \end{aligned} \right\} \tag{1.27}$$

provided the functions p and q are absolutely small and admit absolutely small derivatives. If in (1.27) p and q contain also $\ddot{x}, \ddot{y}$, or some still higher derivatives, i.e.

$$\begin{aligned}\dot{x}(t) &= P(x,y) + p(x,y,\dot{x},\dot{y},\ddot{x},\ddot{y},\dots), \\ \dot{y}(t) &= Q(x,y) + q(x,y,\dot{x},\dot{y},\ddot{x},\ddot{y},\dots), \end{aligned} \right\} \tag{1.28}$$

then in general the solutions of (1.13) and (1.28) will differ qualitatively even though p, q and all their partial first-order derivatives are absolutely as small as desired. This fact is very important in control engineering, because there exist well-behaved physical systems, or successful practical designs, which are described precisely by equations of form (1.28).

As an elementary illustration consider the one time-constant multivibrator (Fig. 2a), nominally described by the first-order system

$$\frac{dv}{dt} = \frac{1}{R_1 C}u, \quad E - v - R_2\phi(u) - \left(1 + \frac{R_2}{R_1}\right)u = 0 \quad (t \geqslant t_0). \tag{1.29}$$

If it is required that the voltage $u(t)$ be a continuous function of time, then, taking into account the shape of the composite tube characteristic (Fig. 2b), it was shown by Andronov in 1929 that (1.29) does not admit any non-constant periodic solution. Such a mathematical result is contrary to physical evidence, because the one time-constant multivibrator is known to oscillate with a continuous periodical waveform $u(t)$. In the discussion of this paradox, which followed between Andronov and

Mandelshtam (summary given in [A5], pp. 463–7), the two now famous alternatives were formulated: (a) either the nominal model (1.29) is not appropriate to describe the practical multivibrator, or (b) it is not being interpreted in a physically significant way.

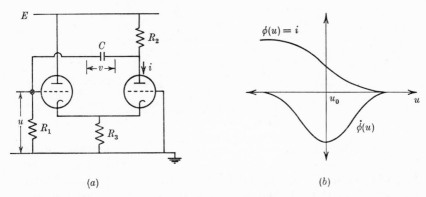

(a)   (b)

Fig. 2. (a) Nominal multivibrator circuit. (b) Composite valve characteristics in the presence of R_2 and R_3.

Andronov has shown later (see for example [A2]) that either alternative may be used to resolve the paradox, provided the space U of the admissible solutions $u(t)$ is properly defined. In fact, specifying that $u(t)$ must be continuous and continuously differentiable leads to the conclusion that (1.29) is inappropriate on physical grounds, because the real multivibrator possesses several small 'parasitic' elements, for example L_1, L_2 and C_0, shown in Fig. 3, which have been neglected in Fig. 2a. For $C_0 = 0$ the more refined circuit of Fig. 3 leads to the improved nominal model

$$\left. \begin{aligned} \frac{dv}{dt} &= \frac{1}{R_1 C} u, \\ E - v - R_2 \phi(u) - \left(1 + \frac{R_2}{R_1}\right) u &= \left[\frac{L_1 + L_2}{R_1} + L_2 \dot{\phi}(u)\right] \frac{du}{dt}. \end{aligned} \right\} \quad (1.30)$$

The system (1.30) is of order two and it possesses an admissible stable periodic solution $u(t)$. The stability conditions are for $C_0 \geqslant 0$:

$$R_1 + R_2[1 + R_1 \dot{\phi}(u_0)] < 0, \quad R_1 R_2 C_0 + L_1 + L_2[1 + R_1 \dot{\phi}(u_0)] < 0.$$

The system of equations (1.30) and the above inequalities constitute a rather unsatisfactory practical result, because the existence and stability of the required periodic solution $u(t) \in U$ of (1.30) depends not only on the *presence* of parasitic parameters, which are difficult to measure in practice, but also on their *relative magnitudes*.

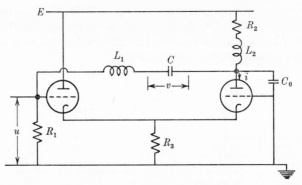

Fig. 3. Refined multivibrator circuit.

Andronov has shown[A 5],[A 2] that the strong dependence on parasitic elements can be alleviated by means of the second alternative, and in particular, by means of a generalization of the set U of admissible functions $u(t)$. In fact, if U is defined as consisting of not only continuous and continuously differentiable but also of piecewise continuous and piecewise differentiable functions $u(t)$, and the first-order differential system (1.29) is supplemented by some appropriate 'jump' conditions, say of the form

$$g(u(t_i^-), u(t_i^+)) = 0, \quad t_i = \text{points of discontinuity of } u(t), \quad (1.31)$$

permitting to join the various pieces of $u(t)$, then (1.29) admits a piecewise continuous periodic solution. In order to compare this essentially discontinuous periodic solution to the observed continuous multivibrator waveform it is necessary to define two functions $u_1(t)$ and $u_2(t)$ as close when they satisfy the inequality

$$\int_{t_0}^{t} |u_1(\tau) - u_2(\tau)| \, d\tau < \epsilon \quad (t \geqslant t_0), \tag{1.32}$$

for some sufficiently small $\epsilon > 0$. Using the metrization of U defined implicitly by (1.32), it is found that the discontinuous periodic solution of (1.29) and the observed continuous multivibrator waveform are indeed quite close. The jump conditions (1.31) are now generally called the Mandelshtam conditions. For the particular case of (1.29) they reduce to the statement that v should be a continuous function of t, and thus also of $u = u(t)$. Physically this means that the electrical current charging the condenser C should always be finite. In general, the physical meaning of the Mandelshtam conditions (1.31) is not so simple.

The preceding Andronov–Mandelshtam argument has produced a noteworthy result: both the second-order differential system (1.30) and the first-order differential system (1.29), supplemented by (1.31), can be considered as a priori models of the one time-constant multivibrator. The difference between these two models consists in the space U of admissible functions $u(t)$, and in the metrization of this space. In the case of equations (1.30) the metric is defined implicitly by the inequality

$$|u_1(t) - u_2(t)| < \epsilon \quad (t \geqslant t_0). \tag{1.33}$$

Continuous and discontinuous a priori models of a physical system need of course not necessarily be based on differential equations.

When analysing practical systems which admit more than one a priori model there arises the problem of choosing the best one. This choice is of course immaterial from the designer's point of view, because it is rather indifferent to him which mathematical entities the theoretician chooses for his work, provided the resulting practical consequences are the same. The theoretician will of course have a tendency to work with a model which suits his personal preferences, or which is likely to result in a more manageable mathematical problem. Discontinuous models have appeared recently in the study of optimal controllers (see for example [P 11]), but it is doubtful whether these models are generally of the a priori type.

To examine some details of the relation between continuous and discontinuous models consider a differential system, of order $n = m + s \geqslant 2$,

$$\dot{y}(t) = G(x, y), \quad \mu \dot{x}(t) = F(x, y) \quad (t \geqslant t_0), \tag{1.34}$$

where the vectors x and y have s and m components, respectively. Let $\mu > 0$ be a small parameter representing the parasitic elements and let $F(x, y)$ and $G(x, y)$ be bounded and continuous for all points (x, y) of the domain Ω of interest. If $F(x, y)$ and $G(x, y)$ depend also on μ, then they should admit finite limits as $\mu \to 0$.

If $\mu = 0$, (1.34) reduces to a system of equations, of order $m < n$,

$$\dot{y}(t) = G(x, y), \quad F(x, y) = 0. \tag{1.35}$$

The possible states of (1.35) are located in some subspace $\bar{\Omega}$ of the 'full' n-dimensional space Ω. This subspace is defined by the algebraic equations $F(x, y) = 0$. The problem consists in determining when the higher derivatives in (1.34) are qualitatively unimportant, i.e. when the 'small' term $\mu \dot{x}$, representing the effect of the parasitic elements, is negligible. In the terminology of non-linear mechanics this problem can be reformulated as follows: If μ is sufficiently small, in what case is the motion, described by (1.34) on the full n-dimensional space Ω, confined to a small neighbourhood of the subspace $\bar{\Omega}$? In other words, when is the motion described by (1.34) sufficiently close to the motion described by (1.35), so that it can be represented by the solution of (1.35) having only $m < n$ dimensions?

To determine this an argument due to Pontryagin can be used[A 5],[P 9], [T 1],[V 3]. Consider the domain located outside of a small $O(\mu^a)$-neighbourhood of the subspace $\bar{\Omega}$. By a $O(\alpha(\mu))$-neighbourhood of $\bar{\Omega}$ is meant the set of points whose distance from $\bar{\Omega}$ is less than a quantity of order $\alpha(\mu)$. Let $0 < a < 1$ and suppose that the neighbourhood $O(\mu^a)$ of $\bar{\Omega}$ reduces to $\bar{\Omega}$ as $\mu \to 0^+$. Outside of $O(\mu^a)$ the following relations hold:

$$|F(x,y)| \geqslant O(\mu^a), \quad |\ddot{x}(t)| \geqslant O(\mu^{a-1}). \tag{1.36}$$

Consequently the velocity $\dot{x}(t)$ is extremely large in that domain, i.e. the motion on the phase trajectories is fast. In the limit $|\ddot{x}(t)| \to \infty$ as $\mu \to 0^+$. This domain is therefore called the domain of fast motion, or more simply the fast domain[A 6],[P 9].

Since in the fast domain the functions $\dot{y}_j(t) = G_j(x,y)$, $j = 1, 2, ..., m$, are bounded by hypothesis for $\mu \to 0^+$, and the phase velocities

$$dy_j/dx_i \quad (i = 1, 2, ..., s),$$

satisfy the inequalities

$$\left|\frac{dy_j}{dx_i}\right| = \mu \left|\frac{G_j(x,y)}{F_i(x,y)}\right| \leqslant O(\mu^{1-a}) \quad \text{as} \quad \mu \to 0^+, \tag{1.37}$$

the variables y change only by quantities of order μ^{1-a}, or smaller, when t is confined to short intervals $\Delta t \leqslant O(\mu^{1-a})$. Consequently, the phase trajectories in the fast domain are located near the s-dimensional subset Ω_1, defined by $y = \lambda$, λ being a constant. Hence, for sufficiently small μ it is possible to approximate the motion outside of the $O(\mu^a)$-neighbourhood of $\bar{\Omega}$ by means of instantaneous jumps of the variables x, the variables y remaining constant. In other words, for sufficiently small values of μ (1.34) can be approximated by

$$y = \lambda, \quad \mu\dot{x}(t) = F(x,y). \tag{1.38}$$

Two cases are possible, depending on the behaviour of fast trajectories near the m-dimensional subspace $\bar{\Omega}$:

(a) For t increasing all fast trajectories enter some small neighbourhood A of $\bar{\Omega}$. If there are no trajectories leaving A, then once in A the motion will remain there. Inside A the motion will be relatively slow and will be well approximated by (1.35). By analogy with fast motions, this type of motion is called slow, and the domain where it occurs is called the slow domain[A 6],[P 3]. Hence, subject to the conditions stated above, μ will be a negligible parameter, provided the motion starts from initial conditions compatible with (1.35). If the initial point is not near $\bar{\Omega}$, then the motion will follow a fast trajectory, enter the neighbourhood A of $\bar{\Omega}$, and will then remain in this neighbourhood. In other words, the approximate equations (1.35), inappropriate initially, become valid after a short time interval $\Delta t \leqslant O(\mu \ln 1/\mu)$.

To determine analytical criteria for the parameter μ to be negligible, it is sufficient to note that the points of the subspace $\bar{\Omega}$ are points of equilibrium for the fast trajectories defined by (1.38). Hence, the behaviour of the fast trajectories in the $O(\mu^a)$-neighbourhood of $F(x,y) = 0$, $0 < a < 1$, is determined by the Liapunov stability of these points of equilibrium. Let $\tau = (1/\mu)t$ be the 'fast' time, then (1.38) becomes

$$y = \lambda, \quad \dot{x}(\tau) = F(x,y). \tag{1.39}$$

Consider a point $(\bar{x}, y) \in \bar{\Omega}$ and let $\xi_i = x_i - \bar{x}_i$, $i = 1, 2, ..., s$, then the corresponding variational equations are

$$\dot{\xi}_i(\tau) = \sum_{j=1}^{s} \frac{\partial F_i}{\partial x_i} \xi_j. \tag{1.40}$$

Designating the derivatives $\partial F_i/\partial x_j$ by p_{ij}, the characteristic equation of (1.40) becomes

$$|p_{ij} - \alpha\delta_{ij}| = 0, \tag{1.41}$$

where δ_{ij} is the Kronecker symbol. For asymptotic stability of the equilibrium points $\xi_i = 0$ all roots of (1.41) must satisfy the inequality $\Re\alpha < 0$. According to a well-known criterion this can be written

$$\left.\begin{array}{l} \dfrac{\partial F}{\partial x} < 0 \quad \text{for} \quad s = 1, \\[2mm] \sigma = \dfrac{\partial F_1}{\partial x_1} + \dfrac{\partial F_2}{\partial x_2} < 0, \quad \Delta = \begin{vmatrix} \dfrac{\partial F_1}{\partial x_1} & \dfrac{\partial F_2}{\partial x_2} \\[2mm] \dfrac{\partial F_2}{\partial x_1} & \dfrac{\partial F_2}{\partial x_2} \end{vmatrix} > 0 \quad \text{for} \quad s = 2, \\[4mm] \quad\cdots\cdots\cdots\cdots \end{array}\right\} \tag{1.42}$$

(b) There exists a part B of the subspace $\bar{\Omega}$ such that the points of B constitute unstable equilibria for the fast trajectories. Consequently, at least one root of (1.41) will satisfy the inequality $\Re\alpha > 0$ in B, and in the full n-dimensional space Ω there exist fast trajectories which leave any $O(\mu^a)$-neighbourhood of $\bar{\Omega}$, no matter how small μ is chosen. The small parameter μ is therefore non-negligible, i.e. the solutions of (1.34) and (1.35) are qualitatively different, even if the associated initial conditions are mutually compatible.

It may happen, as it does for example in the case of (1.29) and (1.30), that in some part of A of $\bar{\Omega}$ the equilibrium points of (1.38) are stable, and for some other part B of $\bar{\Omega}$ they are unstable. Hence, fast trajectories enter $\bar{\Omega}$ at A and they leave $\bar{\Omega}$ at B. In such a case A and B constitute the domain of discontinuity points which can be related by the Mandelshtam conditions. In the simplest case the Mandelshtam conditions can thus be written[A 6],[P 9] (real root of (1.41) changes sign):

$$F(x(t_i^-), y(t_i^-)) = 0, \quad F(x(t_i^+), y(t_i^+)) = 0,$$
$$y(t_i^-) = y(t_i^+), \quad \Delta(x(t_i^-), y(t_i^-)) = 0, \tag{1.43}$$

where $\Delta(x, y)$ designates the Jacobian of F with respect to x. If (1.34) admits a continuous periodic solution, then there will also exist a piecewise continuous periodic solution of (1.35), (1.38) and (1.43), which will be close for $0 < \mu \ll 1$ to the corresponding solution of (1.34), in the sense of the metrization (1.32).

If in addition to stable and unstable equilibria the equations (1.39) admit some other steady states, for example periodic oscillations, then the motion on the fast trajectories can have a non-vanishing duration, and the relation between the solutions of (1.34) and (1.35) can no longer be deduced by means of (1.43) and (1.38), where λ is a constant. More elaborate methods have been devised to deal with such cases (see for example [A 6],[P 9]).

4.4 Sensitivity to small delay

Similarly to the case of high-order derivatives, a system of differential equations may lose the property of inertness when a small delay is taken into account. The introduction of a pure delay into the equations may become necessary because the physical system does possess delay, or because some purely mathematical considerations suggests it as a simplifying artifice. The latter case arises when a physical system involves a large number of relatively small time-constants which are difficult to keep track of individually, as for example in the study of so-called artificial transmission lines. Whatever the reason for the introduction of pure delay, it is advantageous to know how properties of a priori models are affected by this circumstance.

Control systems with delay were studied as early as 1899[P 6], but it is probably Minorsky who first called attention to the fact that small delays may be the cause of some otherwise inexplicable periodic oscillations[M 21]. Such periodic oscillations would of course lead to a loss of system inertness. A concrete example of a differential equation with pure delay, i.e. of a differential-difference equation admitting a periodic solution *independent* of initial conditions was given by Minorsky in 1948[M 22]. Such periodic solutions are analogous to periodic solutions of the limit-cycle type known to occur only in ordinary non-linear differential equations. The notion of phase space is of course not defined unambiguously for differential-difference equations, because the solutions of the latter are not functions but *functionals* of the initial conditions, which include arbitrary functions. Consequently periodic solutions of differential-difference equations independent of initial conditions may not be identified with limit cycles in a phase-space, unless this particular phase space has been properly specified [G 15].

The particular differential-difference equation considered by Minorsky in [M 22] was

$$\ddot{x}(t) + p\dot{x}(t) + q^2 x(t) + r\dot{x}(t-\tau) - s\dot{x}^3(t-\tau) = 0, \qquad (1.44)$$

where p, q, r are real and s, τ small positive constants. Equation (1.44) possesses obvious similarities with the Van der Pol equation, and its periodic solutions were determined essentially by the Van der Pol method (see for example [A 6],[M 23]). The latter method is a particular case of an asymptotic expansion studied in detail by Krylov, Bogoliubov and Mitropolsky [B 15]. About ten years later (p. 176 [L 7]) Minorsky suggested that if a small delay is introduced into the 'elastic term' of the Duffing equation

$$\ddot{x}(t) + a\dot{x}(t) + x(t) + bx^3(t) = 0, \qquad (1.45)$$

where a is a positive and b a real constant, then contrary to (1.45), the resulting differential-difference equation

$$\ddot{x}(t) + a\dot{x}(t) + x(t-\tau) + bx^3(t-\tau) = 0 \quad (\tau > 0), \qquad (1.46)$$

will also admit a non-constant periodic solution independent of the initial conditions. This periodic solution does indeed exist, and for $|b|$ sufficiently small it can be written as power series in b [G 14].

Let a be close to a value $\alpha > 0$, such that the characteristic equation of the linear part of (1.46),

$$s^2 + \alpha s + e^{-\tau s} = 0, \qquad (1.47)$$

admits a conjugate pair of purely imaginary roots, i.e. let

$$a = \alpha + \beta b, \quad s_{1,2} = \pm j\omega \quad (j^2 = -1), \qquad (1.48)$$

where ω is a positive and β a real constant. Choosing the origin of t so that $\dot{x}(0) = 0$ and considering b as an independent variable, an application of the well-known Krylov–Bogoliubov transformation [K 25]

$$\left. \begin{aligned} \omega t &= \theta \left(1 + \sum_{i=0}^{\infty} h_i b^i \right), \\ x(t) &= \sum_{i=0}^{\infty} x_i(\theta) b^i, \end{aligned} \right\} \qquad (1.49)$$

yields for the first terms of (1.49)

$$\left. x_0(\theta) = A_0 \cos \theta, \quad A_0^2 = \frac{4\beta\omega(\tau \sin \omega\tau + 2 \cos \omega\tau)}{3\Delta \cos \omega\tau}, \quad h_1 = \frac{\beta\omega}{\Delta}, \right.$$
$$\left. \Delta = (\tau \sin \omega\tau + 2 \cos \omega\tau)\,tg\omega\tau - \alpha\omega - \tau \cos \omega\tau + 2 \sin \omega\tau. \right\} \qquad (1.50)$$

For small delay, i.e. for $0 < \omega\tau \ll 1$ the expressions for A_0 and h_1, simplify into

$$A_0^2 = \frac{8\beta\omega}{3\Delta_0}, \quad h_1 = \frac{\beta\omega}{\Delta_0}, \quad \Delta_0 = 4\tau^2(1-\omega) - \alpha\omega. \tag{1.51}$$

An asymptotically stable periodic solution, described by (1.50), exists for

$$b > 0, \ \beta < 0, \ \Delta_0 > 0 \quad \text{and for} \quad b < 0, \ \beta > 0, \ \Delta_0 > 0. \tag{1.52}$$

When $\beta b = a - \alpha < 0$, the equilibria solution $x(t) \equiv 0$ of (1.46) is obviously unstable, whereas this is not the case for the corresponding solution of (1.45).

As a third example consider a differential-difference equation with non-constant delay

$$\dot{x}(t) + x(t - \tau) = 0, \quad \tau = \tfrac{1}{2}\pi + \mu[1 - \tfrac{1}{3}x^2(t)], \tag{1.53}$$

where μ is a parameter. If $\mu = 0$ then (1.53) admits a family of periodic bifurcation solutions, depending on two arbitrary constants A and B:

$$x(t) = A\cos t + B\sin t. \tag{1.54}$$

If $\mu \neq 0$ then (1.53) admits a unique periodic solution independent of the choice of the function $\phi(t)$ appearing in the initial condition

$$x(t) = \phi(t) \quad \text{for} \quad -\tau_0 \leqslant t \leqslant 0, \quad \tau_0 = \tfrac{1}{2}\pi + \mu[1 - \tfrac{1}{3}\phi^2(0)] > 0. \tag{1.55}$$

This periodic solution was originally found by means of a numerical method[G 12], and for $0 < |\mu| \ll 1$ it turned out to have an almost sinusoidal form. By means of the Krylov–Bogoliubov transformation it was then relatively straightforward to determine its analytic expression[G 13]:

$$x(t) = \sum_{i=0}^{\infty} x_i(\theta)\mu^i = A_0\cos\theta + \mu(\tfrac{1}{2}\sin\theta - \tfrac{1}{8}\pi\cos\theta - \tfrac{1}{6}\sin 3\theta) + \ldots$$
$$(\theta = t(1 - \tfrac{1}{4}\mu^2 + \ldots), \quad A_0 = 2). \tag{1.56}$$

For $\mu < 0$ this periodic solution is unstable, whereas for $\mu > 0$ it is asymptotically stable.

The existence of a periodic solution of (1.53) of type (1.56) does not depend critically on the form of τ. In fact, if for example τ in (1.53) is replaced by

$$\tau = \tfrac{1}{2}\pi + \mu[\delta_0 + \delta_1 x(t) + \delta_2 x^2(t) + \delta_3 x^3(t) + \delta_4 x^4(t)], \tag{1.57}$$

where δ_0 to δ_4 are real constants, then the value $A_0 = 2$ in (1.56) is simply replaced by

$$A_0 = +\sqrt{\left[\frac{3\delta_2}{5\delta_4}\left(1 \pm \sqrt{\left\{1 - \frac{40}{9}\frac{\delta_0\delta_4}{\delta_2^2}\right\}}\right)\right]}. \tag{1.58}$$

The qualitative properties of (1.53) are thus quite similar to those of the Van der Pol equation.

Because it is well known that a first-order linear differential-difference equation with constant coefficients and a constant delay c_0

$$\dot{x}(t) = ax(t) + bx(t - c_0) \tag{1.59}$$

can be transformed by the transformation

$$x(t) = e^{at}\xi(\tau), \quad \tau = -bt\,e^{-ac_0}, \quad \delta = -bc_0\,e^{-ac_0} \tag{1.60}$$

to the one-parameter form

$$\dot{\xi}(\tau) + \xi(\tau - \delta) = 0 \tag{1.61}$$

(see for example p. 631[K 3]), the physical delay c_0 in (1.59) and the effective delay δ in (1.61) need not be small simultaneously. It may even happen that the effective 'delay' δ turns out to be negative. Consequently the equation (1.59), considered as an a priori model of a system, possesses essentially different properties for $c_0 = 0$ and $c_0 \neq 0$. This is true a fortiori if the delay c_0 is non-constant. A differential-difference equation of the form (1.59), with an almost-constant delay c_0, is known to describe the behaviour of an operational transistor amplifier in the vicinity of its stability boundary[G 11].

4.5   Sensitivity to discretization

Consider a macroscopic physical system whose a priori model is given by the equations

$$\dot{y}(t) = G(x,y), \quad \mu\dot{x}(t) = F(x,y) \quad (t \geqslant t_0), \tag{1.34}$$

discussed in §4.3. If the independent variable t is discretized, i.e. if t is assumed to take only discrete values $t_k = t_0 + kh$, $h > 0$, $k = 0, 1, 2, \ldots$, and the derivatives $\dot{y}(t)$ and $\dot{x}(t)$ are replaced by suitable finite differences, for instance by the linear forms with real-valued coefficients

$$\left.\begin{aligned}
\dot{y}(t) &= \sum_{i=-i_1}^{i=i_2} \alpha_{k+i}(h)\,y(t_{k+i}), \\
\dot{x}(t) &= \sum_{i=-i_3}^{i=i_4} \beta_{k+i}(h)\,x(t_{k+i}) \quad (i_j \geqslant 0, \quad j = 1, 2, 3, 4),
\end{aligned}\right\} \tag{1.62}$$

then (1.34) is converted into a system of difference equations

$$\sum_{i=-i_1}^{i=i_2} \alpha_{k+i}y_{k+i} = G(x_k, y_k), \quad \mu \sum_{i=-i_3}^{i=i_4} \beta_{k+i}x_{k+i} = F(x_k, y_k), \tag{1.63}$$

where for brevity of notation $x(t_k), y(t_k)$ have been shortened into x_k, y_k.

For some physical systems, containing for example mechanical relays or electronic switching circuits, it is possible to arrive at the discrete model (1.63) directly, without the necessity of first formulating the continuous equations (1.34). If however (1.63) is to be an a priori model, and if it is to be equivalent to (1.34), then both (1.34) and (1.63), the latter eventually supplemented by an appropriate interpolation formula, must constitute representative elements of a dense set of 'close' equations $\bar{L}(x, y) = 0$. In other words, the discretization (1.63) of (1.34) must be simply the result of a structure perturbation of (1.34) which leaves the qualitative properties of $x(t)$ and $y(t)$ invariant. In particular, the solutions of (1.34) and (1.63) must have the same stability and sensitivity properties.

In order to compare the solutions of (1.63) to those of (1.34), and in particular to the essentially continuous measurements carried out on the physical system, it is necessary to use an interpolation algorithm on the former, or a discretization algorithm on the latter. In macroscopic control systems all measurements are essentially continuous, no matter whether the *displays* of these measurements are continuous or discrete. The above-mentioned interpolation or discretization algorithms constitute therefore a fundamental stage in the process of comparison of theoretical and experimental data, and they define directly or indirectly the norm to be used for the metrization of the space of admissible solutions.

Similarly to the case of differential equations, it would be useful to have some results concerning the structure perturbation properties of difference equations. Unfortunately neither the Andronov–Pontryagin theory of inertness, nor the Pontryagin–Andronov–Tikhonov theory of parasitic parameter influence have as yet been extended to discrete systems. Stability properties of constant solutions of (1.63) may be studied by an extention of the general Liapunov method (see for example [K 19],[H 4]), and in particular by an extension of the theorem on variational equations. The latter theorem permits to determine local stability of solutions of (1.63) from the asymptotic stability of solutions of the corresponding linear approximation, and thus finally from the roots of certain characteristic polynomials. Most control engineering results on discrete systems are based on this approach, or on the determination of Liapunov V-functions. There exist, however, a few results based on the theory of iterations (see for example [M 27],[G 20],[M 26]) and on certain applications of iterations in numerical analysis (see for example [X 1],[B 5]).

In the preceding argument (1.63) was considered as a discretized version of the continuous system (1.34). There exists however an alternate interpretation of (1.63) not involving a loss of continuity. Suppose that t_0 in (1.34) is replaced by $t_0 + \theta$, $0 \leqslant \theta \leqslant h$, the precise value of the parameter θ being immaterial. If x_k and y_k are interpreted as continuous functions of the new independent variable θ, then (1.63) becomes simply a recurrence relation between a certain number of points of the continuous solutions $x(t)$, $y(t)$ of (1.34). It is well known that such a recurrence relation determines the functions $x(t)$ and $y(t)$ completely, and it may thus be considered as fully equivalent to (1.34). Recurrence relations have been used as equivalent definitions of many elementary and transcendental functions, for example, exponentials, Chebyshev polynomials, Bessel functions, etc. Consequently, if both (1.34) and (1.63) represent a priori models of the same physical system, then these equations define essentially the same functions $x(t)$, $y(t)$, i.e. quantitatively the solutions of (1.34) and (1.63) may differ slightly, by an amount determined by the precision of the formulation, but qualitatively they must be the same. It is well known, however, that most discretizations based on the difference schemes (1.62) do not achieve this result.

4.6 Sensitivity to noise

The formulation of a nominal mathematical model $Ly = 0$ of a practical control system, or even more generally, the formulation of a nominal physical law, appears to imply that this system or this law are assumed to operate in a completely isolated universe, free from any deterministic or stochastic perturbations. Such a narrow interpretation of $Ly = 0$ is highly unrealistic, because the successful formulation of a mathematical model, and a fortiori the successful formulation of a physical law, constitutes the culmination of a complex process of perturbation and error analysis. This perturbation and error analysis takes into account, directly or indirectly, all deterministic and stochastic effects, acting on the practical system under consideration. As was pointed out before, a nominal mathematical model $Ly = 0$, if successful, is merely a representative member of a dense set of close models $\bar{L}y = 0$. The closeness between $L \in \bar{L}$ and $L_1 \in \bar{L}$ is defined implicitly by the restrictive assumptions under which the nominal equations $Ly = 0$ are understood to apply, and by the essentially known precision of the solutions of $Ly = 0$.

Suppose first that the operator L and the dense set of operators $\bar{L}$ can be related by a linear 'superposition principle', i.e. by the equation

$$\bar{L}y = (L + \epsilon_L)y = 0, \tag{1.64}$$

where ϵ_L is the set of 'error' operators, defined implicitly by the precision of y and the validity assumptions of $Ly = 0$. Let us recall now the obvious fact that if a nominal system of equations $Ly = 0$ is to be considered as a physical law, the validity of $Ly = 0$ must have been confirmed by a certain number of *repeated* and *independent* experiments. This repeatability and independence of the experimental confirmation, which may be direct or indirect, involves a systematic comparison of measured and calculated results, and it ensures that the error term $\epsilon_L y$ is small in some sense compared to the nominal term Ly. At the same time this systematic comparison of measured and calculated results provides bounds on the maximal and probable deviations of the variables y. Hence, if a nominal model $Ly = 0$ possesses the a priori property, then it contains adequate information on both the deterministic and stochastic perturbations acting on the corresponding practical system. Unfortunately in many cases little is known explicitly about the details of the actual perturbations, beyond the fact that $Ly \in \bar{L}y$ and $L_1 y \in \bar{L}y$ are 'close', and that the maximal error of y is bounded by a certain constant. The precise meaning of the word close, i.e. the metrization of the space of admissible solutions y, is defined implicitly by the essentially *physical* argument leading to the nominal model $Ly = 0$. It can hardly be stressed enough that because the metric properties of the admissible solutions y are defined by an essentially physical argument, they *cannot* be fixed on purely mathematical grounds. Unfortunately the latter course of action is not entirely unknown in control theory in general, and in the study of optimal systems in particular. As a tentative method it is of course possible to make various ad hoc assumptions, like for example, that $Ly = 0$ and $\bar{L}y = 0$ are related by the equation (1.64), but the resulting consequences must be confronted with experimental facts directly or indirectly.

Let us illustrate now briefly some of the ad hoc assumptions usually made to study the effect of noise. Probably the simplest assumption consists in saying that the noise, i.e. the unavoidable 'external' perturbation, acts instantaneously at a given time $t = t_0$ and then disappears entirely. If the instant $t = t_0$ is chosen as the initial time and the effect of noise is sought on a specific mode of operation, for instance described by the particular solution $x(t) \equiv 0$ of

$$\dot{x}(t) = X(t, x), \quad x(t_0) = x_0 = 0 \quad (t \geqslant t_0), \tag{1.65}$$

where the vector function X is continuously differentiable with respect to its arguments, then only the value of the initial condition x_0 will be affected. The problem of sensitivity to noise is thus reduced to the problem of sensitivity to a parameter variation, which can be handled by means of ordinary deterministic or probabilistic methods. This point of view, adopted essentially in the Liapunov stability theory [L 15], leaves much to be desired, because noise exists at all times and not just for $t = t_0$. Furthermore, the nature of noise is generally unknown in detail, and only its statistical characterization might be available.

To take into account that noise is a constantly acting perturbation, it is convenient to make the assumption of linear superposition, expressed by (1.64), which amounts to replace (1.65) by

$$\dot{x}(t) = X(t, x) + R(t, x), \quad x(t_0) = x_0 \quad (t \geqslant t_0), \tag{1.66}$$

where the vector function $R(t, x)$, describing the noise, is continuous and such that both (1.65) and (1.66) admit unique solutions in the domain of initial conditions under consideration. Since contrary to X the form of R is unknown in principle, it is only possible to assume that R is sufficiently 'small'. The dependence of R on t being unspecified in detail, the reference solution $x(t) \equiv 0$ of (1.65) will generally not be a solution of (1.66).

According to Malkin[M 1],[M 2], the solution $x(t) \equiv 0$ of (1.65) is called 'stable in the presence of constantly acting perturbations' if for every $\epsilon > 0$, no matter how small, there exist two numbers $\eta_1(\epsilon)$ and $\eta_2(\epsilon)$, such that every possible solution of (1.66) satisfies the inequality

$$|x(t)| < \epsilon, \quad \text{provided} \quad |R(t,x)| < \eta_1 \quad \text{and} \quad |x(t_0)| < \eta_2. \quad (1.67)$$

At first sight it might seem that the concept of Liapunov stability is rendered obsolete by the above definition. This is, however, not so, because asymptotic stability in the Liapunov sense turns out to be a necessary condition for the existence of stability in the presence of constantly acting perturbations[M 2]. At least for periodic and for constant steady states it is even a sufficient condition[M 10],[K 19]. The following theorem was proven by Malkin[M 1],[M 2]:

If for (1.65) there exists a positive definite Liapunov function $V(t,x)$ such that its total derivative

$$\dot{V} = \frac{\partial V}{\partial t} + \sum_{i=1}^{n} \frac{\partial V}{\partial x_i} X_i, \quad (1.68)$$

taken along the trajectories of (1.65), is negative definite, and in addition all partial derivatives $|\partial V/\partial x_i|$ are bounded in the x-domain under consideration, then the solution $x(t) \equiv 0$ of (1.65) is stable in the presence of constantly acting perturbations.

Stability in the presence of constantly acting perturbations implies a certain kind of asymptotic stability. In fact, for every η_1, there exists a positive number $\epsilon_1(\eta_1)$ such that for $t > t_0$ sufficiently large the solutions of (1.66) satisfy the inequality $|x(t)| < \epsilon_1$. Furthermore, $\epsilon_1 \to 0$ as $\eta_1 \to 0$. Compared to asymptotic stability in the Liapunov sense, stability in the presence of constantly acting perturbations requires that in addition to the existence of the V-function the partial derivatives $|\partial V/\partial x_i|$ be bounded. This further restriction constitutes indirectly a constraint on the admissible form of (1.65) in the direction of inertness.

Malkin's theory of stability in the presence of constantly acting perturbation is based on an additive interaction between the noise and the nominal model. It is quite obvious that a large number of other, equally reasonable assumptions, can be made about this interaction. For example, it can be assumed that the interaction mechanism is described by the Fokker–Planck equation[A 3].

Suppose for simplicity that the system (1.65) is autonomous and of order one, i.e.

$$\dot{x}(t) = X(x), \quad (1.69)$$

and let $p(x, \tau, y)\, dy$ be the probability that during the time interval τ noise will move the point x into the interval $(y, y + dy)$. If the noise has no preferred direction and the probability of large perturbations diminishes rapidly as $\tau \to 0$, then the probability density $f(x, t) \geqslant 0$ of the x-values defined by (1.65) will be given by a solution of the following boundary-value problem[A 3]:

$$\frac{\partial}{\partial t} f(x, t) + \frac{\partial}{\partial x}[X(x)f(x, t)] - \frac{1}{2}\frac{\partial^2}{\partial x^2}[b(x)f(x, t)] = 0, \qquad (1.70)$$

$$f(x, 0) = g(x), \quad \int_{-\infty}^{+\infty} f(x, t)\, dx = 1 \quad (t \geqslant 0), \qquad (1.71)$$

where $g(x)$ is a specified function, and

$$b(x) = \lim_{\tau \to 0} \frac{1}{\tau} \int_{-\infty}^{+\infty} p(x, \tau, y)\, (y - x)^2\, dy.$$

By analogy with the random walk problem (1.70) and (1.71) can be interpreted as the random swim problem in a canal having a regular current pattern.

It may happen that for sufficiently large t the probability density $f(x, t)$ reaches a stationary form independent of the initial distribution $g(x)$, i.e. $\lim_{t \to \infty} f(x, t) = f(x)$. In such a case, equations (1.70), (1.71) reduce, with suitable continuity conditions, to the stationary form

$$\frac{d}{dx}[X(x)f(x)] - \frac{1}{2}\frac{d^2}{dx^2}[b(x)f(x)] = 0, \quad \int_{-\infty}^{+\infty} f(x)\, dx = 1. \qquad (1.72)$$

It is known that under relatively mild conditions the maxima of the stationary probability density $f(x)$ coincide with the asymptotically stable constant and periodic states of the nominal model (1.65), whereas the minima coincide with the unstable constant or periodic states[A 3]. Because of computational difficulties the relation between $x(t)$ and $f(x, t)$ was not examined for more complex steady states, such as for instance those described by almost-periodic functions of t. In spite of the fact that the Andronov paper[A 3] was published more than thirty years ago, no fundamental progress seems to have been reported since, except for some particular applications[B 2],[M 16], and an occasional review of (1.70) and (1.72) from a purely statistical point of view (see for example [L 9]). Automatic computers being now generally available, the scarcity of particular solutions, which was an insurmountable difficulty in Andronov's time, should no longer constitute a serious obstacle in the study of (1.65) and (1.70), (1.71).

As a fourth ad hoc interaction assumption let the model $Ly = 0$ be such that some of the y may be considered as 'inputs' and others as 'outputs'. Then the noise properties of $Ly = 0$ may be studied by the method of signal transformation developed in communication theory (see for example [L 2],[M 17]). This method may be extended to the dense set of models $\bar{L}y = 0$ by supposing that $\bar{L}$ is equivalent to some particular stochastic operator L_s (see for example §§ 98–108[P 12]).

No matter what kind of interaction is assumed to exist between the nominal model $Ly = 0$ and the noise, the sensitivity to noise must be low if $Ly = 0$ is to be an a priori model suitable for immediate design purposes. This requirement can only be dispensed with when the assumed noise-model interaction is found to be physically non-realistic.

§ 5. GENERAL PROPERTIES OF A PRIORI MODELS CONTAINING FUNCTIONALS EXPLICITLY

In recent years a disproportionateley large number of publications in control engineering were devoted to the study of optimal system and of optimization methods. This statistical fact might create an impression that in the past systems were designed and analysed without any effort to render their properties optimal. Nothing could be further from the truth.

Any practical system, whether it is the product of a systematic design or of a more or less accidental discovery, is optimal, because it is the end-result of a choice with respect to other 'close' but less satisfactory systems. The optimality criterion on which this choice is based is extremely complex as a rule, and it takes into account not only technical, but also economical, social, and sometimes even aesthetical factors. In view of its complexity, such an optimality criterion can rarely, if ever, be expressed in an analytical form, but the existence of such a criterion can hardly be questioned. In spite of this rather self-evident argument Feldbaum found it necessary to prove that without an implicit or explicit optimality criterion no design whatever could be accomplished (p. 9 [F 1]). A similar reasoning applies of course to any problem selected for theoretical investigations.

The subject which has attracted the attention of designers and theoreticians in control engineering in recent years is not the study of optimal systems as such, but the study of systems for which the optimality criterion is expressed *analytically* in the form of an implicit or explicit functional. In what follows we will interpret all functionals in the sense of Volterra [V 5]. Since the perusal of mathematical literature has quickly shown that very little is known about implicit functionals (see for example [B 5]), practically all control engineering work has been concentrated on systems described by models which contain explicit functions of a rather primitive kind, such as for instance

$$I(y, u) = \int_{t_0}^{t_1} F(t, y(t), \dot{y}(t), \ldots, y^{(n)}(t), u(t), \dot{u}(t), \ldots, u^{(m)}(t)) \, dt, \quad (1.73)$$

where m, n are positive integers, the vector $y(t)$ represents the 'constrained' state variables and the vector $u(t)$ the 'free' control variables.

Compared to a traditional overall optimality criterion, assuring a technical and commercial success of a practical design, the explicit functional (1.73) constitutes at most an extremely coarse approximation, but compared to the analytical criteria actually used in control theory, this functional is extremely complex. In fact it contains not only the control vector $u(t)$, but also its derivatives.

We have shown in § 4 that a nominal model of a control system must possess the a priori property if it is to be useful for immediate design purposes. More specifically, this nominal model must be correctly set in the sense of Hadamard, and it must possess low sensitivity to various perturbations of its structure, at least if these perturbations are sufficiently small. The same restriction must apply a fortiori to nominal models intended for optimization purposes. In view of the obvious fact that functionals of form (1.73) are inherently very crude approximations of practical optimality, the requirement of low sensitivity to structure perturbation becomes even more acute. The optimal value of $I(y, u)$ should not only be little affected by a small change of the integration limits, t_0 and t_1, but also by a small variation of the function F appearing under the integral sign. The resulting optimal functions $y(t)$ and $u(t)$ should also differ very little. If the nominal model, intended for optimization purposes, does not possess low sensitivity properties, then the corresponding practical design will be difficult, or even impossible to realize. Should this design turn out to be realizable, then it will be unlikely to operate reliably.

Unfortunately, there does not exist yet a systematic sensitivity analysis of functionals which could be applied directly to nominal models intended for optimization purposes. It will be shown in §§ 6 and 7 that models containing functionals explicitly can often be transformed into forms discussed in § 4. Consequently the methods described in § 4 can be used indirectly for a sensitivity analysis of models intended for optimization purposes. The details of this approach will be worked out in chapters 2 and 3. For the purpose of conciseness we will frequently contract the circumlocution 'model intended for optimization purposes' into 'optimization model'.

§ 6. EQUIVALENT FORMULATION OF EXTREMAL PROBLEMS

From the works of such classical authors as Lagrange, Hamilton and Helmholtz it is known that models of physical systems, intended for optimization or any other purpose, can be obtained by two formally different kinds of reasoning. In the first kind the sequence of events is determined from a global teleological consideration and in the second from a local causal one. Let us recall that 'teleological' means roughly 'directed toward an objective or toward an end'. These two kinds of reasoning are well illustrated by the study of conservative systems in mechanics. It is indeed well known that the motion of a conservative mechanical system, with lumped or with distributed parameters, can either be described by means of the Hamilton principle or by means of the Lagrange equations (see for example [S 5]). The former can be written in the form of a functional of type (1.73)

$$I(t_0, t_1) = \min_{q \in Q} \int_{t_0}^{t_1} L(t, q, \dot{q}) \, dt, \quad L = T - U, \tag{1.74}$$

where the vectors $q(t), \dot{q}(t)$ are the state variables, belonging to the admissible set of motions $Q(q)$, $T = T(t, q, \dot{q})$ is the kinetic and $U = U(q)$ the potential energy of the mechanical system; and the latter in the form of a homogeneous system of differential equations

$$\frac{d}{dt}\left(\frac{\partial L}{\partial \dot{q}}\right) - \frac{\partial L}{\partial q} = 0. \tag{1.75}$$

The formulation (1.74) derives its value from the fact that it is totally independent of the choice of the state variables $q(t)$, $\dot{q}(t)$, and that, if necessary, it can provide the boundary conditions to be imposed on $q(t)$ and $\dot{q}(t)$. It is noteworthy, however, that contrary to the case of equations (1.75), the sequence of events in (1.74) is not determined by an instantaneous cause-and-effect relationship, but by a consideration of all system states, both present and future. This paradox is only apparent, because it has been proved that with suitable restrictions on the space of admissible motions Q, the relations (1.74) and (1.75) are completely equivalent. Consequently, if an optimization problem is described by a functional of form (1.74), then its sensitivity to structure perturbations can be studied on the basis of equations (1.75). Unfortunately (1.74) describes the properties of conservative systems, which are generally incompatible with the a priori property.

When trying to extend the equivalence between (1.74) and (1.75) to non-conservative mechanical systems several mathematical difficulties are encountered. In particular, there no longer exists a potential function

$V(q)$ from which forces acting on the mechanical system can be derived by means of the relation $F = -\operatorname{grad} V$, because that relation is only valid for the conservative components of forces. Since, at least for macroscopic phenomena, dissipation of energy is a universal experimental fact, it is indispensable to generalize (1.74) by adding to it a suitable corrective term. Using essentially a physical argument, which disregards some rather important mathematical aspects, it is generally assumed that the generalized Hamilton principle, applicable to non-conservative mechanical systems, can be written in the form

$$I(t_0, t_1) = \min_{q \in Q} \int_{t_0}^{t_1} (L + W)\, dt, \tag{1.76}$$

where $q, \dot{q}, Q, L$ have the same meaning as in (1.74), and where the function $W = W(t, q, \dot{q})$, which *must not* be considered as a state variable, represents work done by the non-conservative force components F_n (see for example [S 5]). The Lagrange equations analogous to (1.75) become

$$\frac{d}{dt}\left(\frac{\partial L}{\partial \dot{q}}\right) - \frac{\partial L}{\partial q} = F_n. \tag{1.77}$$

Repeating the above argument for Helmholtz's principle, the same formal results are obtained, except that (1.76) and (1.77) will now apply to non-conservative electrical systems.

If the equivalence between (1.76) and (1.77) is accepted in spite of the lack of a mathematically convincing existence proof, then (1.77) differs only in notation from the systems of equation discussed in §4. Consequently, either (1.76) or (1.77) may be interpreted as a priori models provided they turn out to possess the required low sensitivity to small structure perturbations.

In establishing models suitable for optimization purposes it is common practice to arrive at formulations which contain equations of form (1.77) *and* of form (1.76). Such a mixed formulation, based on both causal and teleological arguments, can easily be transformed into the purely causal (see for example [P 11]) or into the purely teleological form (see for example §9, vol. I [C 15]). The control engineering distinction between the 'constrained' state variables, determined by equations of form (1.77), and the 'free' control variables, determined by equations of form (1.76), appears therefore to be merely a matter of expediency and has nothing to do with fundamental principles.

Mixed formulations constitute a partial return to the teleological viewpoint, with the ensuing loss of easy to interpret cause-and-effect relationships. The rather sudden popularity of teleological arguments is not motivated in control theory by the desire to calculate precisely some particular 'optimal' solutions, which in most cases would turn out to be of dubious practical value, but by the desire to determine the *fundamental structure* of 'optimal' solutions (see for example p. 7[B 4]). Probabilistic methods in general, and the theory of games in particular, become useful tools in such a context. But if the mathematical model used for such a purpose is not of the a priori type, then the resulting solution structure will also be practically meaningless. It is noteworthy that in many control engineering problems the ultimate objective remains an explicit cause-and-effect relationship, and the mixed causal-teleological formulation is merely a more or less necessary detour. For example, in the synthesis of optimal controls the main result sought is an instantaneous relation between the values of the optimizing controls and the values of the constrained state variables (see for example the end of §5[P 11]).

Let us recall that the equivalence between (1.76) and (1.77) was deduced by a physical argument, which cannot be accepted as a mathematically convincing proof. In fact, it has been known for a considerable time that the inverse problem of the calculus of variation, i.e. the determination of a functional

$$I(t_0, t_1) = \min_{q \in Q} \int_{t_0}^{t_1} F(t, q, \dot{q}, \ldots)\, dt, \tag{1.78}$$

corresponding to a specified differential equation

$$f(t, q, \dot{q}, \ddot{q}, \ldots) = 0, \tag{1.79}$$

has a straightforward solution only when (1.79) is of order two and $\partial f/\partial \ddot{q} \neq 0$ (see for example no. 604[D 3] and pp. 37–39[B 16]). If (1.79) is of order three or higher, there exist certain difficulties whose nature is not yet completely understood (see for example [D 4],[K 7],[H 13]). The equivalence between (1.78) and (1.79) depends not only on the form of the functions F and f, on the set of admissible functions $Q(q)$, but also on the precise meaning of the 'symbolic' operator $\min_{q \in Q}$. Many problems of contemporary control theory, and in particular the problem of existence of optimal controls, can be traced to this circumstance. Fortunately, at least some of the difficulties presently encountered in control theory happen to parallel the difficulties encountered in the theory of partial differential equations before the turn of the century, and at least some of the latter were resolved in 1900 by Hilbert, in a famous paper on the Dirichlet problem[H 8]. This topic will be discussed in §7 and it will be resumed in chapter 2.

§7. TRANSFORMATION OF EXTREMAL PROBLEMS INTO EQUIVALENT FORMS

The fundamental distinction between a priori and a posteriori models was derived in §4 from the Hadamard distinction between correctly and incorrectly set boundary-value problems associated with partial differential equations. Another of Hadamard's ideas which can be carried over into control theory is the method of descent (proposed in [H 3], and described for example in vol. II[C 15]). In this method a problem with $m \geqslant 1$ independent variables is considered as a special case of the same problem with $m + k, k \geqslant 1$, variables. If something more is known about the latter problem, for instance the properties of a particular solution, then this additional information can be transposed to the former problem by a process of reduction of independent variables, i.e. by a process of descent from an $(m + k)$-dimensional to an m-dimensional space.

Keeping in mind the Hadamard method of descent we will now consider some relations between the problem of minimizing a functional and the problem of solving a partial differential equation with prescribed boundary conditions. Keeping also in mind Halphen's criticism of inessential generality, whenever possible, we will limit our exploration to two-dimensional problems of potential theory.

It is well known that the potential distribution $u(x, y)$ in a region G of the x, y-plane, bounded by the continuous contour Γ, can be described either by a solution of the variational problem

$$u = g(x,y) \text{ on } \Gamma, \quad I(g) = \min_{u \in U} \iint_G \left[\left(\frac{\partial u}{\partial x} \right)^2 + \left(\frac{\partial u}{\partial y} \right)^2 \right] dx \, dy, \quad (1.80)$$

where U is the set of admissible functions $u(x, y)$ and $g(x, y)$ is a specified function, or by the solution of the Dirichlet-type boundary-value problem

$$u = g(x, y) \text{ on } \Gamma, \quad \Delta u = \frac{\partial^2 u}{\partial x^2} + \frac{\partial^2 u}{\partial y^2} = 0 \text{ in } G. \quad (1.81)$$

The functional $I(g)$ in (1.80) may be thought of physically as the energy stored in the electric field produced by the potential distribution $u(x, y)$. When a solution of (1.81) is sought by means of (1.80), then it is said that it is sought by means of the Dirichlet Principle. This terminology is mainly due to Riemann.

The problems (1.80) and (1.81) are merely particular cases of (1.76) and (1.77) when the potential problem is formulated by means of Helmholtz's principle, but what is remarkable about them is that they arouse immediately some doubts about their equivalence. In fact, if (1.81) is to be an identity, then $u(x, y)$ must admit continuous second-order derivatives, whereas if (1.80) is to be an identity, the existence of piecewise continuous first-order derivatives of $u(x, y)$ seems to be sufficient. This discrepancy in the continuity requirements was partially explained by Haar (see for example p. 174, vol. I[C 15]), who has proved that if $u(x,y)$ makes (1.80) into an identity, then it has 'automatically' sufficient continuity to satisfy (1.81). The inverse statement is however not true. In fact, there exist numerous particular cases where a solution of (1.81) exists, but (1.80) is not an identity, because the integral in (1.80) fails to converge (see for example p. 155, vol. I, and chapter 7, vol. II[C 15]). Furthermore, Lebesgue and Wiener have shown that in some cases there may exist a meaningful solution of $\Delta u = 0$ which does not assume on Γ the prescribed values $u = g$ in a strict sense, but only in the sense of a certain mean ([W 10],[W 11], and for example pp. 272–274, vol. II[C 15]). In the teleological representation (1.80) there is no room for such a mechanism if the operator min is understood as the selection of a bilateral minimum of $I(g)$, $u \in U$ in the sense of ordinary calculus.

Without the benefit of the later work of Lebesgue and Wiener, Hilbert has shown that if (1.80) and (1.81) are to be equivalent, then the set of $u \in U$ has to be suitably enlarged, say to $u \in \overline{U}$, and the operator min $u \in \overline{U}$ should be interpreted either as a selection of a minimum or as a selection of a lower limit of the functional in (1.80). That a distinction between a minimum and a lower limit is usually necessary was already pointed out by Todhunter (p. 2, no. 3[T 3]). In the case of (1.80) such a distinction is essential, because if $\overline{U}$ is the set of all functions for which the integral in (1.80) converges, then the values of this integral admit a lower limit, which is not necessarily a minimum. Since the integrand

$$(\partial u/\partial x)^2 + (\partial u/\partial y)^2$$

is positive definite, the lower limit of the integral is non-negative. According to Hilbert the problem of equivalence between (1.80) and (1.81) should therefore be replaced by the problem of equivalence between (1.81) and the limit of the following sequence of problems

$$u_n = g_n(x,y) \text{ on } \Gamma_n, \quad I(g_n) = \min_{u_n \in U} \iint_{G_n} \left[\left(\frac{\partial u_n}{\partial x} \right)^2 + \left(\frac{\partial u_n}{\partial y} \right)^2 \right] dx\, dy,$$

$$(1.82)$$

where $G_n \to G$, $\Gamma_n \to \Gamma$, $g_n \to g$, $I(g_n) \to I(g)$ and $u_n \to u$ as $n \to \infty$. For simplicity it can be assumed that G_n approaches G monotonically from the 'inside', i.e. $G_n \subset G_{n+1} \subset G$ for all n.

The study of test cases has shown that (1.82) is a very significant generalization of (1.80), but that it possesses a fundamental weakness: The convergence of the sequences G_n, Γ_n, g_n and $I(g_n)$ does not necessarily imply the convergence (in the ordinary sense) of the sequence u_n, and even if u_n does converge, some or all of the sequences $\partial u_n/\partial x$, $\partial u_n/\partial y$, $\partial^2 u_n/\partial x^2$, $\partial^2 u_n/\partial y^2$ may diverge. Consequently, there can be no equivalence between (1.82) and (1.81) unless it is possible to define in $\overline{U}$ a suitable convergence process and to select from u_n a suitable subsequence v_n, such that the limits of v_n and its derivatives satisfy (1.81). Hilbert has shown that the required convergence processes and subsequences do indeed exist and that (1.82) and (1.81) are equivalent under far less restrictive conditions that (1.80) and (1.81) (see for example chapter 7, vol. II[C 15]). The sequences $I(g_n)$, u_n and v_n are called minimal sequences, and the use of minimal sequences is called a direct method of the calculus of variations. Other direct methods of the calculus of variations, for instance the method of an infinite number of variables, or the method of Ritz, are only particular variants of the possible use of minimal sequences. The direct methods of the calculus of variations constitute at present the most powerful tool for the investigation of boundary-value problems, no matter whether they are formulated in the form (1.76) or (1.77).

It is noteworthy that some recent papers in control theory report the rediscovery of some of Hilbert's results, but unfortunately without any reference to Hilbert (see for example [K 20],[K 23],[K 24],[B 20],[Z 3]). This is perhaps the reason why the authors of these papers fall into some pitfalls successfully avoided by Hilbert. The Lebesgue–Wiener result, that meaningful extremals cannot always satisfy the prescribed boundary conditions in a strict sense, has also been rediscovered in control theory (see for example p. 109 and §40, pp. 167–171[P 4]).

If the relation between (1.80) and (1.82) is examined from a physical point of view, then for a fixed value of n, (1.82) may turn out to be an equally acceptable or even a better model of a concrete potential problem. The situation is similar to that of the one time-constant multivibrator discussed in §4. The convergence process permitting to select v_n from the generalized sequence $u_n \in \overline{U}$ is analogous to the selection of a continuous model of the one time-constant multivibrator, containing some but not all possible parasitic parameters, and the corresponding limit of (1.82) is analogous to the selection of the discontinuous model. The choice between two alternate models can be carried out on the basis of a physical or of a mathematical reasoning. In the first case the reasoning involves the introduction of some parasitic elements which increase the amount of continuity, and in the second case it involves an appropriate generalization and metrization of the space of admissible solutions.

The research of Hilbert on the Dirichlet principle has been continued by several authors (see for example chapter 5[L 1]), and if the Hadamard method of descent is taken into account, some of the. results obtained have a direct bearing on the development of control theory. As an illustration consider the slightly more general problem

$$u(x) = g(x) \text{ on } \Gamma, \quad I(g) = \min_{u \in U} \int_G F(x, u, u_x)\, dx, \qquad (1.83)$$

where u_x stands for $\partial u/\partial x$, and x is a vector. Let p designate the vector u_x. Provided $F(x, u, p)$ is twice differentiable with respect to its arguments, the boundary-value problem corresponding to (1.83) is

$$u = g \text{ on } \Gamma,$$

$$\sum_{i=1}^{n}\sum_{j=1}^{n}\frac{\partial^2 F}{\partial p_i \partial p_j}u_{x_i x_j} + \sum_{i=1}^{n}\frac{\partial^2 F}{\partial u \partial p_i}u_{x_i} + \sum_{i=1}^{n}\frac{\partial^2 F}{\partial x_i \partial p_i} - F_u = 0 \text{ in } G.$$

Let $F(x, u, p)$ be convex in G and p and $\overline{p}$ two admissible values of u_x, then the following inequalities hold in G:

$$E(x, u, p, \overline{p}) = F(x, u, p) - F(x, u, \overline{p}) - \sum_{i=1}^{n} F_{p_i}(x, u, \overline{p})\,(p_i - \overline{p}_i) > 0$$
$$(p \neq \overline{p}),$$
$$\sum_{i=1}^{n}\sum_{j=1}^{n} F_{p_i p_j}(x, u, p)\,\xi_i \xi_j \geqslant \nu(u, p)\sum_{i=1}^{n}\xi_i^2 \qquad (\nu(u, p) > 0).$$
$$(1.84)$$

These inequalities can either be interpreted as two well-known necessary conditions for the existence of a minimum of $I(g)$, or as the conditions of ellipticity of the partial differential equation. If in addition to the above inequalities the following conditions are satisfied[L 1]

$$x \in G, \quad u \in (-\infty, +\infty), \quad p_i \in (-\infty, +\infty) \quad (i = 1, 2, ..., n),$$
$$F(x, u, p) \geqslant \lambda \left[\sum_{i=1}^{n} p_i^2\right]^{\frac{1}{2}m} \quad (m \geqslant 1, \lambda = ct > 0),$$
$$u \text{ and } p \text{ admit the metric } \|v\| = \left[\int_G (v^2)^{\frac{1}{2}m}\,dx\right]^{1/m},$$
$$(1.85)$$

then there exists at least one function $u \in U$ which minimizes $I(g)$. The conditions (1.85) insure that the lower limit of the integral in (1.83) is finite and that there exist minimal sequences converging to this lower limit. At the same time at least one limit function $u = \lim_{n\to\infty} u_n$ will have the necessary continuity properties to satisfy (1.84). This existence theorem holds under less restrictive conditions than the most general existence theorem known in control theory[F 2].

§ 8. IMBEDDING PROCESSES AND SINGULAR TRANSFORMATIONS OF EXTREMAL PROBLEMS

The direct methods of the calculus of variations make it possible to replace the nominal problem (1.83) by a converging sequence of problems

$$u_n = g_n \text{ on } \Gamma_n, \quad I(g_n) = \min_{u_n \in \overline{U}}\int_{G_n} F(x, u_n, p_n)\,dx, \qquad (1.86)$$

where $p_n = \partial u_n/\partial x$, the symbols G_n, Γ_n, g_n and $u_n \in \overline{U}$ having the same meaning as in equation (1.82) of §7. It is characteristic of (1.86) that the function F under the integral sign is kept fixed. It is of course possible to generalize (1.83) still further and replace F in (1.86) by a converging sequence F_n. The nominal problem becomes thus imbedded in a parametric family of close problems. If (1.83) is an a priori model, then the replacement of F by F_n, at least for n sufficiently large, will not produce a qualitative change in the properties of (1.83). The quantity $F - F_n$ can thus be considered as a small perturbation of the structure of (1.83), and the difference between the corresponding solutions can be used as a sensitivity test.

It is known since the publication of the researches of Todhunter[T 3] that low sensitivity to structure perturbations is a rather exceptional property of functionals, and that very small changes in the form of F may have a very strong effect on the existence and the properties of the corresponding extremal solutions. Far ahead of his contemporaries, Todhunter was studying variational problems admitting discontinuous solutions, which we would call now 'generalized' or 'weak' solutions, and which in their nominal form were inconsistent with the classical theory of continuous functions. As a part of contemporary criticism he received the advice to replace the offending variational problems by other very 'close' ones, which behaved perfectly regularly as far as the theory of continuous functions was concerned (see for example p. 191[W 12]).

Specific illustrations of imbedding processes, i.e. of the replacement of F by F_n in (1.86) will be given in chapters 2 and 3. At present it is sufficient to mention a recent paper of Gamkrelidze[G 5], where a modification of the structure of a nominal model is proposed in order to obtain by a straightforward application of the maximum principle[P 11] some otherwise non-existent optimal solutions, i.e. the so-called sliding regimes reported by Krotov[K 20],[K 24]. The method of Gamkrelidze is not essentially different from a method reported earlier by Swinnerton-Dyer[S 7].

Modifications of the structure of an optimal problem are legitimate steps in the refinement of a deductive model, but they seem out of place in the mathematical process of solving a specific problem. If the nominal model is inductive, the necessary mathematical modifications should be carried out during the formulation of the problem, in conformity with the physical circumstances, and not as a sort of mathematical afterthought. This remark applies particularly to the inequality constraints so popular in contemporary control theory. In many cases inequality constraints are introduced as a substitute for the law of evolution of a completely determined state variable; for example the continuous torque–load curve of a motor is replaced by an inequality involving the maximal torque and load values, or the continuous amplitude–gain curve of an amplifier is replaced by an inequality involving the amplitude–gain curve asymptotes. The replacement of a continuous relation by a discontinuous one is generally inadvisable, because it is likely to produce a qualitative effect on the nature of the closeness between two functions, and thus on the metric properties of the set of functions admissible for optimization purposes.

It happens quite frequently that the inequalities introduced more or less artificially during the formulation of the problem are transformed back into equalities during the solution of the problem, for example by using penalty functions[K 15],[K 18], or the artifice proposed by Valentine[V 1]. The use of inequality constraints was made widely known by the publications of Leitman[L 8] and Miele[M 18]. Since equality–inequality–equality transformations can hardly be beneficial to the process of establishing a priori models, the use of inequality constraints should be limited to cases where they are inherently necessary. In the vast majority of cases inequality constraint approximations do not turn out to be labour-saving devices.

If a specific formulation of an optimization problem appears to be inconvenient for some reasons, then it is of course advisable to transform it into another form. Formally it is possible to consider a large variety of different transformations, but unless a transformation is known to leave the a priori property invariant, it should be used with precautions. If the qualitative properties of optimal solutions are not preserved by a transformation, then this transformation will be called singular. It will be shown later that certain transformations are capable of transforming discontinuous functions into continuous ones, and thus prevent the existence of a unique inverse transformation. Such transformations are of course singular. Transformations which affect the structure of a nominal model, such as for example the discretization of one or more state or control variables, are also singular as a rule. This is true no matter whether Euler's method (see for example p. 151, vol. I[C 15]) or Bellman's dynamic programming (see for example[B 4],[B 5],[B 6]) are used as a basis for the discretization. The use of discretization must therefore be accompanied by a direct or indirect proof that no change of qualitative properties is involved.

Since the precise scope of validity of dynamic programming has not yet been established, and it is known that recurrence processes of the type used in dynamic programming lead generally to a gradual loss of continuity properties[H 2], dynamic programming should be used with a lot of circumspection. The problem of preservation of continuity in recurrence processes occurs in connection with Huyghens principle, and it has occupied many investigators (see for example [F 8]). Preservation of the a priori property is far more important in control theory than the realization of an apparent mathematical simplicity.

CHAPTER 2

FUNDAMENTAL PROPERTIES OF
ELEMENTARY EXTREMAL PROBLEMS

§ 9. SOME PROPERTIES OF NOMINAL
EXTREMAL PROBLEMS

If a theoretical optimization problem is to be meaningful for design purposes, then, as was shown in chapter 1, it must possess the a priori property. Suppose for example that this optimization problem has been obtained by means of a mixed causal-teleological reasoning and that it can be written in the form

$$\left.\begin{aligned} I(t_0, t_1) &= \min_{u \in U} F(y, u, t), \\ G(y, u, t) &= 0, \\ L(y, u, t) &\leqslant 0, \quad y \in Y, \quad t_0 \leqslant t \leqslant t_1, \end{aligned}\right\} \quad (2.1)$$

where F is a functional of the point-functions $y(t)$, $u(t)$, G is a continuous point-function, and L is either a functional or a point-function of the same arguments, representing the optimality criterion and the constraints, respectively. Suppose furthermore that the interval $t_0 \leqslant t \leqslant t_1$ and the admissible sets U, Y have been defined in such a way that I and L are both unambiguous and single-valued.

If the nominal optimization problem (2.1) is to be of the a priori type, then it must be correctly set in the Hadamard sense and it must be relatively insensitive to small variations of its structure. The variations of structure considered should correspond to all types of perturbations likely to occur in practice. Similarly, to the procedure followed in § 4, we will study (2.1) in stages, starting with a very simple problem and then adding various features till a physically realistic level is reached.

In conformity with the first condition of (1.6), the very least requirement to be imposed on (2.1) is that there should exist at least one solution. This requirement signifies in particular that the following conditions should hold:

(a) U and Y are completely and unambiguously defined, $\left.\vphantom{\begin{aligned}a\\b\end{aligned}}\right\}$ (2.2)
(b) all admissible $u(t)$, $y(t)$ can be tested by means of F and L.

In spite of the fact that the conditions (2.2) are self-evident, and thus appear to be trivial, let us recall that their importance has not been fully recognized until Hilbert has published his results on the Dirichlet principle[H 8]. In order to show that the importance of (2.2) should not be underestimated let us examine a very simple illustrative example.

Let U be a group of papers submitted to a symposium, say on control theory, and suppose that the optimization problem consists in selecting one paper from U which communicates an original result in the most concise form. The conciseness should be judged, for example, by the number of equations used to establish the result in question. Suppose furthermore that the maximal length of all acceptable papers is limited to a certain number of pages of some standard format. It is then quite obvious that unless the exact number of acceptable symposium papers is known, and each paper is actually available for examination, it is impossible to make the required selection, even if a completely unambiguous criterion of what constitutes an original result is available.

When interpreting the practical meaning of results obtained by means of a nominal problem of form (2.1), a complication sometimes arises from the fact that the definition of the set, or sets, of admissible functions is not necessarily the same in theory and in practice. We have seen in §4.3 that in some physically meaningful problems the closeness of two functions is defined by the integral inequality (1.32), whereas in contemporary control theory (see in particular [B 9] and [H 6]) the closeness of two functions is defined by the algebraic inequality (1.33). Hence, the corresponding admissible sets cannot be identical.

Returning to the preceding illustrative example, let us note that the selection of the best symposium paper will be completely devoid of practical significance if it has been obtained from a hypothetical list of papers which is not identical with the actual list. This conclusion remains valid even if the hypothetical list is a 'very general' one and if it contains the actual list as a particular case. Expressed in other words: a solution of a nominal optimization problem is completely devoid of practical significance if it has been obtained on the basis of some hypothetical set of admissible functions not identical with the practical set, no matter the generality of the former.

It is of course permissible to replace a specified admissible set U by a more general set $\overline{U}$ which contains U as a subset. But if a practically meaningful result is required, then it is necessary to show how to deduce an optimal solution valid in U from an optimal solution valid in $\overline{U}$. In the case of the Hilbert method of dealing with the Dirichlet principle this was done by means of the definition of special convergence processes (cf. §7). In general, the process of deduction of a solution of (2.1), valid in U, from a corresponding solution valid in $\overline{U}$, $U \subset \overline{U}$, constitutes essentially an interpretation algorithm. It is obvious that this interpretation algorithm increases in complexity as the admissible set $\overline{U}$ increases in generality. In some cases the required interpretation algorithm is more difficult to establish than the optimal solution defined in $\overline{U}$.

In the illustrative example involving symposium papers the set U had a finite number of elements. In more complex problems the admissible set may have an enumerable or a non-enumerable number of elements, but the latter circumstance does not invalidate the conditions (2.2). Quite to the contrary, it makes them even more important. We will show that in some cases the conditions (2.2) furnish the principal explanation of an otherwise baffling behaviour of extremal solutions.

Unless the contrary is stated, we will always assume that the optimal problem is of the form (2.1). The omission of the case where a maximum of a functional F is sought does not constitute a loss of generality, because the latter case can be deduced from the obvious relationship

$$\max_{u \in U} F = -\min_{u \in U}(-F). \tag{2.3}$$

The operator $\min_{u \in U}$ in (2.1) and (2.3) will be used to designate three mathematically distinct notions: an 'ordinary' minimum, a lower limit and a least value. An ordinary minimum, or more briefly a minimum, will be related to a two-sided (bilateral) local neighbourhood and a lower limit to a one-sided (unilateral) local neighbourhood. The idea of a local neighbourhood is irrelevant for a least value. The three notions will be defined more precisely in §10, and the differences between them will be illustrated by examples.

Before examining optimization problems of form (2.1), where F is a functional, we will discuss first the simple case when (2.1) defines a minimal value of a single point-function. We will suppose in succession that the admissible sets U and Y, which are then only point-sets, have a finite, an enumerable, and a non-enumerable number of elements. Although the results thus obtained will not be directly applicable to the case when F is a functional, they will furnish a convenient base for the comparison of analogous extremal properties.

§ 10. EXTREMA OF ORDINARY FUNCTIONS WITHOUT CONSTRAINTS

10.1 Functions defined on a finite point-set

Consider the problem (2.1) and suppose that F is a single-valued point-function of a discrete m-dimensional variable u, $m < \infty$, and that the constraints $G = 0$ and $L \leqslant 0$ are absent. Then (2.1) reduces to the form

$$I = \min_{u \in U} F(u), \tag{2.4}$$

where U is a specified point-set in a m-dimensional Euclidean space. If the set U contains only a finite number of elements u_i, $i = 1, 2, ..., n < \infty$, then the solution of (2.4) can be found in a straightforward manner, which is independent of the *internal structure* of U, i.e. of the order in which the u_i are arranged in U. In fact, for every u_i it is sufficient to calculate the corresponding values $I_i = F(u_i)$, and then to arrange the numbers I_i into a non-increasing sequence $\bar{I}_j, j = 1, 2, \dots$. Since there are at most n different numbers I_i, this ordering is always possible and the last element $\bar{I}_n$ of $\bar{I}_j$ will be one solution of (2.4). This solution is not necessarily unique, because from a certain rank k all elements of the sequence $\bar{I}_j, j = k, k+1, \dots, n$, may turn out to be identical. If $k = 1$, then the problem (2.4) is degenerate in the sense that all admissible elements u_i are equally optimal. The operator $\min_{u \in U}$ in (2.4) is thus seen to define a comparison algorithm of a finite number of stages, and the end-product $\bar{I}_n$ of this algorithm will be called the least value of $F(u)$ in U.

10.2   Functions defined on an enumerable point-set

If U in (2.4) contains an enumerable number of elements $u_i, i = 1, 2, \dots$, then the possibility of solving (2.4) depends both on the nature of the function F and on the internal structure of the admissible set U, i.e. on the order in which the u_i occur in U. If the u_i occur in such a way that the corresponding sequence of numbers $I_i = F(u_i)$ shows no regular pattern, no matter how large $i < \infty$ is chosen, then *for practical purposes* the problem (2.4) has no solution. If the problem (2.4) is to be practically meaningful, with or without the aid of automatic computers, then it should be possible to determine the least value of $F(u)$ in U by means of a *finite* number of operations. This conclusion leads to the perhaps unexpected result that practically meaningful extremal problems should not involve the use of Zermello's axiom of choice[Z4].

If starting from some sufficiently large $i < \infty$ the numerical sequence $I_i = F(u_i)$ does show a regular pattern, then (2.4) does admit a practically meaningful solution. Let us consider the simplest possible type of regular behaviour by supposing that the sequence I_i is bounded from below, i.e. that there exists a finite real number M_1 such that $F(u_i) > M_1$ for all $u_i \in U$. According to the Bolzano–Weierstrass theorem the numerical sequence $F(u_i)$ admits then at least one accumulation value. Let M be the largest lower bound and $F(\bar{u})$ an accumulation value of the sequence $F(u_i)$. The values $F(\bar{u})$ are usually called local lower limits of $F(u)$ in U and they are designated by various distinctive symbols. We will use the notation
$$F(\bar{u}) = \liminf_{u \in U} F(u).$$

Suppose that there exists only one local lower limit $F(\bar{u})$. Two cases are then possible depending on whether M coincides with $F(\bar{u})$ or not. If $F(\bar{u}) \neq M$, then necessarily $F(\bar{u}) > M$, and (2.4) admits a solution which can be found by replacing the enumerable set U by a sufficiently large but finite subset $u_1, u_2, \dots, u_n$. The solution of (2.4) is therefore a least value.

As an illustration consider the example

$$F(u_1) = 1, \quad F(u_2) = 2, \quad F(u_3) = 1, \quad F(u_4) = 0,$$

$$F(u_5) = -1, \quad F(u_i) = 1/i \quad (i > 5).$$

The largest lower bound of this sequence is $M = -1$, whereas its lower limit is $F(\bar{u}) = \lim_{i \to \infty} F(u_i) = 0$. The solution of (2.4) is therefore the least value $I = F(u_5) = -1$.

If $F(\bar{u}) = M$, then the operator $\min_{u \in U}$ in (2.4) can be interpreted as a lower limit. Two cases are possible depending on whether $F(\bar{u})$ is an element of the sequence $F(u_i)$, $i = 1, 2, ...$, or not, i.e. depending on whether one of the following relations holds:

$$F(\bar{u}) = F(\lim_{i \to \infty} u_i) \quad \text{or} \quad F(\bar{u}) = \lim_{i \to \infty} F(u_i). \tag{2.5}$$

Using the terminology due to Hadamard[H 1], in the first case the problem (2.4) is said to admit a solution in *the strict sense*. In the second case there is no solution in the strict sense, because the value $F(\bar{u})$ is not attained on any element $u \in U$, but there exist elements $u_j \in U$ such that the difference $|F(u_j) - F(\bar{u})|$ can be rendered as small as desired. In control engineering terminology the problem (2.4) can then be said to admit suboptimal solutions defined, for example, by the interpretation algorithm

$$|F(u_j) - F(\bar{u})| < 1/j \quad (j = 1, 2, ...). \tag{2.6}$$

Let us illustrate the two possibilities by examples.

Example 1. $F(u_i) = 1/i$, $i = 1, 2, ...$ Since $M = F(\bar{u}) = \lim_{i \to \infty} F(u_i) = 0$, the operator $\min_{u \in U}$ in (2.4) designates a lower limit. If $u_\infty \in U$, the limit value $F(\bar{u}) = 0$ is attained on the sequence u_i for the fixed value $i = \infty$. The problem (2.4) has therefore an optimal solution in the strict sense.

Example 2. $F(u_i) = -(1 + [1/i])^i$, $i = 1, 2, ..., u_\infty \in U$. Since

$$M = F(\bar{u}) = \lim_{i \to \infty} F(u_i) = -e,$$

the operator $\min_{u \in U}$ in (2.4) designates again a lower limit, but the limit value $F(\bar{u}) = -e$ is not attained for any fixed value of i. For $i = \infty$ the expression $F(u_i)$ is an indetermination of the form $-(1^\infty)$. The problem (2.4) has therefore no optimal solution, but it admits suboptimal solutions of any accuracy.

The distinction between optimal and suboptimal solutions of a point-function problem is mathematically important but physically completely meaningless. In fact values of physical constants must not be interpreted as rigorously fixed, but only as representative elements of some corresponding dense sets (cf. §4.1). This is so because physically there is no qualitative difference between the numbers $F(u_j)$ and $F(\overline{u})$ in (2.5). Unfortunately, the same conclusion does not necessarily apply to the case when F is a functional, because $\overline{u} = \lim_{i\to\infty} u_i$ and $F(\overline{u}) = \lim_{i\to\infty} F(u_i)$ may become qualitatively different from u_i and $F(u_i)$, respectively.

To illustrate this difficulty on a somewhat artificial point-function problem consider again example 2, with the added restriction that both $u_i = i$ and $F(u_i) = -(1 + [1/i])^i$ should be finite rational numbers, and that the same restrictions should apply to the validity of the relations (2.5) and (2.6). Since $\overline{u} = \lim_{i\to\infty} u_i$ is not finite and $F(\overline{u}) = \lim_{i\to\infty} F(u_i)$ is not rational, the above problem has neither optimal nor suboptimal solutions under the conditions stated. Suboptimal solutions become only possible when the interpretation algorithm (2.6) is 'generalized' to admit irrational numbers $F(\overline{u})$, or when (2.6) is replaced for example by

$$|F(u_j) - F(u_{j+k})| < 1/j \quad (k > 1,\, j = 1, 2, \ldots).$$

Since contrary to the case of points in a Euclidean space and the convergence of numerical sequences the distance between two functions and the convergence of functional sequences can be defined in a variety of ways (cf. §§4.2 and 4.3), the functions $\overline{u}$, u_j and the functionals $F(\overline{u})$, $F(u_j)$ need not be of the same qualitative type, respectively. Even if $\overline{u}$ and u_j are of the same qualitative type, they need not satisfy an inequality of the form (2.6) for any value of j. The closeness of $\overline{u}$ and u_j depends in general on the precise definition of distance in the admissible set U, and the closeness of $F(\overline{u})$ and $F(u_j)$ depends on the precise definition of the notion of convergence. These aspects will be examined more thoroughly in §12.

10.3   Functions defined on a non-enumerable point-set

Discontinuous functions

Consider again the problem (2.4), but suppose that F is a discontinuous single-valued point-function of the continuous m-dimensional variable $u \in U$, $m < \infty$. If F is so strongly discontinuous that the set H of the values of $F(u)$ shows no regular pattern whatever, then without Zermello's axiom of choice the problem (2.4) admits no solution. It is of course also reasonable to expect that the increase of the number of elements of U will require a more stringent condition on the orderliness of $F(u)$ to ensure a meaningful solution of (2.4). Contrary to the case of an enumerable point-set U, the condition that the set H of the F-values be bounded from below does not generally insure the existence of a solution of (2.4) in a strict sense. This can be shown by means of the following one-dimensional example:

$$F(u) = \begin{cases} \tfrac{1}{2} & \text{for} \quad u = 0 \quad \text{and} \quad u = 1, \\ u & \text{for} \quad 0 < u < 1, \end{cases}$$

where U is the closed interval $0 \leqslant u \leqslant 1$ and H is the open interval $0 < F(u) < 1$.

The largest lower bound of H is $M = 0$, and it coincides with the lower limit $F(\bar{u})$ of $F(u)$. The value $F(\bar{u})$ is however not attained on any element u of U. The above problem admits therefore only suboptimal solutions defined, for example, by the interpretation algorithm (2.6).

Let us recall here that the failure to distinguish between extreme values of a function and its upper and lower limits has produced many famous mathematical errors (see for example the summary given in [L6]). Such a distinction is especially important in the case of functionals.

The least restrictive condition which will insure the existence of a strict solution of (2.4) is that H be both bounded and closed from below. A set H is said to be closed from below if it contains as elements the lower limits of all its infinite sequences h_i, $i = 1, 2, ...$, where all values $h_i \in H$.

Consider (2.4) where the non-enumerable set U is such that the set H of F-values is bounded and closed from below. If the largest lower bound M of H is not known, then the determination of a strict solution of (2.4), although possible in principle, may turn out to be quite laborious. In fact it will be necessary to calculate M by iteration, for example, by choosing a certain number of enumerable subsets of U, calculating the corresponding subsets of H, and then determining the largest lower bounds M_i and the lower limits $\bar{h}_i$ of the latter. The iteration process is stopped when an improvement of the values of M_i and $\bar{h}_i$ no longer occurs. Since a regular behaviour of the sequences M_i and $\bar{h}_i$ is *not* a consequence of the property that H is bounded and closed from below, extrapolation is usually not possible and the speed of convergence of the iteration process cannot be established in advance.

Strongly discontinuous point-functions are of course of negligible practical interest by themselves, but as we will see in § 12, this conclusion does not apply when they occur in functionals.

Continuous and piecewise continuous functions

Suppose now that the function F in (2.4) is single-valued and continuous in a specified region U of an m-dimensional Euclidean space, $m < \infty$. Since all metrizations are known to be equivalent in a Euclidean space of a finite number of dimensions, it can be assumed without loss of generality that the distance between two points of U is defined by the sum of the absolute differences of their coordinates. Let the same definition of distance apply in the one-dimensional space H of the values of $F(u)$.

Let us recall that $F(u)$ is said to be continuous in U if for any point $u_0 \in U$ and any $\epsilon > 0$ there exists a $\delta > 0$ such that

$$|u - u_0| < \delta \quad \text{implies} \quad |F(u) - F(u_0)| < \epsilon \quad (u \in U, \quad F(u) \in H). \quad (2.7)$$

The definition (2.7) is essentially possible because the sets U and H have the same quantity of elements, i.e. they have the same cardinal number. Unfortunately such a property is generally absent in the case of functionals.

Consider a fixed point $u_0 \in U$. The function F is said to admit a relative or a local minimum at u_0 if for any u, located in a sufficiently small bilateral neighbourhood of u_0, defined by the first inequality in (2.7), $F(u_0)$ satisfies the inequality

$$F(u) - F(u_0) \geqslant 0. \quad (2.8)$$

If instead of (2.8) $F(u_0)$ satisfies the somewhat stronger inequality

$$F(u) - F(u_0) > 0 \quad (u \neq u_0),$$

then, according to Hadamard[H 1], the local minimum of F is said to be strict.

With a slight restriction on the set U, the problem of existence of a solution of (2.4) for continuous functions $F(u)$ can be considered as completely solved. In fact, Weierstrass has proved that if the set U is bounded and closed (from below and from above), then the set H is bounded (from below and from above), and it contains both the smallest lower limit and the largest upper limit of $F(u)$. For single-variable problems of the form (2.4) the Weierstrass theorem can be reworded as follows: If $F(u)$ is single-valued and continuous in a finite closed interval $U: -\infty < a \leqslant u \leqslant b < +\infty$, then the values of $F(u)$ are located inside a closed interval $H: -\infty < c \leqslant F(u) \leqslant d < +\infty$ and there exist two points u_1, u_2 in U such that

$$F(u_1) = c \quad \text{and} \quad F(u_2) = d.$$

The points u_1 and u_2 are located either in the interior or on the boundary of U. In the former case $F(u_1) = c$ represents the smallest local minimum and in the latter case the smallest lower limit of $F(u)$.

The Weierstrass theorem insures the existence of a solution of (2.4), but it does not provide any method to determine it. A problem satisfying the conditions of the Weierstrass theorem may admit a large number of local minima and at most two lower limits which do not coincide with local minima. Once the set $F(u_i)$ of isolated local minima and the non-trivial lower limits of $F(u)$ is known, then the determination of a solution of (2.4) is reduced to the determination of the least value of the set $F(u_i)$, which is at most enumerable. The solution of (2.4) can then be determined by the procedure described in §§ 10.1 or 10.2.

If instead of being continuous the function $F(u)$ is only piecewise continuous in U, then to determine a solution of (2.4) the above method can be combined with the iterative method described earlier.

Piecewise differentiable functions

Let us recall first that continuity of a function does not imply anything about its differentiability. This statement is quite evident from an examination of the inequalities (2.7), which stipulate nothing about the order of magnitude of ϵ, considered as a function of δ. In other words, the inequalities (2.7) stipulate nothing about the $O(\epsilon(\delta))$-neighbourhood (cf. §4.3) of the value $F(u_0)$. For instance, if $F(u_0)$ possesses only an $O(\delta^a)$-neighbourhood, $0 < a < 1$, then the ratio

$$\frac{F(u) - F(u_0)}{u - u_0} = O(\delta^{a-1}), \tag{2.9}$$

and it can be made as large as desired by choosing u sufficiently close to u_0. Examples of continuous functions which are not differentiable at any point of their domain of definition were first given by Weierstrass. The following one-dimensional example is particularly simple: the function

$$F(u) = \sum_{k=0}^{\infty} \frac{\cos{(k!)}\,u}{k^2}, \tag{2.10}$$

is continuous but not differentiable for any value of u in $-\infty < u < +\infty$, because the series

$$-\sum_{k=0}^{\infty} \frac{k!}{k^2} \sin{(k!)}\,u,$$

which represents formally the derivative $F'(u)$, diverges for all values of u.

Suppose that $F(u)$ in (2.4) is continuous and piecewise differentiable in some region U of an m-dimensional Euclidean space, $m < \infty$, and that U contains only an enumerable number of points $\bar{u}_i$ where the total differential

$$dF = \sum_{j=1}^{m} \frac{\partial F}{\partial u_j}\,du_j$$

fails to exist. u_j are the components of the vector u. Since the points of non-differentiability $\bar{u}_i$ are necessarily isolated, they can be surrounded by a sufficiently small hypervolume (hypercube in the case of a metric defined by the first inequality in (2.7)), inside of which $F(u)$, $u \neq \bar{u}_i$, is differentiable. Consequently the values of the partial derivatives of $F(u)$

can be explored in these hypervolumes, and the following criterion can be formulated:

> $F(\overline{u}_i)$ is a local extremum (minimum or maximum) if and only if the partial derivatives $\partial F/\partial u_j, j = 1, 2, ..., m$, have different signs for $\overline{u}_i + \epsilon$ and $\overline{u}_i - \epsilon$, where $\epsilon > 0$ is a constant m-dimensional vector of a sufficiently small length.   (2.11)

As an illustration of the use of the above criterion consider two single-variable examples.

Example 1. $F(u) = |u|$. The point $u = 0$ is a local minimum because $F'(0^-) = -1$ and $F(0^+) = +1$.

Example 2. $F(u) = u^{\frac{2}{3}}$. The point $u = 0$ is a local minimum because for $\epsilon > 0$, $F'(\epsilon) = \frac{2}{3}\epsilon^{-\frac{1}{3}} > 0$ and $F'(-\epsilon) = -\frac{2}{3}\epsilon^{-\frac{1}{3}} < 0$.

When the points $\overline{u}$ where $F(u)$ is not differentiable are non-enumerable, then there exists no general criterion for a minimum, and each case must be handled on its own merits. In many cases the problem (2.4) can be reduced to a problem of the same form but of a smaller number of dimensions.

Continuously differentiable functions

When the function $F(u)$ in (2.4) is continuously differentiable in U, then, by definition, for any $u_0 \epsilon U$ $F(u_0)$ has a $O(\delta^a)$-neighbourhood, $a \geqslant 1$, and the ratio (2.9) has a finite limit as $u \to u_0$. Comparing (2.9) to (2.8) it is obvious that every partial derivative of $F(u)$ must change its sign as u is made to traverse the point u_0 in a suitable direction. Since by definition these partial derivatives are continuous, they must vanish at $u = u_0$. In other words, a necessary, but not sufficient condition for the existence of a local extremum of $F(u)$ at $u = u_0$ is that

$$\frac{\partial F}{\partial u_j} = 0 \quad \text{for} \quad u = u_0 \quad (j = 1, 2, ..., m), \qquad (2.12)$$

where as before u_j are the components of the vector u. A point at which (2.12) is satisfied is said to be a stationary point of $F(u)$.

A necessary and sufficient condition for a local extremum can be obtained by combining (2.12) and (2.11), the point $\overline{u}_i$ in the latter criterion being replaced by the stationary point u_0. Since the combination (2.11), (2.12) involves a bilateral neighbourhood of $u = u_0$, it is only applicable to interior points of U. An element u of U is said to be an interior element if it cannot be expressed as a limit of elements, which do not belong to U. Consequently the combined criterion (2.11), (2.12) does not apply if u_0 is a boundary point of U. In the latter case the extremum of $F(u)$ will be given by a local lower or a local upper limit.

Functions admitting higher-order derivatives

If the function $F(u)$ in (2.4) is twice continuously differentiable in U, then in the neighbourhood of every interior point of U it can be approximated by the first two terms of its Taylor series. This Taylor series can be used to explore the values of $F(u)$ and of its first-order partial derivatives in the neighbourhood of any stationary interior point $u = u_0$, and thus to determine whether $F(u_0)$ is a local minimum or a local maximum. Indeed, near a stationary point

$$F(u) - F(u_0) = \frac{1}{2} \sum_{j=1}^{m} \sum_{k=1}^{m} F_{u_j u_k}(u_0) \, (u_0 - u_{0j}) \, (u_k - u_{0k}) + \dots, \quad (2.13)$$

where the subscripts of F stand for partial differentiations with respect to the components u_j of the vector u. Comparing (2.13) to (2.8), it is obvious that for a local minimum the second member of (2.13) must be non-negative. The following criterion can therefore be formulated:

If the associated quadratic form

$$M_2 = \sum_{j=1}^{m} \sum_{k=1}^{m} F_{u_j u_k}(u_0) x_j x_k \quad (2.14)$$

is positive definite, i.e. $M_2 > 0$ for all real values of the m-dimensional vector variable $x \neq 0$, then the stationary point $u = u_0$ is a local minimum of $F(u)$.

If some of the second-order partial derivatives of $F(u)$ vanish at $u = u_0$ and the associated quadratic form M_2 becomes degenerate, then the criterion (2.14) fails. The exploration of the neighbourhood of $F(u)$ must then be carried out by determining $F(u)$ or its first derivatives in the neighbourhood of $u = u_0$.

If $F(u)$ admits derivatives up to the order $n = 2k, k > 1$, and all partial derivatives up to the order $n - 1$ inclusive vanish at $u = u_0$, then the sign of the difference $F(u) - F(u_0)$ will be determined by the nth-order term of the Taylor series. Similarly to (2.14), $F(u)$ will be a local minimum if n is an even integer and the associated nth-order form M_n is positive definite.

§ 11. EXTREMA OF CONSTRAINED POINT-FUNCTIONS

Consider the problem (2.1) and suppose that F is a single-valued function of two vector variables u and v of m and k dimensions, respectively. Let G and L be two sets of k point-functions $G_i(u, v)$ and $L_i(u, v)$, then the point-function version of (2.1) can be written in the form

$$I = \min_{u \in U} F(u, v), \quad G_i(u, v) = 0, \quad L_i(u, v) \leqslant 0 \quad (v \in V), \quad (2.15)$$

where U and V are specified regions of corresponding Euclidean spaces. The vector variables u and v may be either discrete or continuous.

The problem (2.15) can be reduced immediately to a problem of form (2.4) by interpreting the constraints $G_i = 0$, $L_i \leqslant 0$, $v \in V$, as a part of the complete definition of the admissible set U. In other words, only those elements u are considered to be admissible for the minimization of $F(u, v)$ which, in addition to $u \in U$, satisfy $G_i = 0$, $L_i \leqslant 0$, and $v \in V$. The problem (2.15) can therefore be written

$$I = \min_{u \in U_1} F_1(u), \tag{2.16}$$

where $F_1(u) = F(u, \bar{v})$, and $\bar{v}$, U_1 are point-sets in the region $G_i = 0$, $L_i \leqslant 0$, $v \in V$, $u \in U$.

The problem (2.16) is therefore not essentially different from the problem (2.4), and it can be treated by the methods discussed in §10. The only difference is that the definitions of $F_1(u)$ and U_1 are implicit and from the point of view of (2.4) are given in a somewhat unusual form. At least when the variables u and v are discrete, (2.16) may turn out to be more convenient than (2.15), and this observation is claimed as one of the advantages of dynamic programming (see for example chapter 1, §17[B6]).

All methods of solving (2.15) are based essentially on the fact that constraints of the form $L_i < 0$ increase neither the number of local minima nor the number of attained local lower limits of F. For conciseness let us designate by 'minimal point' a point where F admits either a local minimum or a local lower limit. The sole effect of the constraints $L_i < 0$ is then to render inadmissible certain minimal points of the problem

$$\min_{u \in U} F, \quad G_i = 0.$$

Contrary to $L_i < 0$, the constraints $L_i = 0$ may, and as a rule do produce new minimal points of F. These additional minimal points may be either local minima or non-trivial local lower limits. The procedure of solving (2.15) reduces therefore to the determination of two sets of minimal points from which the least value of F is chosen by means of a comparison algorithm. The first set consists of the minimal points of the less constrained problem $\min_{u \in U} F$, $G_i = 0$, from which minimal points inconsistent with the constraints $L_i < 0$, have been omitted. The second set consists of the minimal points of the problem

$$I = \min_{u \in U} F(u, v), \quad G_i(u, v) = 0, \quad L_i(u, v) = 0 \quad (v \in V). \tag{2.17}$$

The determination of the second set of minimal points is considerably simplified when the functions G_i or L_i are such that the system of k simultaneous equations $G_i(u, v) = 0$ or $L_i(u, v) = 0$ can be transformed into the explicit form $v = g(u)$. Substituting $v = g(u)$ into $F(u, v)$, the problem (2.17) is then transformed immediately into the form (2.16).

Let the constraints $L_i = 0$ be absent in (2.17). If *only non-degenerated interior local minima* of the simplified problem (2.17) are of interest and if the $k + 1$ functions F, G_i are all continuously differentiable with respect to their arguments, then the determination of the explicit solution $v = g(u)$ of $G_i = 0$ can be circumvented by means of an artifice due to Lagrange. Introducing the notation

$$x_j = u \quad (j = 1, 2, ..., m),$$

$$x_j = v \quad (j = m + 1, ..., m + k = n) \quad (x \in X),$$

the simplified problem (2.17) can be interpreted as a particular case of the problem

$$I = \min_{x \in X} F(x), \quad G_i(x) = 0, \quad x = (x_1, x_2, ..., x_n).$$

Consider the two matrices

$$\left(\frac{\partial F}{\partial x_j} \frac{\partial G_i}{\partial x_j} \right), \quad \left(\frac{\partial G_i}{\partial x_j} \right) \quad (i = 1, 2, ..., k, \quad j = 1, 2, ..., n).$$

If the rank of these matrices are $k + 1$ and k, respectively, then there exists a non-vanishing constant vector $\lambda = (\lambda_1, \lambda_2, ..., \lambda_k)$ such that the stationary values of $F(x)$, subject to the constraints $G_i(x) = 0$, coincide with the stationary values of the auxiliary function

$$\bar{F}(x, \lambda) = F(x) + \sum_{i=1}^{k} \lambda_i G_i(x). \tag{2.18}$$

The components λ_i of the vector λ are called Lagrange multipliers. The auxiliary function $\bar{F}(x, \lambda)$ is interpreted as a function of $m + k$ independent variables. Supposing in this particular case that the operators max and min stand only for the determination of non-degenerated interior local maxima and minima, respectively, the following relations are known to hold (see for example §§ 186–7[C6] and vol. 1, chapter 4, § 9[C15])

$$I = \min_{x, \lambda} \bar{F}(x, \lambda) = \max_{\lambda} \min_{x} \bar{F}(x, \lambda). \tag{2.19}$$

The two relations in (2.19) are very useful in numerical calculations, because they permit to estimate the value of I both from below and from above.

It should be noted at this point that because partial derivatives of $F(x)$ and $G_i(x)$ are used in the proof of (2.18) and (2.19), these relations are inappropriate for the determination of local minima of non-continuously differentiable functions (cf. § 10.3). The same conclusion applies to the determination of non-trivial local lower limits, because it is obvious that a point where a function $F(x)$ attains a local lower limit need not be a stationary point of F.

Consider now the artifices due to Valentine[V 1] which permits to replace a system of k inequalities $L_i(x) \leqslant 0$ by a system of k simultaneous equations

$$y_i^2 + L_i(x) = 0, \qquad (2.20)$$

where $y = (y_1, y_2, \ldots, y_k)$ is an auxiliary vector variable. Since this artifice has no effect on the continuity properties of the functions $F(x)$ and $L_i(x)$, its use in minimization problems constitutes neither a generalization of the proof, nor an extension of the scope of validity of the relations (2.18) and (2.19). Consequently, the usefulness of the artifice of Valentine does not go beyond a certain *notational* simplification. A similar conclusion will be shown to apply when F is a functional.

§ 12. SOME PROPERTIES OF FUNCTIONALS

12.1 The field of a functional

Consider the nominal extremal problem (2.1), where F is no longer a point-function but a functional (cf. §5). It is convenient to know under what condition (2.1) can be transformed into an equivalent problem involving only point-functions, for example into a problem of the form (2.15) or (2.16). A transformation of the required kind can obviously not exist, unless (2.1) and (2.15) involve essentially the same type of continuity. Since the continuity of functionals is a far more complex notion than the continuity of point-functions, a functional will be equivalent to a point-function only in exceptional circumstances.

Let us now discuss some definitions and results of the theory of functionals which are useful for the study of extremal problems. A list of some elementary definitions is given in the appendix.

Consider the single-variable functional

$$I = F(u(t)), \quad u(t) \in U, \quad t \in (a, b),$$

where $u(t)$ is a single-valued point-function defined for every t in the interval $a \leqslant t \leqslant b$. U is called the field of definition of the functional F. Let H designate the set of values of $I = F(u(t))$, corresponding to $u \in U$. For example, the field of definition of the functional

$$I = \int_a^b u(t)\, dt \qquad (2.21)$$

is the set of functions $u(t)$ which are integrable in the closed interval $a \leqslant t \leqslant b$, and H is the set of all real numbers $-\infty < I < +\infty$. The precise meaning of the term 'integrable' depends on the definition of the integral in (2.21). In problems of physical origin (2.21) may be either a line integral or an integral in the Riemann sense. In some cases, for instance in Fourier analysis, an integral of type (2.21) may stand for a Cauchy principal value, for a Borel limit, or even for a finite part (definitions of these concepts are given, for example, in [B 17] and [H 3]). In more mathematical problems the integral in (2.21) may be understood in the Lebesgue, or in the Lebesgue–Stieltjes sense. It is important to note that from the point of view of conditions (2.2), the field U of (2.21) is *not* completely defined unless the meaning of the integral has been fully specified.

Suppose that the functional $F(u(t))$ is equivalent to some point-function $G(t)$, defined in a given region of a Euclidean space. Then the corresponding sets U and H of F and G must have the same quantity of elements, respectively, i.e. they must have the same cardinal numbers. But the sets U and H associated with a non-degenerated point-function have the same cardinal number, which is at most that of a continuum. Unfortunately, this property does not generally hold for functionals. In fact, the cardinal number of the set U of all single-valued functions $u(t)$, defined in a finite interval $a \leqslant t \leqslant b$, is known to be *larger* than the cardinal number of a continuum. There exist thus *more* elements $u \in U$ than there are distinct values of $F(u(t))$.

Because of this intrinsic incompatibility very few functionals are studied in the *totality* of their field of definition. To insure that a functional F be single-valued it is necessary to restrict its total field of definition to some subfield U_1 (for example to some metric space), and to obtain some analogy between the properties of functionals and the properties of point-functions, it is necessary to restrict U_1 to some subfield $U_2 \subset U_1$. Unfortunately, these restrictions can be made in a variety of ways, depending more on the subjective preferences of each investigator than on the objective requirements of the subject-matter. The end-result is generally that the final restrictions are too severe and that a strict solution of the problem $I = \min_{u \in U} F(u(t))$ fails to exist in the subfield U_2. There can be thus no functional analogue to the Weierstrass theorem on the extrema of continuous point-functions, except when some very special conditions are satisfied by F and U.

It is known that at least a particular class of functionals can be considered as limits of point-functions of n independent variables as $n \to \infty$ (see for example [V 5]). This property can be used to classify at least some functionals according to their dimensionality, by interpreting n as a cardinal number. It is obvious that functionals of an enumerable dimensionality will form only a particular subclass of the class of all functionals. It is also obvious that if a field U is to be adequate for the investigation of extremal properties of a functional F, then in general the dimensionality of U should not be smaller than the dimensionality of F. Having made these preliminary remarks let us now determine the dimensionality of some commonly used sets of functions.

The set of analytic functions $u(t)$, i.e. functions admitting a converging Taylor series, has an enumerable dimensionality, because an analytic function is fully defined by the coefficients of its Taylor series.

The set of functions $u(t)$ which can be represented by a convergent series of orthogonal functions is enumerable in dimensionality, because any $u(t)$ is completely defined by the coefficients of the corresponding orthogonal series.

The set of linearly independent functions is enumerable in dimensionality because these functions can be orthogonalized (see for example [C 15]).

The dimensionality of the set of continuous functions $u(t)$ is enumerable, because a continuous function is completely defined when its value for rational values of t are known.

The dimensionality of the set of piecewise continuous functions $u(t)$, admitting an enumerable number of points of discontinuity, is enumerable, because each $u(t)$ is described by an enumerable set of continuous functions.

The set of Lebesgue-measurable functions $u(t)$, which are bounded almost everywhere in an interval $a \leqslant t \leqslant b$, is enumerable in dimensionality. In fact, according to a theorem of Borel, for any two numbers $\epsilon > 0$ and $\delta > 0$ there exists a continuous function $v(t)$, such that $|u(t) - v(t)| \geqslant \delta$ only on a subset of $a \leqslant t \leqslant b$ of measure less than ϵ. Furthermore, according to Luzin[L 18], there exists in $a \leqslant t \leqslant b$ a perfect set P on which $u(t)$ is continuous and the measure of P differs from the measure of $a \leqslant t \leqslant b$ by less than ϵ.

As far as extremal properties of functionals are concerned, the set of Lebesgue measurable functions appears therefore as essentially not more general than the set of piecewise continuous functions. This is true a fortiori in the content of control theory where by definition quantities of zero measure have no physical significance. Mathematically, none of the above-mentioned sets of functions are adequate for an exhaustive study of extremal properties of functionals, and extremal solutions will exist inside these sets only under exceptional circumstances. An adequate set of functions must therefore contain 'strongly' discontinuous functions, like for instance the distributions of Schwartz[S 1], but it is not clear yet whether such a set of generalized functions is unique for a given functional, or a given extremal problem. Fortunately, function-sets of enumerable dimensionality are sufficient for the study of many extremal problems encountered in contemporary control theory. At present only few exceptions are known (see for example [S 7],[G 5]), but such cases are likely to multiply in the future. This problem will be discussed more thoroughly in § 14.4.

12.2   Functional spaces and the continuity of functionals

Let us now discuss some aspects of the theory of abstract functional spaces so far as they are necessary for the study of optimization problems of form (2.1). For this limited purpose it is sufficient to state the various definitions and theorems in a considerably 'de-generalized' form. Readers interested in an exhaustive treatment are referred to more specialized literature (see for example [S 2]).

A functional $F(u)$ is said to be single-valued in a set U if for every $u(t) \in U$ there exists only one value of $I = F(u)$.

A set U is said to be a Frechet space, provided:

(a) a notion of convergence has been defined in U;

(b) this notion of convergence is such that if an infinite sequence $u_1(t), u_2(t), \ldots$ converges to $\bar{u}(t)$, then every infinite subsequence of this sequence converges to $\bar{u}(t)$;

(c) a sequence composed of identical elements $u(t)$ converges to $u(t)$.

A set U is said to be a metric space if U is a Frechet space into which a metric has been introduced, i.e. if U admits the notion of distance, defined by some norm function $N(u_1(t), u_2(t))$ satisfying the conditions

$$\left. \begin{array}{ll} (a) & N(u_1, u_2) = N(u_2, u_1) \geqslant 0, \\[2mm] (b) & N(u_1, u_2) = 0 \quad \text{if and only if} \quad u_1(t) \equiv u_2(t), \\[2mm] (c) & N(u_1, u_2) \leqslant N(u_1, u_3) + N(u_3, u_2). \end{array} \right\} \quad (2.22)$$

Since a Frechet space possesses already a definition of convergence, the norm N is of no use for the transposition of properties of numerical sequences to functional sequences $u_i(t)$, $i = 1, 2, \ldots$, unless it satisfies the additional condition:

(d) $\bar{u}(t)$ is a limit of a sequence $u_1(t), u_2(t), \ldots$ if

$$N(\bar{u}, u_i) \to 0 \quad \text{as} \quad i \to \infty. \quad (2.23)$$

It is noteworthy that all Frechet spaces do not admit a norm N satisfying the conditions (2.22), (2.23). For example, the space of all functions $u(t)$, defined in a finite interval $a \leqslant t \leqslant b$, for which convergence is the ordinary convergence of the values of $u(t)$, is inconsistent with (2.22), (2.23). It is for this reason that ordinary convergence is rarely used in the study of functionals. The notion of uniform convergence is more convenient although considerably less general.

A set U is said to be a complete space if U is a metric space and it admits the Cauchy criterion as a necessary and sufficient condition of convergence. If it applies, the Cauchy criterion states that a sequence $u_1(t), u_2(t), \ldots$ is convergent provided $N(u_i, u_{i+k}) < \epsilon$ for any $\epsilon > 0$, no matter how small, and i and k sufficiently large. It should be noted that the validity of the Cauchy criterion is generally a necessary but not a sufficient condition for a metric space to be complete. For example, the metric space composed of rational numbers admits the Cauchy criterion but is not complete, because limits of its sequences may turn out to be irrational numbers. If a specific metric space U is not complete with a given norm N_0, then it can be either 'completed', by including in U some other elements (irrational numbers in the case of the preceding example), or it can be made complete by replacing the norm N_0 by another norm N_1.

A set U is said to be closed if it contains the limits of all its possible infinite sequences.

A set U is said to be compact if it has either a finite number of elements, or if all its infinite sequences admit at least one limit element. The notion of compactness was not needed in the study of extrema of point-functions, because there is no difference between a bounded and a compact set of points belonging to a Euclidean space.

A set U is said to be equally bounded in the interval $a \leqslant t \leqslant b$, if there exists a constant $M > 0$ such that $|u(t)| < M$ for all t in $a \leqslant t \leqslant b$ and for all $u(t) \in U$.

(2.24): A set U is said to be equally continuous in an interval $a \leqslant t \leqslant b$ if for any $\epsilon > 0$, no matter how small, there exists a $\delta > 0$ such that $|t_2 - t_1| < \delta$ implies $|u(t_1) - u(t_2)| < \epsilon$ for all values t_1, t_2 in $a \leqslant t \leqslant b$ and for all $u(t) \in U$.

(2.25): Let U be a Frechet space and let $F(u(t))$ be a functional defined and single-valued for all $u \in U$. F is said to be continuous for $u = u_0$ if the convergence of a sequence $u_1(t), u_2(t), \ldots$ implies the convergence of the sequence $F(u_1), F(u_2), \ldots$, and if in addition $\lim_{n \to \infty} u_n(t) = u_0(t)$ implies $\lim_{n \to \infty} F(u_n) = F(u_0)$. If the Frechet space in which F is continuous is also a metric space, then for every $\epsilon > 0$, no matter how small, there exists a $\delta > 0$, such that

$$N(u_0, u_n) < \delta \quad \text{implies} \quad |F(u_0) - F(u_n)| < \epsilon.$$

(2.26): The functional $F(u(t))$ is said to be uniformly continuous in a domain D of U if the inequalities of the definition (2.25) are valid for a fixed value of ϵ and any two elements $u_0(t)$ and $u_n(t)$ of D. It should be noted that a functional which is continuous for every element $u \in D$, is not necessarily uniformly continuous in D. This property has no counterpart in the theory of point-functions. The definitions of continuity (2.25) and (2.26) depend on the definition of convergence used in connection with a Frechet space and on the definition of the norm N used in connection with a metric space. By a change of the definition of convergence or of N some functionals can be made either to gain or to lose the property of continuity.

(2.27): A functional $F(u(t))$, which is continuous for every element $u(t)$ of a set U, is uniformly continuous in U if U is compact and closed.

(2.28): A sequence of equally bounded and equally continuous functions admits a limit function which is an element of the sequence. This theorem is due to Ascoli, and it can be used occasionally to deduce the existence of an extremum of a functional. There exists however a theorem due to Frechet, which is closer to the Weierstrass theorem on the extrema of point-functions.

(2.29): A single-valued functional $F(u(t))$ which is continuous in the sense of the definition (2.25) in a compact and closed set U is bounded for every $u \in U$, and there exist two elements $u_1(t)$ and $u_2(t)$ in U such that

$$\liminf_{u \in U} F(u) = F(u_1) \quad \text{and} \quad \limsup_{u \in U} F(u) = F(u_2).$$

Unfortunately most functionals occurring in realistic optimization problems satisfy neither the conditions of theorem (2.28) nor those of theorem (2.29). This is due to the property that the assumed or implied definitions of convergence and distance are incompatible with the continuity definitions (2.24) and (2.25). There exist therefore two possibilities to obtain a solution: (a) modifying the nominal optimization model so that either (2.28) or (2.29) apply (cf. §2); and (b) showing that there exists a meaningful 'generalized' solution for the specific nominal optimization model chosen. Both approaches are full of pitfalls and require a complete re-thinking of the optimization problem. The present trend in control theory appears however to be slightly in favour of possibility (b).

12.3   Unilateral continuity of functionals

The definitions of continuity (2.24) and (2.25) are not indispensable for the existence of extremal values of functionals and they do not preclude the existence of other kinds of continuity. In Hilbert's treatment of the Dirichlet principle neither (2.24) not (2.25) is used to deduce the existence of a lower limit of a functional. Hilbert's success has prompted Tonelli to formulate a definition of continuity of functionals which would be particularly appropriate for the study of lower and upper limits. Baire had previously introduced a rather unproductive notion of unilateral continuity of point-functions, which he called semi-continuity. By adapting Baire's idea to simple functionals Tonelli has found[T 6] that the direct method used in Hilbert's argument could be applied to a variety of extremal problems containing functionals of the form (1.73). The result was a systematic theory of extremal problems without any recourse to the theory of differential equations. Let us now give a short account of Tonelli's reasoning, the full details of which can be found in [T 6].

Let U be a metric space with the norm N, and $F(u(t))$ a single-valued functional defined for every $u(t) \in U$.

(2.30): F is said to be semi-continuous from below for an element $u_0 \in U$ if for any $\epsilon > 0$, no matter how small, there exists a $\delta > 0$, such that

$$N(u_0, u) < \delta \quad \text{implies} \quad F(u) > F(u_0) - \epsilon.$$

(2.31): Similarly, F is said to be semi-continuous from above for $u_0 \in U$ if

$$N(u_0, u) < \delta \quad \text{implies} \quad F(u) < F(u_0) + \epsilon.$$

Like continuity, semi-continuity is a *local* property of a functional and its presence or absence depends on the norm N used for the metrization of the set U. By combining (2.30) and (2.31) it is obvious that if for some element $u_0 \in U$ a functional $F(u)$ is simultaneously semi-continuous from below and from above, then, with the same norm N, F is continuous for $u_0 \in U$ in the sense of (2.25).

(2.32): If a single-valued functional $F(u)$, defined for every element $u(t)$ of a set U, attains in U a lower limit for $u_0 \in U$, then $F(u)$ is semi-continuous from below for $u = u_0$. A similar result holds for an upper limit.

(2.33): If the functional $F(u)$ is semi-continuous from below in a compact and closed set U, then $F(u)$ is bounded from below and there exists at least one element $u_0(t) \in U$ such that

$$\liminf_{u \in U} F(u) = F(u_0).$$

A similar result holds for an upper limit.

Like theorem (2.29), theorem (2.33) is also analogous to the Weierstrass theorem on the existence of extrema of point-functions, but it is less restrictive than theorem (2.29).

If the definition of semi-continuity is applied to functionals of the form (1.73) then these functionals can be classified into certain standard types. For some of these standard types it is possible to establish the existence of extremal solutions, and to determine the properties of the latter[T6]. It should be stressed at this point that Tonelli's approach constitutes merely one particular direct method of solving extremal problems. Like all direct methods, it requires a certain insight and ingenuity to be applied successfully. From a practical point of view direct methods become indispensable when indirect methods, based on the theory of differential equations, run into fundamental difficulties, and they become advantageous when the indirect methods become unwieldy. Fundamental difficulties associated with one indirect method, based on Euler's equations, will be discussed in §§ 14.3 and 14.4.

12.4 Continuity of functionals in the sense of Hadamard

A definition of continuity of functionals which is particularly convenient in the physical interpretation of mathematical results obtained on the basis of idealized models has been introduced by Hadamard[H 3]. In order to present Hadamard's ideas on this subject we will simplify the nature of the admissible elements $u(t)$. This can be done without loss of generality

Consider a set U composed of one-dimensional functions $u(t)$, defined in a finite interval $a \leqslant t \leqslant b$. Suppose that the functions $u(t)$ are single-valued and continuous and that the convergence of a sequence of elements of U is understood as uniform convergence. In order to metricize U let

$$N(u_1(t), u_2(t)) = |u_1(t) - u_2(t)|. \qquad (2.34)$$

The above norm function N is first used to characterize a local neighbourhood of a function $u_0(t) \in U$. Similarly to a definition used by Kneser[K 17], a subset U_1 of U is said to be a neighbourhood of order zero in h, $h > 0$, of $u_0(t)$, if

$$|u(t) - u_0(t)| < h \quad \text{for all} \quad u(t) \in U_1. \qquad (2.35)$$

Consider a single-valued functional $I = F(u(t))$ defined in U_1. If $F(u)$ is continuous in U_1 in the sense of the definition (2.25), then $F(u)$ is said to be continuous of order zero in h in U_1.

Example 1. Let

$$F(u) = \int_a^b g(t, u(t)) \, dt \qquad (2.36)$$

be a convergent Riemann integral, where $g(t, u)$ is a continuous point-function for $a \leqslant t \leqslant b$, $-\infty < u < +\infty$. If the inequality (2.35) holds, then the functional $F(u)$ is continuous of order zero in h in U_1, as can be easily verified by substitution.

Example 2. Suppose that the functions $u(t) \in U_1$ are not only continuous but also differentiable. Let $a = 0$, $b = 1$ and $F(u) = \lim \sup_{u \in U_1} \dot{u}(t)$. The functional $F(u)$ is not continuous in U_1 of order zero in h. In fact, let $u_1(t) = k$, $k = ct$, $u_2(t) = k + \delta \sin(2\pi/\delta)t$, $\delta > 0$. The functions $u_1(t)$ and $u_2(t)$ are located in a neighbourhood of order zero in h, provided $h \geqslant \delta$, because $|u_1(t) - u_2(t)| = |\delta \sin(2\pi/\delta)t| \leqslant \delta$. Furthermore,

$$\lim_{\delta \to 0} u_2(t) = u_1(t).$$

Since $F(u_1) = 0$ and $F(u_2) = 2\pi$, $|F(u_1) - F(u_2)|$ cannot be rendered less than a pre-assigned $\epsilon > 0$ by any choice of $\delta > 0$, no matter how small.

It is noteworthy that functionals which are continuous of order zero in h in their whole field of definition are highly exceptional, and they occur rather rarely in practical optimization problems. Such functionals are not essentially different from point-functions. Continuity of order zero in h appears therefore as a sufficient condition that a problem of form (2.1) be for all practical purposes equivalent to a point-function problem of form (2.15). More precisely, the following theorem holds:

Functionals which are continuous of order zero in h in a field U_1 can be expressed in this field as limits of continuous point-functions of n independent variables, as n is made to increase indefinitely. This theorem is due to Gateaux.

A neighbourhood of order zero in h of a function $u_0(t)$ is too general for some optimization problems. One way to restrict it consists in requiring that not only $u(t) \in U_1$ but all derivatives of $u(t)$ up to the order $k \geqslant 1$ satisfy inequalities of type (2.35), i.e. that

$$\left. \begin{array}{l} |u(t) - u_0(t)| < h, \quad |\dot{u}(t) - \dot{u}_0(t)| < h, ..., \\[2mm] |\overset{(k)}{u}(t) - \overset{(k)}{u_0}(t)| < h \quad \text{for all} \quad u(t) \in U_1. \end{array} \right\} \qquad (2.37)$$

If for a given function $u_0(t)$ there exist functions $u(t)$ such that the inequalities (2.37) are satisfied, then $u_0(t)$ is said to possess a neighbourhood U_1 of order k in h.

If a functional $F(u)$ is continuous at $u = u_0$ in the sense of (2.25) when the $u(t)$ are restricted to a neighbourhood of order k in h, then $F(u)$ is said to be continuous at $u = u_0$ of order k in h.

The continuity of a functional of order $k \geqslant 1$ in h is a less restrictive property than continuity of order zero in h. In fact many functionals occurring in contemporary optimization problems are locally continuous of a finite order $k \geqslant 1$ in h.

From an examination of (2.35), (2.36) and (2.37) it is obvious that local continuity of order $k \geqslant 0$ in h of a functional can exist only if its field of definition U consists exclusively of continuous functions. When discontinuous optimization models are considered (cf. §4.3), then U will contain necessarily at least some discontinuous functions and the norm (2.34) becomes inappropriate. The following norms have been used successfully:

$$N(u_1, u_2) = \int_a^b |u_1(t) - u_2(t)| \, dt, \qquad (2.38)$$

$$N(u_1, u_2) = \left[\int_a^b (u_1(t) - u_2(t))^m \, dt \right]^{1/m} \quad (m > 0). \qquad (2.39)$$

By analogy with (2.35) and (2.37) the norm (2.38) can be said to define a neighbourhood and a local continuity of order $k = -1$ in h. The norm (2.39) gives rise to a neighbourhood and a local continuity 'in the mean', and it is especially convenient when the integral is interpreted in the Lebesgue sense. In control theory it is used mostly when $m = 2$. Still more general neighbourhoods and continuities can be easily devised in theory and in practice (see for example vol. 2, chapter 7[C 15] and [G 10]), but the scope of their usefulness in control theory has not yet been established.

If an optimization problem possesses the a priori property, then the functional $F(u)$ contained therein must admit some kind of continuity, because otherwise the relation between $F(u)$ and $u(t)$ would appear to an observer as entirely random (cf. §4.1). Concrete optimization problems studied so far seem to indicate that local continuity of an order $k \geqslant -1$ in h meets this practical need.

12.5   The least value of a functional

Up to the present we have discussed mainly local minima and local lower limits of functionals, but in many control engineering problems only the least possible value of the functional is of interest. This least value may be attained on an isolated function or it may coincide either with a local minimum or with a local lower limit, and the latter need not necessarily be strict. To illustrate the necessity of distinguishing between local minima, local lower limits and the least value consider a functional $F(u)$ defined in a field U, the elements of which can be represented by the imbedding process

$$g(t, u(t), \lambda) = 0, \tag{2.40}$$

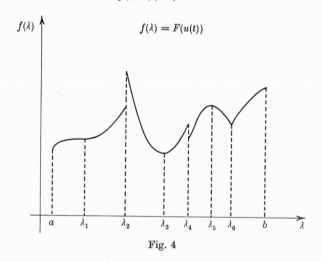

Fig. 4

where λ is a real-valued parameter. Suppose that (2.40) defines a single function $u(t) \in U$ for each value of λ in the finite interval $a \leqslant \lambda \leqslant b$. If instead of $u(t)$ the parameter λ is considered as an independent variable of $F(u)$, then this functional can be interpreted as a point-function of λ, say $f(\lambda)$. Suppose that $f(\lambda)$ is a piecewise continuous function of the form shown in Fig. 4. For brevity let us designate by lower limit of $f(\lambda)$ a non-trivial lower limit, i.e. a lower limit which does not coincide with a minimum of $f(\lambda)$. The function $f(\lambda)$ is seen to possess local minima for $\lambda = \lambda_3$ and $\lambda = \lambda_6$, and local lower limits for $\lambda = a$ and $\lambda = \lambda_4$. The least value of $f(\lambda)$, and thus also of $F(u)$, is a local minimum, and it occurs for $\lambda = \lambda_3$. The largest value of $F(u)$ is a local upper limit, and it occurs for $\lambda = \lambda_2^+$. $f(\lambda)$ admits a local maximum for $\lambda = \lambda_5$ and local upper limits for $\lambda = \lambda_2^-$, $\lambda = \lambda_4$, and $\lambda = b$. The point $\lambda = \lambda_1$ is an inflection point of $f(\lambda)$.

From an examination of Fig. 4 it is obvious that the least value of a functional $F(u)$ is a meaningful notion even if $F(u)$ lacks overall continuity in its field of definition U. We have deliberately avoided the use of the term *absolute minimum* as a designation of the least value, because the term absolute minimum is frequently used as a synonym for a local lower limit (see for example [T 6] and [K 23]).

§ 13. STUDY OF FUNCTIONALS BY DIRECT METHODS

13.1   Functionals defined on a discrete set

Consider the single-variable extremal problem

$$I(x_0, x_1) = \min_{y \in Y} I(y) = \min_{y \in Y} \int_{x_0}^{x_1} F(x, y(x), y(\dot{x}))\, dx$$

$$(y(x_0) = y_0, \quad y(x_1) = y_1), \quad (2.41)$$

where the sense of the integral has been defined, F is a single-valued integrable point-function of three independent variables, and Y is a set of piecewise continuous and piecewise differentiable functions $y(x)$, defined and single-valued in the finite interval $x_0 \leqslant x \leqslant x_1$.

Let us suppose first that the admissible set Y is discrete and that it can be represented by means of a sequence $y_i(x)$, $i = 1, 2, \dots$. Similarly to the point-function problem (2.4), the solution of (2.41) depends both on the nature of the point-function F and on the structure of the sequence $y_i(x)$. The problem (2.41) will admit a meaningful solution only if F and $y_i(x)$ are such that the numerical sequence $I(y_i)$ shows a regular pattern starting by some sufficiently large value of i. If, for instance, this sequence is bounded from below by a constant M_1, then it admits a lower limit $I(\bar{y})$ and a largest lower bound M. The least value of (2.41) is $I(x_0, x_1) = M$, and it occurs either for some finite value of i or as $i \to \infty$. In the latter case the solution $I(x_0, x_1) = I(\bar{y})$ can be either strict or not, depending on whether $\bar{y}(x) = \lim_{i \to \infty} y_i(x)$ is an element of the admissible set Y or not. The closeness of two admissible functions $y_i(x), y_j(x)$ is then defined by the closeness of their indices i, j. In some cases it may even happen that the convergence of the sequence $I(y_i)$ does not occur simultaneously with the convergence, in the ordinary sense, of the sequence $y_i(x)$, that is $\bar{y}(x)$ may be a generalized function such as, for instance, a distribution[S 1],[S 7]. The admissible sub-optimal curves $y_j(x), j \gg 1$, are then not defined without an interpretation algorithm, which permits to select from the sequence $y_i(x)$ a suitable subsequence $y_j(x)$. If (2.41) is a correctly set problem in the sense of Hadamard, then the formulation of an interpretation algorithm is always possible, either by redefining Y on physical grounds, or by conserving Y and changing the definition of convergence on mathematical grounds. Both procedures amount essentially to an improvement of the nominal optimization model (2.41) (cf. § 2). For a discrete admissible set Y the optimization problem (2.41) is therefore not essentially different from the point-function problem (2.4). Hence, if a correctly set problem of form (2.1) can be reduced to the form (2.41) with a discrete set Y, then its solution presents no longer any fundamental obstacles. Such an objective is accomplished by the use of direct methods.

13.2 General properties of functionals defined on a non-discrete set

Consider again (2.41) and suppose that the set Y is non-enumerable, i.e. that all the admissible piecewise-differentiable functions $y(x) \in Y$ cannot be represented by means of one sequence $y_i(x)$, $i = 1, 2, \dots$. Let Y and F be such that the functional $I(y)$ is single-valued for all $y \in Y$. We have seen in § 12 that if the problem (2.41) is to admit a solution within the framework of a limited theory of functionals, then some additional restrictions must be imposed on F and Y.

Except for special cases, the least restrictive conditions of the existence of a solution of (2.41) appears to be the local boundedness of $I(y)$ from below, i.e. the boundedness of $I(y)$ from below when the comparison functions $y(x)$ are located in a sufficiently small neighbourhood Y_1 of a fixed function $y_0(x)$. This requirement is met, for instance, at the abscissae a, λ_3, λ_4 and λ_6 of Fig. 4. With a suitable definition of convergence and of the norm N, $I(y)$ may turn out to be semi-continuous from below at $y = y_0$ (cf. § 12.3). Then according to (2.33) $I(y)$ will admit a local lower limit in a strict sense in Y. If this local lower limit is the only one possible in Y, then it will be the required solution of (2.41); if not, then it is necessary to find all other local lower limits of $I(y)$, and afterwards to select the least value of $I(y), y \in Y$, by means of a comparison algorithm.

In some cases the local or even the 'global' semi-continuity of $I(y)$ can be deduced from the properties of the function $F(x, y, \dot{y})$ appearing in (2.41). Let the norm (2.35) define a zero-order neighbourhood in h for an element $y \in Y$ and let F be continuous with respect to its three arguments and differentiable twice with respect to $\dot{y}$. Suppose that the integral in (2.41) is a line integral and that C designates an admissible curve $y = y(x)$ joining the prescribed points (x_0, y_0) and (x_1, y_1). According to Hilbert and Tonelli[T6]:

(2.42): $I(C)$ is said to be regular negative (positive) if

$$F_{\dot{y}\dot{y}} < 0 \ (>0) \text{ on } C,$$

(2.43): $I(C)$ is said to be quasi-regular negative (positive) if

$$F_{\dot{y}\dot{y}} \leqslant 0 \ (\geqslant 0) \text{ on } C.$$

Tonelli[T6] has established numerous results concerning regular and quasi-regular functionals $I(C)$. Two of these results are particularly significant for control theory:

(2.44): A necessary condition that $I(C)$ be semi-continuous from below at $C = C_0$ is that $I(C)$ be at least quasi-regular positive in a zero-order neighbourhood in h of $C = C_0$.

(2.45): Suppose that $I(C)$ is quasi-regular positive for $C = C_0$ and let $\alpha > 0$, $\beta > 0$ and r real be three constants such that

$$F(x, y, \dot{y}) \geqslant \beta |\dot{y}|^{1+\alpha} + r, \quad |\dot{y}| < \infty,$$

in a sufficiently small zero-order neighbourhood in h of C_0. Then in this neighbourhood there exists a curve $C = \bar{C}$ on which $I(C)$ attains its lower limit.

The proof of theorem (2.45) has been slightly generalized by Ewing and Morse [E3]. It should be stressed, however, that some optimization problems of form (2.41) admit meaningful solutions even if the conditions of theorem (2.45) are not satisfied.

13.3   Continuous functionals

Let us now examine some simple examples of continuous functionals. Consider the problem (2.41), where the line integral

$$I(y) = \int_{x_0}^{x_1} F(x, y(x))\, dx \tag{2.46}$$

coincides with the functional (2.36) discussed in example 1 of §12.4. Since F does not contain $\dot{y}(x)$, $F_{\dot{y}\dot{y}} \equiv 0$ and $I(y)$ is quasi-regular (both positive and negative) for all curves joining the points (x_0, y_0) and (x_1, y_1). If there exists a real constant r such that $F(x, y) \geqslant r$ for some admissible curve C, and this curve admits a zero-order neighbourhood in h contained in Y, then according to theorem (2.45), $I(y)$ admits at least one lower limit $I(C_0)$. Since under the above conditions $I(C)$ is semi-continuous from above and from below, it is continuous in the sense of definition (2.25).

According to Gateaux's theorem the problem (2.41), where $I(y)$ is given by (2.46), is therefore not essentially different from a point-function problem.

Tonelli has established the rather obvious but previously unknown fact that the most general function of form (1.73), which is semi-continuous both from below and from above, must necessarily have the form

$$I(y) = \int_{x_0}^{x_1} [M(x,y) + \dot{y}N(x,y)] \, dx. \tag{2.47}$$

If the above integral is a line integral and the functions $M(x,y)$ and $N(x,y)$ are such that $F = M + \dot{y}N$ satisfies theorem (2.45) on some admissible curves having a zero-order neighbourhood in h, then $I(y)$ admits at least one lower limit in this neighbourhood. The problem (2.41), where $I(y)$ is given by (2.47), is also not essentially different from a point-function problem.

The lower limit of (2.47), and also of (2.46), because the latter is a particular case of the former, can be calculated by means of successive approximations. If the successive approximations are used to construct a minimal sequence, then the problem (2.41), where $I(y)$ is given by (2.47), reduces to the determination of the least value of $I(y)$ defined on a discrete set. A degenerated case occurs when $M + \dot{y}N$ is a total differential, i.e. $M \, dx + N \, dy = dG$, because $I(y) = G(x_1, y_1) - G(x_0, y_0)$, and the value of $I(y)$ is independent of the shape of the curve $y = y(x)$ joining the points (x_0, y_0), (x_1, y_1), $G(x, y)$ being singled-valued.

If M and N are differentiable functions, then guided by the reasoning of Tonelli[T6], the extremal values of the line integral (2.47) can be determined by successive applications of Green's theorem. In fact let G be a finite closed domain in the x, y-plane and let C_1 and C_2 be two curves in G joining the points (x_0, y_0), and (x_1, y_1). Suppose for simplicity that the curves C_1 and C_2 are non-intersecting except at the limit points (x_0, y_0) and (x_1, y_1). Evaluating (2.47) on C_1 and C_2, and considering $I(C_1)$ as a reference value, gives

$$\Delta I(C_1) = I(C_1) - I(C_2) = \oint_{C_1 - C_2} [M(x,y) \, dx + N(x,y) \, dy]$$

$$= \iint_{G_1} \left(\frac{\partial N}{\partial x} - \frac{\partial M}{\partial y} \right) dx \, dy, \tag{2.48}$$

where $G_1 \subset G$ is the area enclosed by the curves C_1 and C_2. The sign of the increment ΔI, which depends on the sign of the point-function

$$v(x, y) = \frac{\partial N}{\partial x} - \frac{\partial M}{\partial y}, \tag{2.49}$$

can be used to determine the shape of C_2 so that $I(C_2) < I(C_1)$. Taking $I(C_2)$ as a reference and repeating the process, a curve C_3, located inside G, can be determined so that $I(C_3) < I(C_2)$. Continuing this process a curve C_0 will be found on which $I(y)$ attains its lower limit. Since the domain G is finite and closed, the limit curve C_0 exists and is the limit of the minimal sequence of curves C_i. The limit curve C_0 consists generally of pieces of the boundary of G, pieces of the curve $v(x, y) = 0$, if the latter is located partly inside G, and eventually of two vertical segments which join the points (x_0, y_0), (x_1, y_1) either to the boundary of G or to the curve $v(x, y) = 0$. When (x_0, y_0) and (x_1, y_1) are located on the boundary of G, the details of the above method have been worked out by Miele[L 8].

Example. Consider the extremal problem

$$\min_{y \in Y} I(y) = \min_{y \in Y} \int_0^1 [-(x+y)^2 + \dot{y} \sin y]\,dx \quad (y(0) = 0, y(1) = 0), \quad (2.50)$$

where the integral is a line integral and G is the rectangle $0 \leqslant x \leqslant 1$, $-1 \leqslant y \leqslant 0$. Y is the set of continuous piecewise differentiable curves located inside G. Since $v(x, y) = -2(x+y)$, a minimal sequence for (2.50) is

$$y_n(x) = \begin{cases} -x, & 0 \leqslant x \leqslant 1-(1/n) \\ (n+1)(x-1), & 1-1/n \leqslant x \leqslant 1 \end{cases} \quad (n = 1, 2, \ldots). \quad (2.51)$$

Substituting (2.51) into (2.50) yields

$$I(\bar{y}) = \lim_{n \to \infty} I(y_n) = 0,$$

where

$$\bar{y}(x) = \lim_{n \to \infty} y_n(x)$$

is a 'discontinuous' curve consisting of the straight line $y(x) = -x$, $0 \leqslant x \leqslant 1$, and of the vertical segment joining the points $(1, -1)$, $(1, 0)$. Since $\bar{y}(x)$ is not an element of Y, the problem has only suboptimal solutions, some of which are given by the elements of the sequence (2.51).

13.4   Functionals continuous of order $k \neq 0$ in h

The functional (2.47) is highly exceptional because it is the only functional of form (1.73) which, with moderate restrictions on the form of M and N, is known to be continuous of order zero in H for all curves C of its field of definition Y. Other functionals of form (1.73) are either discontinuous with the norm (2.34), or continuous only of some order $k \neq 0$ in h. For purposes of control engineering the knowledge of the type of continuity of a functional is extremely important, because the type of continuity characterizes the type of perturbations of $y(x)$ which will produce a small change in the value of the functional $I(y)$. In other words, the nature of the continuity properties of the functional determines the sensitivity properties of the corresponding optimization model, and hence the practical usefulness of the latter. As an illustration of various types of continuity let us examine a few elementary examples.

As a first example consider the classical problem of determining the minimum distance between two fixed points in the Euclidean plane, which, without loss of generality, can be written in the form

$$\min_{y \in Y} I(y) = \min_{y \in Y} \int_0^{x_0} \sqrt{[1 + \dot{y}^2(x)]}\, dx \quad (y(0) = 0, \quad y(x_0) = 0), \quad (2.52)$$

where the integral is a line integral and Y is the set of continuous piece-wise differentiable functions $y(x)$. Since $F = \sqrt{[1 + \dot{y}^2]} > 0$ and $F_{\dot{y}\dot{y}} > 0$ for all $y \in Y$, the functional $I(y)$ in (2.52) is semi-continuous from below. Since, furthermore, F satisfies the conditions of theorem (2.45), the problem (2.52) admits a lower limit in the strict sense. This lower limit can be determined by means of any properly chosen minimal sequence, and if $x_0 = 1$, for instance, by either

$$y_n(x) = \frac{1}{n} x(x-1) \quad \text{or} \quad y_n(x) = -\frac{1}{n} x(x-1) \quad (n = 1, 2, \ldots). \quad (2.53)$$

In fact, substituting (2.53) into $I(y)$, yields

$$I(y_n) = \int_0^1 \sqrt{[1 + \dot{y}_n^2(x)]}\, dx = \frac{2}{n} \int_0^1 \sqrt{\left[x^2 - x + \frac{n^2 + 1}{4} \right]}\, dx$$

$$= \frac{1}{2} \sqrt{\frac{n^2 + 1}{n^2}} + \tfrac{1}{4} n \log \frac{\sqrt{[n^2 + 1]} + 1}{\sqrt{[n^2 + 1]} - 1}, \quad (2.54)$$

and $\lim\limits_{n \to \infty} I(y_n) = 1$. Because $I(y_n)$ is a numerical sequence and the limit function

$$\bar{y}(x) = \lim_{n \to \infty} y_n(x) \equiv 0$$

possesses a bilateral neighbourhood, the lower limit $I(\bar{y})$ is a strict minimum (cf. §10.3), i.e.

$$I(y_n) - I(\bar{y}) > 0 \quad \text{for all } y_n \text{ in (2.53).} \quad (2.55)$$

Let us now prove that the functional $I(y)$ in (2.52) is not continuous of order zero in h. In fact, consider the following classical geometrical argument, which shows that the closeness of two admissible functions $y_1(x)$ and $y_2(x)$ does not insure the closeness of $I(y_1)$ and $I(y_2)$. Consider the isosceles triangle with a unity base and two sides of length m, $2m > 1$, shown in Fig. 5. If the height of this triangle is halved and the sides of length m are folded by moving the summit A_0 to the point B_0 of abscissa $x = \frac{1}{2}$, then the length of the curve $OA_1 B_0 A_1^1 C$ is the same as the length of the curve $OA_0 C$, and the length of the latter is equal to $2m$. Halving the height of the isosceles triangles $OA_1 B_0$ and $B_0 A_1^1 C$ and repeating the folding of the sides of length $\frac{1}{2}m$ yields a sawtooth curve

$$OA_2 B_1 A_{\frac{1}{2}}^1 B_0 A_{\frac{1}{2}}^2 B_1^1 A_{\frac{1}{2}}^3 C$$

of the unchanged length $2m$. Continuing the halving of heights and the folding process n times yields a sawtooth curve $y_n(x) \geqslant 0$ whose maximum value does not exceed $2^{-n}\sqrt{[m^2 - \frac{1}{4}]}$. For n sufficiently large the curve $y_n(x)$ will be contained inside any zero-order neighbourhood in h of the function $\bar{y}(x) \equiv 0, 0 \leqslant x \leqslant 1$, but its length will remain equal to $2m$.

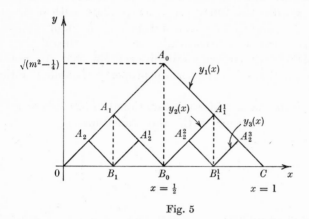

Fig. 5

Consequently the length of two curves located inside a zero-order neighbourhood in h can differ by as much as desired. It follows as a corollary that if a sequence $y_n(x)$ converges to a limit function $\bar{y}(x)$, then the sequence of the lengths of $y_n(x)$ does not necessarily converge to the length of $\bar{y}(x)$.

The reason for the discrepancy between the length of $y(x) \equiv 0$ and the length of the sawtooth curves $y_n(x)$ of Fig. 5 might be attributed to the fact that the latter have a discontinuous slope. This objection is unfounded as the following example shows. Consider the problem (2.52) with $x_0 = \pi$, and the minimal sequence

$$y_n(x) = \frac{\sin nx}{n} \quad (n = 1, 2, \ldots) \tag{2.56}$$

composed of continuously differentiable functions. The length of $y_n(x)$ is given by the elliptic integral

$$I(y_n) = \int_0^\pi \sqrt{[1 + \cos^2 nx]}\, dx = \frac{\sqrt{2}}{n} \int_0^{n\pi} \sqrt{[1 - \tfrac{1}{2}\sin^2 x]}\, dx = \frac{\sqrt{2}}{n} E(n\pi). \tag{2.57}$$

Recalling that $E(m\pi + \phi) = 2mE(\tfrac{1}{2}\pi) + E(\phi)$, where m is an integer and $0 \leqslant |\phi| \leqslant \tfrac{1}{2}\pi$, it follows that

$$\lim_{n \to \infty} I(y_n) = 2\sqrt{2} \int_0^{\frac{1}{2}\pi} \sqrt{[1 - \tfrac{1}{2}\sin^2 x]}\, dx > 3 \cdot 82. \tag{2.58}$$

Furthermore

$$\bar{y}(x) = \lim_{n\to\infty} \frac{\sin nx}{n} \equiv 0, \quad I(\bar{y}) = \pi \quad \text{and} \quad \lim_{n\to\infty} I(y_n) > I(\lim_{n\to\infty} y_n).$$

It might now be argued that the two minimal sequences $y_n(x)$, Fig. 5, and $y_n(x)$, equation (2.56), are both non-representative for a physical phenomenon, because the functions $y_n(x)$ oscillate with a frequency which increases indefinitely as $n \to \infty$, and such a behaviour is physically impossible. This argument is intrinsically irrelevant in the case of general functionals, although it confirms our observation that the notion of continuity of a functional is an important property of nominal optimization models.

As a related example consider the functional (2.52), where $I(y)$ is a Riemann integral, together with the functions $y(x) = -\delta(x-1)$, $0 < x \leqslant 1$, $y(0) = 0$ and

$$y_\epsilon(x) = \begin{cases} \dfrac{\delta}{\sqrt{\epsilon}}\sqrt{x} & 0 \leqslant x \leqslant \epsilon \\ -\dfrac{\delta}{1-\epsilon}(x-1) & \epsilon \leqslant x \leqslant 1 \end{cases}, \tag{2.59}$$

where δ and ϵ are two sufficiently small positive constants. The functions $y(x)$ and $y_\epsilon(x)$ are clearly located inside a zero-order neighbourhood in h, provided $\delta < h$; and the functions $y_\epsilon(x)$ consist of only two segments, which are even monotonic. The 'length' of the first segment of $y_\epsilon(x)$ is

$$\lim_{\alpha\to 0} \int_\alpha^\epsilon \sqrt{\left[1 + \frac{\delta^2}{4\epsilon}\frac{1}{x}\right]}\, dx = 2\int_0^{\sqrt{\epsilon}} \sqrt{\left[\frac{\delta^2}{4\epsilon} + v^2\right]}\, dv. \tag{2.60}$$

The second integral is obtained from the first by the rather obvious transformation $x = v^2$. Completing the elementary integration of (2.60) and adding the 'length' of the second segment yields

$$I(y_\epsilon) = \sqrt{[\delta^2 + (1-\epsilon)^2]} + \sqrt{[\tfrac{1}{4}\delta^2 + \epsilon^2]} + \frac{\delta^2}{4\epsilon}\log\left[\frac{2}{\delta}(\epsilon + \sqrt{[\tfrac{1}{4}\delta^2 + \epsilon^2]})\right]. \tag{2.61}$$

No matter how small $\delta > 0$ is chosen, the expression

$$\lim_{\epsilon\to 0} I(y_\epsilon) = \sqrt{[\delta^2 + 1]} + \delta,$$

derived from (2.61), does not coincide with the expression

$$I(\lim_{\epsilon\to 0} y_\epsilon) = I(y) = \sqrt{[\delta^2 + 1]}.$$

The preceding results seem to be inconsistent with the intuitive physical idea of length. This inconsistency is a consequence of the difference of the respective definitions. Contrary to the mathematical definition (2.52), the physical definition of length does not involve any limiting processes performed on infinitesimal quantities, such as those occurring in derivatives and integrals. The significance of the difference between the mathematical and physical definitions of length, the latter describing essentially the outcome of an experiment with 'standard' meters or 'standard' yards, was already recognized by Lebesgue[L5], who has shown that the mathematical length of a curve $y = y(x)$ is a continuous functional of order $k = 1$ in h. The three preceding examples, $y_n(x)$ of Fig. 5, $y_n(x)$ of (2.56) and the solution of (2.59), show merely that, except for isolated families of functions, the continuity of the functional $I(y)$ in (2.52) is not of an order less than $k = 1$.

The minimal solution $\bar{y}(x) \equiv 0$, $I(\bar{y}) = x_0$ of (2.52) is strict and unique. A further property of this solution is that the value of $I(y)$ cannot be made smaller than $I(\bar{y})$ by the use of any admissible comparison functions $y(x)$, located in a $k = 0$ order neighbourhood in h of $\bar{y}(x)$. Such an *unimprovable* minimum is called a *strong* minimum of $I(y)$. In the case of the problem (2.52) the strong minimum coincides with the least value of $I(y)$. In the general case a unique and strict minimum of a functional is of course not necessarily unimprovable. An improvable minimum is called a *weak* minimum. A weak minimum can therefore not coincide with the least value of a functional. Let us illustrate this possibility by means of an example.

Consider the problem

$$\min_{y \in Y} I(y) = \min_{y \in Y} \int_0^1 \dot{y}^3(x)\, dx \quad (y(0) = 0,\ y(1) = 1), \tag{2.62}$$

where Y is the set of continuous piecewise differentiable functions and $I(y)$ is a line integral. It can be easily verified that in a sufficiently small neighbourhood of the curve $y(x) = x$ the functional $I(y)$ is continuous of order $k = 1$ in h. Since the function $F(x, y, \dot{y}) = \dot{y}^3$ satisfies theorem (2.45) for $y(x) = x$, and $F_{\dot{y}\dot{y}} = 6\dot{y} > 0$, $I(y)$ admits a strict local lower limit in a sufficiently small neighbourhood of $y(x) = x$. It can be easily verified that this local lower limit is a local minimum and that the corresponding minimal solution $\bar{y}(x) = x$, $I(\bar{y}) = 1$ is strict and unique. The value $I(\bar{y}) = 1$ is however not unimprovable.

Let $\delta > 0$ be sufficiently small and $0 < \epsilon < \frac{1}{2}$, then the function

$$y_\epsilon(x) = \begin{cases} (1+2\delta)x, & 0 \leqslant x \leqslant \frac{1}{2} - \epsilon \\ \dfrac{(1+2\delta)\epsilon - \delta}{\epsilon}(x - \frac{1}{2}) + \frac{1}{2}, & \frac{1}{2} - \epsilon \leqslant x \leqslant \frac{1}{2} + \epsilon \\ (1+2\delta)x - 2\delta, & \frac{1}{2} + \epsilon \leqslant x \leqslant 1 \end{cases} \epsilon Y, \tag{2.63}$$

and it is clearly located in a zero-order neighbourhood of the function $\bar{y}(x) = x$. Substituting (2.63) into the functional $I(y)$ yields

$$I(y_\epsilon) = (1+2\delta)^2 + \frac{2\delta^2(1+2\delta)}{\epsilon^2}\left(\epsilon - \frac{\delta}{1+2\delta}\right), \tag{2.64}$$

which for any $\delta > 0$ can be rendered less than $I(\bar{y})$ by a suitable choice
of ϵ. In fact, $I(y_\epsilon)$ can be rendered negative and as small as desired.
Hence, $I(y)$ in (2.62) does not have a finite least value. Since with a suit-
able choice of ϵ, $I(y_\epsilon) < I(\bar{y})$, the minimal solution $\bar{y}(x) = x$, $I(\bar{y}) = 1$ constitutes a weak minimum, which is present only when the comparison
functions $y(x) \in Y$ are restricted to a small neighbourhood of order $k = 1$
in h. Weak minima are therefore intimately related to the order of
continuity in h of the corresponding functional. A weak minimum may
occur with respect to any order k of the admissible neighbourhood in h.

As an illustrative example consider the problem

$$\min_{y \in Y} I(y) = \min_{y \in Y} \int_{x_0}^{x_1} [(x-a)^2 \dot{y}^2(x) + (x-a)\dot{y}^3(x)] \, dx$$
$$(y(x_0) = 0, \quad y(x_1) = 0), \quad (2.65)$$

where $x_0 < a < x_1$, $I(y)$ is a line integral, and Y is the set of continuous
piecewise differential functions. Hadamard has shown that the problem
(2.65) has no solution unless Y contains a subset Y_1, the elements of which
admit second-order derivatives[H 1]. For $y(x) \in Y_1$ a minimal solution of
(2.65) is given by $\bar{y}(x) \equiv 0$, $I(\bar{y}) = 0$. This minimal solution is a weak
minimum with respect to comparison functions $y(x)$ located in a neighbourhood of order $k = 2$ in h, with $h < \frac{1}{3}$. To show that the minimum
$I(\bar{y}) = 0$ is a weak one, it is sufficient to consider the continuous piecewise
differentiable function

$$y_\epsilon(x) = \begin{cases} 0, & x_0 \leqslant x \leqslant a-\epsilon \\ \dfrac{\delta}{\epsilon}(x-a+\epsilon), & a-\epsilon \leqslant x \leqslant a \\ -\dfrac{\delta}{\epsilon}(x-a-\epsilon), & a \leqslant x \leqslant a+\epsilon \\ 0, & a+\epsilon \leqslant x \leqslant x_1 \end{cases} \quad (\delta > 0), \quad (2.66)$$

for which
$$I(y_\epsilon) = \delta^2 \left(\frac{\epsilon}{3} - \frac{\delta}{2\epsilon}\right). \quad (2.67)$$

For any fixed $\delta > 0$ it is possible to choose a value of $\epsilon > 0$ such that

$$I(y_\epsilon) < I(\bar{y}) = 0.$$

Consider (2.41) with $F \geqslant 0$ and $F_{\dot{y}\dot{y}} \geqslant 0$. It is obvious that a weak
minimum of $I(y)$ does not necessarily exist under these conditions. If the
integral in (2.41) is a line integral and the conditions of theorem (2.45)
are satisfied, then only the existence of a strict lower limit is assured.
Theorem (2.45) does not apply when the integral in (2.41) is *not* a line integral. To illustrate such a case consider the classical Weierstrass example

$$\min_{y \in Y} I(y) = \min_{y \in Y} \int_{-1}^{+1} x^2 \dot{y}^2(x) \, dx \quad (y(-1) = -1, \quad y(1) = 1), \quad (2.68)$$

where Y is the set of continuous piecewise differentiable functions and the integral is a Riemann integral. Since $F(x, y, \dot{y}) = x^2 \dot{y}^2 \geqslant 0$ in $-1 \leqslant x \leqslant 1$, the functional $I(y)$ is bounded from below. Its lower limit can be found, for example, by means of the minimal sequence

$$y_n(x) = \begin{cases} -1, & -1 \leqslant x \leqslant -\dfrac{1}{n} \\[2mm] nx, & -\dfrac{1}{n} \leqslant x \leqslant \dfrac{1}{n} \\[2mm] +1, & \dfrac{1}{n} \leqslant x \leqslant 1 \end{cases} \quad (n = 2, 3, \ldots), \qquad (2.69)$$

which yields
$$I(y_n) = \frac{2}{3n}.$$

Letting $n \to \infty$ gives $I(\bar{y}) = 0$ and

$$\bar{y}(x) = \lim_{n \to \infty} y_n(x) = \begin{cases} 1, & x > 0 \\ -1, & x < 0 \end{cases}. \qquad (2.70)$$

The function $\bar{y}(x)$ is not an element of the set Y, because it is only piecewise continuous. The problem (2.68) admits therefore only suboptimal solutions, some of which are elements of the minimal sequence (2.69). The minimal sequence is of course not unique. Another example is

$$y_n(x) = \begin{cases} -1, & -1 \leqslant x \leqslant 1/n \\[2mm] \dfrac{n+1}{n-1} - \dfrac{2}{n-1}\dfrac{1}{x}, & 1/n \leqslant x \leqslant 1 \end{cases} \quad (n = 2, 3, \ldots), \qquad (2.71)$$

for which $I(y_n) = 4/(n-1)$. The limit function (2.70) is obviously independent of the choice of the minimal sequence $y_n(x)$.

The set Y in (2.68) does not constitute the total field of definition of the corresponding functional $I(y)$, and this is the principal reason why a strict minimal solution fails to exist. In fact, $I(y)$ is meaningful on the more general set $\bar{Y}$ of piecewise continuous and piecewise differentiable functions, and the discontinuous solution (2.70) can be interpreted as a strict minimum in $\bar{Y}$. The comparison functions $y(x)$, and in particular the elements of the minimal sequences (2.69) and (2.71) are located in a neighbourhood of order $k = -1$ in h of $y = \bar{y}$. The minimum (2.70) is a strong one.

On the basis of the reasoning used by Hilbert in his treatment of the Dirichlet principle, which is merely illustrated by the one-dimensional problem (2.68), it is possible to conjecture that any *interior* local lower limit of a functional $I(y)$, defined in a field Y, can be interpreted as a local minimum of the same functional, defined in a more general field $\bar{Y}$, $Y \subset \bar{Y}$. Such a result would be very attractive because it would eliminate the necessity of distinguishing between minima and interior lower limits. Unfortunately, the above conjecture still lacks a general proof, but there exists a fragmentary result pointing in the right direction.

Consider a minimal problem defined by the Lebesgue integral

$$I(y) = \int_0^1 F(x, y, \dot{y})\, dx \quad (y(0) = 0, \quad y(1) = 1), \tag{2.72}$$

where $F(x, y, \dot{y})$ is a continuous point-function of its three arguments, and the variables x and y are limited to the rectangle $G: 0 \leqslant x \leqslant 1$, $0 \leqslant y \leqslant 1$. Let $y = \phi(x)$ and $\dot{y} = \psi(x)$ be two independent functions. Lavrentiev[L 3] has shown that for a suitable choice of the measurable functions $\phi(x)$ and $\psi(x)$ the functional

$$\int_0^1 F(x, \phi(x), \psi(x))\, dx \quad (\phi(0) = 0, \quad \phi(1) = 1, \quad (x, y) \in G)$$

attains always its lower limit, but that in general $\psi(x)$ is *not* a derivative of $\phi(x)$. Hence, if the functional (2.72) is to possess a strict lower limit, it is first necessary to generalize the field of $I(y)$ so that $\psi(x)$ can be interpreted as a generalized derivative of $\phi(x)$. This can be sometimes accomplished by means of the theory of distributions (see for example [S 6],[S 7]). It is then necessary to generalize the field of (2.72) some more, so that the resulting lower limit coincides with a local minimum.

It should be stressed at this point that the lower limit of a functional $I(y)$ is a quantity which depends on the nature of the admissible field Y. This property, which holds of course also for the minima of $I(y)$, was already known to Hilbert (see for example [H 8]), but until 1926 no illustrative examples of type (2.41) were available. The first example was given by Lavrentiev[L 3], who has constructed a functional $I(y)$ of type (2.72) such that
$$\liminf_{y \in Y} I(y) = 1 \quad \text{and} \quad \liminf_{y \in \overline{Y}} I(y) = 0,$$

where Y is the set of continuously differentiable functions and $\overline{Y}$ is the set of absolutely continuous functions.

The above argument in general, and Lavrentiev's example in particular, have a direct bearing on the optimization of practical control systems. Because several methods based on the so-called 'sufficient condition for an absolute minimum' are based on the hypothesis that the lower limit of a functional $I(y)$ is independent of its field Y, these methods must be considered as basically unsound (see for example [K 24], [B 20], [Z 3], [G 23], and in particular lemma I of [K 23].

§ 14. INDIRECT STUDY OF FUNCTIONALS BY MEANS OF EULER'S EQUATIONS

14.1 Differentials and derivatives of functionals

When an extremal problem of the form

$$I(x_0, x_1) = \min_{y \in Y} I(y) = \min_{y \in Y} \int_{x_0}^{x_1} F(x, y(x), \dot{y}(x))\, dx \quad (y(x_0) = y_0,\ y(x_1) = y_1)$$

$$(2.41)$$

is studied by means of a direct method, the admissible elements $y(x) \in Y$ are first subjected to an ordering process, which consists, for example, in selecting a subset $Y_1 \subset Y$ and arranging certain elements of Y_1 into a minimal sequence $y_i(x)$, $i = 1, 2, \ldots$. A direct method is only successful when the subset Y_1 constitutes an admissible local neighbourhood of the solution of (2.41). The selection and ordering of Y_1 can be carried out in many different ways and the details of this procedure constitute a 'method' of solving (2.41). In some cases it is convenient to choose as Y_1 a parametric imbedding of the form (2.40), because the problem (2.41) can then be interpreted as an equivalent point-function problem (cf. § 12.5).

To illustrate the use of a parametric imbedding consider a particularly simple case. Suppose that the function $F(x, y, \dot{y})$ in (2.41) is twice differentiable with respect to its three arguments and that the solution $y = \bar{y}(x)$ of (2.41) is a local minimum which admits a neighbourhood Y_1 of order $k \geqslant 0$ in h. Suppose, furthermore, that in this neighbourhood there exists a linear parametric family

$$\left. \begin{array}{l} y(x) = \bar{y}(x) + \lambda \eta(x), \quad \eta(x_0) = 0, \quad \eta(x_1) = 0, \\[2mm] |\eta(x)| < h, \quad |\dot{\eta}(x)| < h, \quad \ldots, \quad |\overset{(k)}{\eta}(x)| < h, \end{array} \right\}$$

$$(2.73)$$

where λ is a 'small' real parameter. Substituting (2.73) into the functional $I(y)$ of (2.41) yields a point-function

$$f(\lambda) = \int_{x_0}^{x_1} F(x, \bar{y}(x) + \lambda \eta(x),\ \dot{\bar{y}}(x) + \lambda \dot{\eta}(x))\, dx, \tag{2.74}$$

which in the neighbourhood of the point $\lambda = 0$ has by definition the same behaviour as that of the functional $I(y)$ in the neighbourhood of order $k \geqslant 0$ in h of the function $y = \bar{y}(x)$. If $I(y)$ is such that $f(\lambda)$ is differentiable at $\lambda = 0$, and this differentiation can be carried out *under* the integral sign of (2.74), then a necessary condition for $f(0)$ to be a minimum is that

$$f'(0) = \int_{x_0}^{x_1} \left(\frac{\partial F}{\partial y}\, \eta(x) + \frac{\partial F}{\partial \dot{y}}\, \dot{\eta}(x) \right)_{\lambda = 0} dx = \int_{x_0}^{x_1} \left(\frac{\partial F}{\partial y} - \frac{d}{dx}\, \frac{\partial F}{\partial \dot{y}} \right) \eta(x)\, dx = 0$$

$$(2.75)$$

for all $\eta(x)$ of the linear parametric family (2.73). Let us now recall the so-called fundamental lemma of the calculus of variations, which states that if $b(x)$ is a fixed function and $y(x)$ a function belonging to an infinite set E, then

$$b(x) \equiv 0 \quad \text{if} \quad \int_{x_0}^{x_1} b(x) \cdot y(x) \, dx = 0 \quad \text{for all} \quad y(x) \in E. \qquad (2.76)$$

From (2.76) and (2.75) it follows that the parentheses in the second integral of (2.75) vanish identically, or in other words that $\bar{y}(x)$ is a particular solution of the ordinary differential equation

$$\frac{\partial F}{\partial y} - \frac{d}{dx} \frac{\partial F}{\partial \dot{y}} = 0. \qquad (2.77)$$

Equation (2.77) is called the Euler or the Euler–Lagrange differential equation corresponding to the extremal problem (2.41). The solutions of (2.77) are called the extremals of $I(y)$. From a geometrical point of view the left-hand side of (2.77) can be interpreted as a gradient of $I(y)$ in the functional space Y_1.

The stationarity conditions (2.75) and (2.77) of $I(y)$ can be put in a form which is analogous to that for a point-function problem. In fact, letting

$$\lambda \eta(x) = \delta y, \quad \lambda f'(0) = \delta I, \quad \frac{\partial F}{\partial y} - \frac{d}{dx} \frac{\partial F}{\partial \dot{y}} = \frac{\delta I}{\delta y}, \qquad (2.78)$$

these conditions can be written in the form

$$\delta I = \int_{x_0}^{x_1} \frac{\delta I}{\delta y} \, \delta y \, dx = 0 \quad \text{and} \quad \frac{\delta I}{\delta y} = 0. \qquad (2.79)$$

The symbols δy, δI and $\delta I / \delta x$ are called the variation of $y(x)$, the first variation of $I(y)$ and the functional derivative of $I(y)$, respectively. Sometimes δI is also called the functional differential of $I(y)$, because formally it can be interpreted as a limit of the total differential

$$dF = \sum_{i=1}^{n} \frac{\partial F}{\partial y_i} \, dy_i$$

as n, the number of independent variables y_i, becomes non-enumerable.

Volterra has shown[V 5] that the functional derivative of a functional is generally a more complicated functional. It is noteworthy that the functional derivative in (2.79) is *not* a more complicated functional but merely an explicit point-function of the variables x, $y(x)$, $\dot{y}(x)$ and $\ddot{y}(x)$. This property is highly exceptional. It is a consequence of the fact that the functional $I(y)$ in (2.41) has a very special form.

Similarly to the notation (2.78) it is possible to introduce a distinctive symbol for relations involving higher-order derivatives of $f(\lambda)$ at $\lambda = 0$. For example

$$\delta^2 I = \tfrac{1}{2}\lambda^2 f''(0) = \tfrac{1}{2}\lambda^2 \int_{x_0}^{x_1} (F_{yy}\eta^2(x) + 2F_{y\dot{y}}\eta(x)\dot{\eta}(x) + F_{\dot{y}\dot{y}}\dot{\eta}^2(x))\,dx \quad (2.80)$$

is called the second variation of $I(y)$ at $y = \bar{y}(x)$. Using $\delta I, \delta^2 I, \ldots$ it is possible to formulate various criteria for a minimum analogous to the corresponding point-function criteria.

The expressions (2.75) and (2.80) were obtained on the basis of a neighbourhood of order $k \geqslant 0$ in h, defined by the parametric imbedding (2.73). These expressions are therefore inherently related to the norm function

$$N(\bar{y}, y) = |\bar{y}(x) - y(x)|.$$

Similarly to the continuity of a functional, the notions 'functional differential' and 'functional derivative' can also be related to an arbitrary norm function $N(\bar{y}, y)$, which at the same time fixes the admissible neighbourhood of the minimal solution $y = \bar{y}(x)$. This can be done in a variety of ways. For example, δI and $\delta^2 I$ can be defined as solutions of the following two implicit expressions

$$\left. \begin{array}{l} \displaystyle\lim_{\epsilon \to 0} \frac{I(\bar{y} + \delta y) - I(\bar{y}) - \delta I}{N(\bar{y}, \bar{y} + \delta y)} = 0, \\[3mm] \displaystyle\lim_{\epsilon \to 0} \frac{I(\bar{y} + \delta y) - I(\bar{y}) - \delta I - \tfrac{1}{2}\delta^2 I}{[N(\bar{y}, \bar{y} + \delta y)]^2} = 0, \end{array} \right\} \quad (2.81)$$

where $\epsilon = N(0, \delta y)$, and in such a case they are called differentials in the sense of Frechet. If, on the contrary, ϵ is a parameter and δI and $\delta^2 I$ are solutions of the two equations

$$\left. \begin{array}{l} \displaystyle\lim_{\epsilon \to 0} N\left(\frac{I(\bar{y} + \epsilon\,\delta y) - I(\bar{y})}{\epsilon}, \delta I \right) = 0, \\[3mm] \displaystyle\lim_{\epsilon \to 0} N\left(\frac{I(\bar{y} + \epsilon\,\delta y) - I(\bar{y}) - \delta I}{\epsilon^2}, \tfrac{1}{2}\delta^2 I \right) = 0, \end{array} \right\} \quad (2.82)$$

then they are called differentials in the sense of Gateaux. If N defines a neighbourhood of order $k \geqslant 0$ in h, then (2.82) and (2.81) coincide with the explicit definitions of δI and $\delta^2 I$ given in (2.78) and (2.80). The examples (2.81) and (2.82) do not exhaust all possibilities; indeed they constitute merely particular applications of the notion of functional differentiation introduced by Volterra [V4],[V5].

Consider a functional $I(y)$, $y(x) \in Y$, not necessarily of form (2.41), where the $y(x)$ are defined in an interval $x_0 \leqslant x \leqslant x_1$. Let δy be an absolutely integrable function which differs from $y(x) \equiv 0$ only in the interval $\xi - \epsilon \leqslant \xi \leqslant \xi + \epsilon$, $x_0 \leqslant \xi - \epsilon \leqslant \xi \leqslant \xi + \epsilon \leqslant x_1$, $\epsilon > 0$. If

$$\left. \begin{aligned} \lim_{\epsilon \to 0} \sigma_\epsilon &= \lim_{\epsilon \to 0} \int_{\xi - \epsilon}^{\xi + \epsilon} |\delta y| \, dx = 0 \\ \lim_{\epsilon \to 0} \frac{I(y + \delta y) - I(y)}{\sigma_\epsilon} &= I'(y(x), \xi) \end{aligned} \right\} \tag{2.83}$$

and

exist, then $I'(y(x), \xi)$ is called a functional derivative of $I(y)$ at the point $x = \xi$[V 5]. The functional derivative of $I(y)$ is thus a functional with respect to the variable $y(x) \in Y$, and a point-function with respect to the variable ξ, $x_0 \leqslant \xi \leqslant x_1$. In degenerated cases $I'(y, \xi)$ may reduce to a point-function.

From a practical point of view the definitions of N, δI, $\delta^2 I$, etc., should be chosen in such a way that there exists a *maximal* amount of qualitative agreement between the properties of the physical system and the corresponding properties of the nominal optimization model. An arbitrary choice of these definitions can lead to unnecessary complications or it can render the nominal optimization model completely unrealistic.

14.2　Equivalence between direct methods and the method based on Euler's equation

From the way in which the relation $\delta I = 0$ was obtained it is quite clear that the vanishing of the functional derivative $\delta I / \delta y$ is not equivalent to the operation $\min_{y \in Y} I(y)$. Equation $\delta I / \delta y = 0$ implies only that $I(y)$ is stationary at $\bar{y} \in Y_1$, i.e. $I(\bar{y})$ is a minimum, a maximum or an inflection point of $I(y)$. Since by definition $\bar{y} \in Y_1$ admits a bilateral neighbourhood, the possibility that $I(\bar{y})$ is a non-trivial local lower limit of $I(y)$ is thus specifically excluded. Minima where $\delta I / \delta y$ does not exist must also be sought by other methods. This situation arises usually when $F(x, y, \dot{y})$ in (2.41) has poor differentiability properties.

Similarly to the case of point-functions (cf. §§ 10.6 and 10.7), local minima of $I(y)$ can be determined from stationary solutions $y_s(x)$, which satisfy the boundary conditions $y(x_0) = y_0$ and $y(x_1) = y_1$, by taking into account the sign of $\delta^2 I$ at $y = y_s$, or the values of δI in a sufficiently small neighbourhood of $y = y_s$. Similarly, to the case of point-functions, the following criterion can be formulated:

A necessary condition that $I(y_s)$ be a local minimum of $I(y)$ is that $\delta I = 0$ at $y = y_s$ and $\delta^2 I \geqslant 0$ for all admissible $y(x)$ which are sufficiently close to $y = y_s$. The local minimum $I(y_s)$ is called strict if $\delta^2 I > 0$ for all admissible $y \neq y_s$. Several other necessary criteria can be easily evolved from $\delta I = 0$ and $\delta^2 I \geqslant 0$. Two examples were given in § 7 in connection with the functional derivative (1.84), and in the case of (2.41), (2.77) they can be written in the form

$$E(x,y_s,\dot{y}_s,p) = F(x,y_s,p) - F(x,y_s,\dot{y}_s) + (\dot{y}_s - p) F_{\dot{y}}(x,y_s,\dot{y}_s) > 0$$
$$(p \neq \dot{y}_s),$$
$$F_{\dot{y}\dot{y}}(x,y_s,\dot{y}_s) > 0, \qquad (2.84)$$

where p is the value of the derivative $\dot{y}(x)$, and $y(x)$ is a function located in a neighbourhood of order $k \geqslant 0$ in h of $y_s(x)$. The inequalities (2.84) are called the Weierstrass criterion and the Legendre strong condition, respectively.

Consider a one-parameter family D of extremals passing through the point $y(x_0) = y_0$ containing in its interior a stationary solution $y_s(x)$ which passes also through the point $y(x_1) = y_1$. Since the solutions of (2.77) form a two-parameter family, this hypothesis does not involve a loss of generality. It may happen that except for (x_0, y_0) the extremal $y_s(x)$ has no point in common with any other extremal $y(x) \in D$ in $x_0 \leqslant x \leqslant x_1$. In such a case it is said that $y_s(x)$ possesses no point conjugate to (x_0, y_0), or more simply that $y_s(x)$ possesses no conjugate point.

If $y = y_s(x)$ is to furnish a local minimum of $I(y)$, then at least one of the inequalities (2.84) must be satisfied. If both are satisfied, $I(y)$ is interpreted as a line integral with $dx > 0$, and the extremal $y_s(x)$ has no conjugate points on the segment joining the points (x_0, y_0) and (x_1, y_1), then the minimum $I(y_s)$ may be a relatively strong one, i.e. the value $I(y_s)$ cannot be improved by any other local minimum of $I(y_\sigma)$, where $y_\sigma \neq y_s$. From the way conditions (2.84) were evolved it is obvious that they contain no information whatever on minima of $I(y)$ where $\delta I/\delta y$ fails to exist, or on non-trivial lower limits of $I(y)$. A relatively strong minimum should therefore not be confused with the least value of $I(y)$, i.e. with a strong minimum of $I(y)$ as defined in §12.4.

If the least value of (2.41) is a unique local minimum, then (2.41) is equivalent to $\delta I = 0$, $\delta^2 I \geqslant 0$, and in the case of non-uniqueness to $\delta I = 0$, $\delta^2 I \geqslant 0$ and an appropriate comparison algorithm. The ties between (2.41) and the two-point boundary-value problem

$$\frac{\delta I}{\delta y} = 0, \quad y(x_0) = y_0, \quad y(x_1) = y_1 \quad (\delta^2 I \geqslant 0, \quad x_0 \leqslant x \leqslant x_1) \quad (2.85)$$

can be strengthened by various generalizations of the notion of derivative, but the equivalence between (2.41) and (2.85) is essentially incomplete. Fortunately, in many cases of practical interest the partial equivalence between (2.41) and (2.85) is sufficient to determine a meaningful solution of (2.41).

Local lower limits of $I(y)$ can sometimes be considered by separating the interval $x_0 \leqslant x \leqslant x_1$ into subintervals, introducing unilateral variations and replacing $\delta I/\delta y = 0$ in (2.85) by $\delta I/\delta y \geqslant 0$. This method was used with success by Todhunter [T 3], and it was later justified by Bliss and Underhill [B 11],[B 12]. With a different notation it was reintroduced by Pontryagin (see for example [P 11]). Examples of the use of $\delta I/\delta y \geqslant 0$ will be given in §15.

Let us now examine the examples of §13.4 from the point of view of the formulation (2.85).

For the problem (2.52) the Euler equations and the extremals are

$$\frac{d}{dx}\dot{y}(x) = 0, \quad y(x) = Cx + C_1.$$

Taking into account the boundary conditions $y(0) = 0$, $y(x_0) = 0$ yields a unique solution $y_s(x) \equiv 0$, which does not possess any conjugate points. The inequalities (2.84) become

$$E(x, y_s, \dot{y}_s, p) = \sqrt{(1+p^2)} - 1 > 0 \quad \text{for } p \neq 0, \quad \text{and} \quad F_{\dot{y}\dot{y}}(x, y_s, \dot{y}_s) = 1.$$

Hence, $y_s(x) \equiv 0$ may be a relatively strong minimum of $I(y)$. The formulation (2.85) has furnished a correct solution of the problem (2.52), but there is no indication that $I(y)$ is continuous of order $k = 1$ in h, or that $I(y_s) = x_0$ is the least value of $I(y)$.

The problem (2.62) has the same Euler equation and the same extremals as the problem (2.52). This common property produces a misleading impression, because the functionals in (2.52) and (2.62) are entirely different. This stationary solution $y_s(x) = x$, satisfying the boundary conditions $y(0) = 0$, $y(1) = 1$, is unique. Since $F_{\dot{y}\dot{y}}(x, y_s, \dot{y}_s) = 1$ and $E(x, y_s, \dot{y}_s, p) = (p-1)(p^2+p-2) > 0$ only for $p > -2$, the local minimum $I(y_s) = 1$ is not relatively strong. The study of $\delta I = 0$ and $\delta^2 I \geqslant 0$ gives no information that $I(y)$ is not bounded from below.

The Euler equation resulting from the problem (2.65) is

$$\frac{d}{dx}[3(x-a)^2\dot{y}^2 + 2(x-a)\dot{y}] = 0. \tag{2.86}$$

Equation (2.86) admits a first integral

$$\dot{y}(x) = -\tfrac{1}{3}(x-a) \pm \sqrt{\left[\tfrac{1}{9}(x-a)^2 - \frac{C}{x-a}\right]},$$

where C is a constant of integration. Since $y(x)$ becomes imaginary in the neighbourhood of the abscissa $x = a$, $x_0 < a < x_1$, a two-parameter family of extremals fails to exist in the whole interval $x_0 \leqslant x \leqslant x_1$. The isolated stationary solution $y_s(x) \equiv 0$ satisfies (2.86) as well as the boundary conditions $y(x_0) = 0$, $y(x_1) = 0$, but the inequalities (2.84) produce a completely erroneous result. In fact, purely formally

$$F_{\dot{y}\dot{y}}(x, y_s, \dot{y}_s) = 2(x-a)^2 \geqslant 0$$

and

$$E(x, y_s, y_s, p) = p^2(x-a)(p+x-a).$$

Since for p fixed $E(x, y_s, \dot{y}_s, p)$ changes sign at the abscissa $x = a$, the impression is produced that $I(y_s)$ may not possess a minimum. This result is due to the fact that in the case of (2.65) the function $f(\lambda)$, defined by (2.74), may *not* be differentiated under the integral sign. The problem (2.65) was constructed by Scheeffer as an objection to the transition from equation (2.74) to equation (2.75), and is sometimes called Scheeffer's objection (see for example [H 1]). The practical significance of (2.65) consists in illustrating the importance of the fact that the form of (2.77) does not depend on the interval of definition $x_0 \leqslant x \leqslant x_1$ of $I(y(x))$, whereas the minimal value of $I(y)$ obviously does. Other illustrative examples will be given in §14.3.

Consider now the problem (2.68), for which the Euler equation and the extremals are

$$\frac{d}{dx}(x^2\dot{y}) = 0, \quad y(x) = \frac{C_1}{x} + C.$$

Since for $C_1 \neq 0$ $y(x)$ is not bounded in the interval $-1 \leqslant x \leqslant +1$, there exists no continuous extremal joining the points $(-1, -1)$ and $(1, 1)$. This negative result was to be expected, because the least value of $I(y)$ in (2.68) is a lower limit and not a minimum. The formulation (2.85) fails again, but for a different reason than in the case of the problem (2.65).

If instead of the neighbourhood of order $k \geqslant 0$ in h a neighbourhood of order $k = -1$ in h is used to derive (2.85), then

$$\frac{\delta I}{\delta y} = \frac{d}{dx}(x^2\dot{y}) = 0 \tag{2.87}$$

appears to be unchanged, but the derivatives in (2.87) must now be interpreted as generalized derivatives. Recalling that in such a case the number of integration constants may be larger than the order of the differential equation (see for example [S 1]), the (generalized) solution of (2.87) becomes

$$y(x) = C_0 + (C_1/x) + C_2 u(x) + C_3 \delta(x), \tag{2.88}$$

where C_0 to C_3 are real constants, $u(x)$ is the unit-step function and $\delta(x)$ is the Dirac unit-impulse. The solution (2.88) can be easily verified by substitution in (2.87). The four integration constants in (2.88) can be determined by the boundary conditions $y(-1) = -1$, $y(1) = 1$ and by the implied condition that the solution of (2.68) should be a bounded point-function. The latter condition yields $C_1 = 0$, $C_3 = 0$ and the former $C_0 = -1$, $C_2 = 2$; hence the unique solution of (2.68) is

$$y(x) = -1 + 2u(x). \tag{2.89}$$

Except for notation, (2.89) coincides with the limiting solution (2.70), determined by means of a minimal sequence. Since ordinary criteria, like for instance (2.84), are not valid for generalized extremals, the nature of the stationary solution (2.89) has to be examined by evaluating δI in the neighbourhood of (2.89). Such an examination shows that $I(y)$ attains on (2.89) a strong minimum.

14.3 Dependence of the minimal properties of a functional on the boundary conditions

In connection with problem (2.65) and the associated Euler equation (2.86) it was stated that the form of extremals $y(x)$ is independent of the interval $x_0 \leqslant x \leqslant x_1$, whereas, on the contrary, the least value of a functional $I(y)$ is strongly affected by both the location and the size of this interval. Let us now examine some simple examples from this point of view.

Consider the problem

$$I(0, 1) = \min_{y \in Y} I(y) = \min_{y \in Y} \int_0^1 \sqrt{[1 + \dot{y}^2(x)]}\, dx \quad (y(0) = 0, \quad \dot{y}(1) = 1),$$

$$(2.90)$$

which differs from the problem (2.52) only by the nature of the boundary conditions. The Euler equation associated with $I(y)$ being unaffected by the boundary conditions, the extremals are still the straight lines $y(x) = Cx + C_1$. The two boundary conditions $y(0) = 0$ and $\dot{y}(1) = 1$ determine the integration constants C and C_1 uniquely, and the stationary solution of (2.90) is

$$y_s(x) = x, \quad I(y_s) = \sqrt{2}.$$

Substituting $y_s(x)$ into (2.84) yields

$$F_{\dot{y}\dot{y}} = \frac{\sqrt{2}}{4} > 0, \quad E = \sqrt{[1 + p^2]} - \frac{\sqrt{2}}{2}(1 + p) > 0 \quad (p \neq 1),$$

and, hence $I(y_s)$ appears to possess all characteristics of a relatively strong minimum. Unfortunately $I(y_s) = \sqrt{2}$ does not coincide with the least value of $I(y)$, $y \in Y$, because $I(y)$ admits also a non-trivial lower limit. Consider in fact the minimal sequence

$$y_n(x) = \begin{cases} 0, & 0 \leqslant x \leqslant 1 - (1/n) \\ x - 1 + (1/n), & 1 - (1/n) \leqslant x \leqslant 1 \end{cases} \quad (n = 2, 3, \ldots) \quad (2.91)$$

every element of which belongs to Y and satisfies the boundary conditions of (2.90). Since $I(y_n) = 1 + [\sqrt{(2)} - 1]/n$, $I(y_n) < I(y_s)$ for $n > 3$. The limit $\bar{y}(x) = \lim_{n \to \infty} y_n(x) \equiv 0$ of (2.91) does not satisfy the boundary condition $\dot{y}(1) = 1$, and because of this (2.90) admits only suboptimal solutions, some of which are given by the elements of the sequence (2.91).

From the study of the above example it is obvious that the notion of a relatively strong minimum, defined by means of the inequalities (2.84), contains a very small amount of information on the minimal properties of $I(y)$. In the practice of control engineering the statement that a local minimum is relatively strong is usually irrelevant, because what is wanted is a 'minimal' value of $I(y)$ which is unimprovable by other admissible functions $y(x)$, and not just by other local minima of $I(y)$. An additional misunderstanding between designers and theoreticians arises from the careless use of terminology, because many authors do not differentiate between a strong minimum and a relatively strong minimum, using the term strong minimum for both notions.

Consider now another modification of the problem (2.52):

$$I(0, x_1) = \min_{y \in Y} I(y) = \min_{y \in Y} \int_0^{x_1} \sqrt{[1 + \dot{y}^2(x)]} \, dx \quad (y(0) = 0, \, y(x_1) = y_1),$$
$$\tag{2.92}$$

where the values of x_1 and y_1 are not given in advance but are specified implicitly by the condition that the point (x_1, y_1) should lie on a pre-scribed continuous piecewise differentiable curve $g(x, y) = 0$.

In order to deal with floating boundary conditions $y(x_0) = y_0, y(x_1) = y_1$ in (2.85), it is necessary to drop the conditions $\eta(x_0) = 0$ and $\eta(x_1) = 0$ in (2.73). Supposing that the points (x_0, y_0) and (x_1, y_1) are located on the continuous and continuously differentiable curves $\phi(x, y) = 0$ and $\psi(x, y) = 0$, respectively, and repeating the procedure leading to (2.77), it is found that (2.77) should be supplemented by the equations

$$\left. \left[(F - \dot{y} F_{\dot{y}}) \frac{\partial \phi}{\partial y} - F_{\dot{y}} \frac{\partial \phi}{\partial x} \right] \right|_{x=x_0} = 0, \left. \begin{array}{c} \\ \\ \end{array} \right\}$$
$$\left. \left[(F - \dot{y} F_{\dot{y}}) \frac{\partial \psi}{\partial y} - F_{\dot{y}} \frac{\partial \psi}{\partial x} \right] \right|_{x=x_1} = 0, \quad \tag{2.93}$$

called usually the transversality conditions. In a non-degenerate extremal problem the set of four equations $\phi(x, y) = 0$, $\psi(x, y) = 0$, and (2.93) is sufficient to determine the values of the four unknown constants x_0, x_1, y_0, y_1. If the curves $\phi(x, y) = 0, \psi(x, y) = 0$ are given in the explicit form $y = u(x), y = v(x)$, respectively, with $|\dot{y}| < \infty$, then the equations (2.93) simplify into

$$[F + (\dot{u} - \dot{y}) F_{\dot{y}}]_{x=x_0} = 0, \quad [F + (\dot{v} - \dot{y}) F_{\dot{y}}]_{x=x_1} = 0. \tag{2.94}$$

With the hypothesis that the terminal points (x_0, y_0), (x_1, y_1) are both floating, the first variation of $I(y)$ can be written in the more general form

$$\delta I = \int_{x_0}^{x_1} \frac{\delta I}{\delta y} \, \delta y \, dx + \int_{x_0}^{x_1} \frac{d}{dx} [F_{\dot{y}} \, \delta y + (F - \dot{y} F_{\dot{y}}) \, \delta x] \, dx. \tag{2.95}$$

The conditions (2.93) are contained implicitly in $\delta I = 0$.

After the above digression let us now return to the problem (2.92). The extremals passing through the point $y(0) = 0$ are $y = Cx$. Substituting $F = \sqrt{[1 + \dot{y}^2(x)]}$ and $g(x, y)$ into the second equation of (2.93) yields

$$\frac{\partial g}{\partial y} - \dot{y}\frac{\partial g}{\partial x} = 0.$$

Consequently the required extremal is orthogonal to the curve $g(x, y) = 0$ at the point $y(x_1) = y_1$. Suppose that the curve $g(x, y) = 0$ has the form shown in Fig. 6, where the point B is a point of discontinuity of slope.

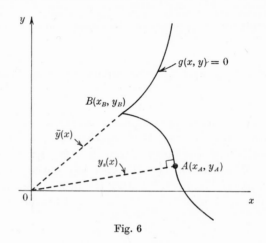

Fig. 6

The unique solution of (2.92), compatible with $\delta I = 0$ and the transversality condition, is therefore $y_s(x) = (y_A/x_A)x$, $I(y_s) = \sqrt{[x_A^2 + y_A^2]}$. From an examination of Fig. 6 it is obvious that $I(y)$ admits also a non-stationary minimum $\bar{y}(x) = (y_B/x_B)x$, $I(\bar{y}) = \sqrt{[x_B^2 + y_B^2]}$. Supposing that Fig. 6 is a true-scale representation of the curve $g(x, y) = 0$,

$$I(\bar{y}) < I(y_s),$$

and the least value of $I(y)$ is attained on an extremal which does not satisfy the transversality condition. By definition, the normal to $g(x, y) = 0$ is not uniquely defined at the point (x_B, y_B). Non-stationary minimal solutions of the type shown in Fig. 6 were already known to Todhunter[T 3].

Unless the boundary conditions associated with a functional $I(y)$ are exceedingly simple, it is generally impossible to tell in advance whether the study of all extremals making $I(y)$ a local minimum will actually furnish the least value of $I(y)$. The study of extremals may therefore be only considered as a *first step* in the solution of an optimization problem. The sufficiency of such a first step requires always a proof.

Let us recall that the criterion for a relatively strong minimum, based on the absence of conjugate points and the inequalities (2.84), holds only with the assumption that $I(y)$ is a rather restricted line integral [G 7], [G 22]. In general this criterion may fail. As an illustration consider the problem

$$I(\tfrac{1}{2}, 1) = \min_{y \in Y} I(y) = \min_{y \in Y} \int_{\tfrac{1}{2}}^{1} x^2 \dot{y}^2(x)\, dx \quad (y(\tfrac{1}{2}) = 2 \quad y(1,1)), \quad (2.96)$$

where $I(y)$ is a Riemann integral and Y is the set of single-valued, continuous piecewise differentiable functions $y(x)$, $\tfrac{1}{2} \leqslant x \leqslant 1$. The problem (2.96) differs from the Weierstrass problem (2.68) only by the interval of definition of x. The extremals of $I(y)$ are hyperbolas and the unique extremal satisfying the boundary conditions of (2.96) is $y_s(x) = 1/x$. The value $I(y_s) = 1$ is a local minimum which coincides with the least value of $I(y)$, i.e. the minimal value $I(y_s)$ cannot be improved by any other $y(x) \in Y$. There are no conjugate points on $y_s(x) = 1/x$, $F_{\dot{y}\dot{y}} = 2x^2 > 0$ and

$$E = x^2 p^2 + 2p + 1/x^2 > 0 \quad \text{all } p \text{ and } x > 0.$$

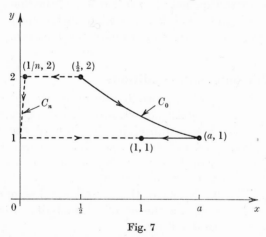

Fig. 7

If the integral $I(y)$ in (2.96) is a line integral, and x is not restricted to $\tfrac{1}{2} \leqslant x \leqslant 1$, then the minimal value $I(y_s) = 1$ can easily be improved. Suppose in fact that the continuous piecewise differentiable functions $y \in Y$ need not be single-valued for a fixed x, and may define curves located inside the rectangle $G: 0 \leqslant y \leqslant b, 0 \leqslant x \leqslant a, a > 1, b \leqslant 2$. The curve C_0 (Fig. 7), defined by

$$y(x) = \begin{cases} 2(1-\alpha) + \dfrac{\alpha}{x}, & \tfrac{1}{2} \leqslant x \leqslant a \\ 0, & 1 \leqslant x \leqslant a \end{cases} \quad \left(\alpha = \dfrac{a}{2a-1} \right), \quad (2.97)$$

is then clearly admissible. Substituting $y(x)$ into $I(y)$ and integrating in the direction shown in Fig. 7 yields

$$I(C_0) = \frac{a}{2a-1} < I(y_s) = 1 \quad \text{for} \quad a > 1.$$

The solution $I(C_0)$ is not the least value of $I(y)$, even if a is allowed to grow indefinitely [G 19]. The value $\lim_{a\to\infty} I(C_0) = \frac{1}{2}$ can be improved by means of the minimal sequence of curves C_n, defined by

$$y_n(x) = \begin{cases} 2, & 1/n \leqslant x \leqslant \frac{1}{2} \\ nx+1, & 0 \leqslant x \leqslant 1/n \\ 1, & 0 \leqslant x \leqslant 1 \end{cases} \quad (n > 1), \tag{2.98}$$

the form of which is shown in Fig. 7. Carrying out the integration in the direction shown in Fig. 7 yields

$$I(C_n) = -1/3n < I(C_0) < I(y_s).$$

The limit solution $I(C_\infty) = 0$ is optimal or suboptimal, depending on whether on the vertical segment $x = 0$, $1 \leqslant y \leqslant 2$, the line integral $I(y)$ is considered meaningful or not. Control engineers favour the former and some mathematicians the latter alternative (see for example [K 20],[K 23] and [G 7]).

14.4 Attainable boundary conditions

When an extremal problem is studied by means of Euler's equation it may happen that there exist no extremals satisfying the prescribed boundary conditions. For instance, in the case of the Weierstrass problem (2.68) there does not exist a continuous curve of the form $y(x) = C_1/x + C$ satisfying the boundary conditions $y(-1) = -1, y(1) = 1$. In such circumstances it is customary to say that the point (x_1, y_1) is not attainable from the point (x_0, y_0), or that the terminal points (x_0, y_0) and (x_1, y_1) are not contained inside the set of attainability of the extremal problem (see for example [M 5], [C 12]).

As it was already shown in the case of the Weierstrass problem (2.68), from the lack of attainability it is not permissible to conclude that an extremal problem has no solution (cf. §14.2), but only that the study of (non-generalized) extremals fails to provide it. Several simple examples of non-attainability have been discussed by Carathéodory [C 6],[C 7].

As a first illustration consider the problem (2.41) written in the parametric form

$$\left. \begin{array}{l} I(s_0, s_1) = \min_{x,\, y \in U} I(x, y) = \min_{x,\, y \in U} \int_{s_0}^{s_1} F(x, y, \dot{x}, \dot{y})\, ds, \\ x(s_0) = x_0, \quad x(s_1) = x_1, \quad y(s_0) = y_0, \quad y(s_1) = y_1, \end{array} \right\} \tag{2.99}$$

where $s_0 \leqslant s \leqslant s_1$, U is the set of continuous and continuously differentiable curves defined by $x = x(s)$, $y = y(s)$, $I(x, y)$ is a line integral, and

$$F(x, y, \dot{x}, \dot{y}) = e^x \sqrt{[\dot{x}^2 + \dot{y}^2]}. \tag{2.100}$$

The extremals of (2.99), (2.100) can be written in the form

$$\cos y + C_1 \sin y + C_2 e^{-x} = 0, \tag{2.101}$$

C_1, C_2 being two integration constants. If $x_0 = 0$, $y_0 = 0$, (2.101) describes a one-parameter family of curves passing through the point $(0, 0)$ and located inside a strip of the x, y-plane bounded by the asymptotes $y = \pi$ and $y = -\pi$. The set of attainability of the point $(0, 0)$ is therefore the strip $-\infty < x < +\infty$, $-\pi < y < \pi$. If $|y_1| \geqslant \pi$ there exists thus no extremal joining the points $(0, 0)$ and (x_1, y_1). The problem (2.99), (2.100) admits obviously suboptimal solutions in the set U, and an optimal solution in a suitably generalized set $\bar{U}$, because $F(x, y, \dot{x}, \dot{y}) \geqslant 0$ and the integral $I(x, y)$ is bounded from below. Carathéodory has shown[C5] that

$$I(s_0, s_1) \geqslant -e^{x_0} + e^{x_1} > 0.$$

As a second illustration consider the problem (2.99) where

$$F(x, y, \dot{x}, \dot{y}) = \sqrt{[\dot{x}^2 + \dot{y}^2]} + 2\,\frac{y\dot{x} - x\dot{y}}{1 + x^2 + y^2}. \tag{2.102}$$

Carathéodory has shown[C6] that the extremals of (2.99), (2.102) consist of two families of spirals approaching asymptotically the circle

$$x^2 + y^2 = 1 \tag{2.103}$$

from the inside and the outside, respectively. The points (x_0, y_0) and (x_1, y_1) belong to the set of attainability of (2.99), (2.102) only if they are both located either inside or outside of the circle (2.103). But since the integral $I(x, y)$ is bounded from below, the problem (2.99), (2.102) admits a meaningful 'non-extremal' solution regardless of the location of the points (x_0, y_0) and (x_1, y_1). To show this let $x = \rho \cos \theta$, $y = \rho \sin \theta$. With this substitution $F(x, y, \dot{x}, \dot{y})$ becomes

$$F(\rho, \theta, \dot{\rho}, \dot{\theta}) = \sqrt{[\dot{\rho}^2 + \rho^2 \dot{\theta}^2]} - \frac{2\rho^2 \dot{\theta}}{1 + \rho^2},$$

and since $\rho\dot{\theta} \leqslant \sqrt{[\dot{\rho}^2 + \rho^2 \dot{\theta}^2]}, \quad 1 - \frac{2\rho}{1 + \rho^2} = \frac{(1 - \rho)^2}{1 + \rho^2},$

$$F(\rho, \theta, \dot{\rho}, \dot{\theta}) \geqslant \frac{(1 - \rho)^2}{1 + \rho^2} \sqrt{[\dot{\rho}^2 + \rho^2 \dot{\theta}^2]} \geqslant 0. \tag{2.104}$$

The value $I(s_0, s_1)$ is therefore non-negative regardless of the location of the points (x_0, y_0) and (x_1, y_1), and $I(s_0, s_1)$ admits a lower limit.

Instead of generalizing the admissible set U in (2.99), a meaningful solution which joins two non-attainable points (x_0, y_0), (x_1, y_1) can also be

obtained by replacing F in (2.99) by a modified function $\overline{F}$, which is close
in some sense to F (see for example [S 7], [G 5] and cf. § 14.6). For the problem
(2.99) thus modified the points (x_0, y_0), (x_1, y_1) become attainable. The
resulting extremal curve $x = \overline{x}(s)$, $y = \overline{y}(s)$ is often called an optimal
sliding regime (cf. [G 5]). This terminology may introduce misunder-
standings between designers and theoreticians, because in control
engineering the term sliding regime is already used with another meaning.
In fact, the term sliding regime is used as a description of a certain
vibratory (periodic or non-periodic) operation of a relay. In control
systems containing relays there may thus exist optimal transitions from
one system state to another which are sliding regimes in the usual
engineering sense without being sliding regimes in the sense of unattainable boundary conditions.

As an illustration consider the following optimization problem

$$I(-1, 1) = \min_{y \in Y} I(y) = \min_{y \in Y} \int_{-1}^{1} (-\dot{y}^2 - xy^2)\, dx \quad (y(-1) = 0, \quad y(1) = 0),$$

$$(2.105)$$

studied by Krotov [K 24], where Y is the set of continuous, piecewise
differentiable curves $y(x)$ satisfying the inequality $|\dot{y}(x)| \leqslant 1$. Krotov
shows that the problem (2.105) admits the optimal sliding regime represented by the limit of the sawtooth curves C_n shown in Fig. 8, as n, the
number of teeth, increases indefinitely. All teeth of C_n have a slope of ± 1,

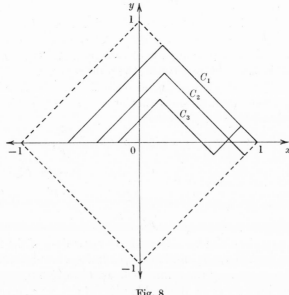

Fig. 8

and the abscissae of slope discontinuities are given by the roots of certain algebraic equations. The optimal sliding regime $\bar{y}(x) = \lim_{n \to \infty} C_n \equiv 0$ of (2.105) is not a sliding regime in the sense of unattainable boundary conditions, because the function $\bar{y}(x) \equiv 0$ satisfies the boundary conditions, $y(-1) = 0$, $y(1) = 0$, as well as the corresponding Euler equation

$$\ddot{y}(x) - xy(x) = 0. \tag{2.106}$$

Let us recall at this point that 'sliding regimes' of the type shown in Fig. 8 are not a new mathematical discovery. One simple example appears in Fig. 5. Another simple example has been reported by Carathéodory in 1906[C2].

Consider the problem of finding the least value of the line integral

$$I(C) = \int_{s_0}^{s_1} \frac{(y^2 + 1)(\dot{x}^2 + \dot{y}^2)}{(2\dot{x}^2 + 2\dot{y}^2)^{\frac{1}{2}} - \dot{x}} ds, \tag{2.107}$$

where $x = x(s)$, $y = y(s)$ is a parametric representation of a continuous piecewise differentiable curve contained inside the closed region G bounded by the straight lines $x = 0$, $x = 1$, $y = 1$ and $y = 2$ (Fig. 9).

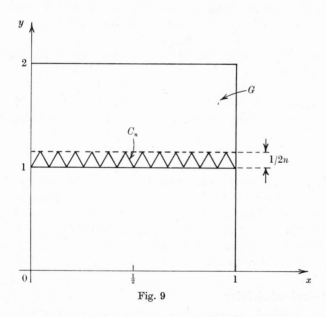

Fig. 9

Let C_n be a sawtooth curve having n equal teeth of slope $\dot{x} = 1/\sqrt{2}$, $\dot{y} = \pm 1/\sqrt{2}$, which joins the points $(0, 1)$ and $(1,1)$. If n is sufficiently large, the curve C_n will be located inside any zero-order neighbourhood in h of the straight line segment C_0, described by $y(x) = 1$, $0 \leqslant x \leqslant 1$ (Fig. 9).

Substituting C_0 and C_n into (2.107) and carrying out the elementary integrations yields

$$I(C_0) = \frac{2}{\sqrt{[2]} - 1} > 4 \cdot 8 \quad \text{and} \quad \lim_{n \to \infty} I(C_n) = 4,$$

respectively. The curve C_n, $n \to \infty$, has all the characteristics of a usual sliding regime, but it is clearly not a sliding regime in the sense of unattainable boundary conditions. The property $I\left(\lim_{n \to \infty} C_n\right) \neq \lim_{n \to \infty} I(C_n)$ signifies only that the functional $I(C)$ is not continuous of order zero in h for $C = C_0$.

From the preceding illustrative examples it is clear that the notion of the set of attainability, based on the study of extremals, is not very significant for control engineering purposes. If for some subsidiary reason the set of attainability is of interest, then it is convenient to interpret Euler's equation as a nominal model of a dynamical system. The set of attainability is then a by-product of the subdivision of the phase-plane into cells of qualitatively identical trajectories (cf. §4). This subdivision is accomplished by determining the bifurcation solutions of the Euler equation with respect to the parameters contained in the boundary condition of the corresponding extremal problem (cf. §4.1).

If the terminal points (x_0, y_0) and (x_1, y_1), defined explicitly or implicitly by the boundary conditions, are both located inside the same cell, then they are mutually attainable, and vice-versa. The case when one of the terminal points is located on the boundary of a cell, i.e. on a bifurcation solution of the Euler equation, needs a special investigation, because its attainability depends on the nature of the bifurcation solution in question.

The determination of the set of attainability is particularly straightforward when one of the terminal points, say (x_0, y_0), is known to describe a stable equilibrium state of the dynamical system $\delta I / \delta y = 0$. In this case the set of attainability of (x_0, y_0) coincides with its zone of influence in the sense of Liapunov (cf. §3), and the latter can, for example, be estimated by various adaptations of the second Liapunov method. The work of several contemporary authors is based on such a simpifying assumption (see for example [M 5]).

14.5   Local additivity

Before studying the extremal properties of a functional $I(y)$ systematically it is advisable to ascertain first whether $I(y)$ possesses some salient particularity, either with respect to the admissible functions $y(x) \in Y$, or with respect to the interval of definition $G: x_0 \leqslant x \leqslant x_1$ of the latter. In the affirmative, this particularity may be the starting-point for the derivation of a special method which permits the determination of the extremal values of $I(y)$.

As an illustration consider the following example. Suppose that the admissible functions $y(x)$ are all single-valued. Choose a constant ξ, $x_0 < \xi < x_1$, permitting to separate the interval $G: x_0 \leqslant x \leqslant x_1$ into two subintervals $G_1: x_0 \leqslant x \leqslant \xi$ and $G_2: \xi \leqslant x \leqslant x_1$. Each $y(x)$ can then be separated into two segments $y_1(x)$ and $y_2(x)$, the first defined for x in G_1 and the second for x in G_2. Let Y_1 and Y_2 designate the sets of $y_1(x)$ and $y_2(x)$, respectively. The separation process just described can be characterized symbolically by the three equations

$$G_1 + G_2 = G, \quad y_1(x) + y_2(x) = y(x), \quad Y_1 + Y_2 = Y. \tag{2.108}$$

If a functional $I(y)$ is such that the fourth equation

$$I(y_1) + I(y_2) = I(y) \tag{2.109}$$

can be added to (2.108), then $I(y)$ is said to be locally additive, or to possess the property of local additivity.

Most functionals encountered in control engineering are expressed by integrals of the form (1.73) and are therefore locally additive. Local additivity is however not a consequence of the fact that a functional $I(y)$ is expressible by integrals. Simple examples can be found in mechanics, where

$$I(y) = \frac{\int_{x_0}^{x_1} x\sqrt{[1 + \dot{y}^2(x)]}\, dx}{\int_{x_0}^{x_1} \sqrt{[1 + \dot{y}^2(x)]}\, dx}, \quad I(y) = \frac{\int_{x_0}^{x_1} x^2\sqrt{[1 + \dot{y}^2(x)]}\, dx}{\int_{x_0}^{x_1} x\sqrt{[1 + \dot{y}^2(x)]}\, dx}$$

and

$$I(y) = \frac{\int_{x_0}^{x_1} x^2\sqrt{[1 + \dot{y}^2(x)]}\, dx}{\int_{x_0}^{x_1} \sqrt{[1 + \dot{y}^2(x)]}\, dx}$$

are used to describe the centre of gravity, the centre of percussion, and the moment of inertia of a one-dimensional mass distribution $y(x)$, respectively. These functionals are obviously not locally additive.

From the presence of local additivity of $I(y)$ it might be conjectured that

$$\min_{y_1 \in Y_1} I(y_1) + \min_{y_2 \in Y_2} I(y_2) = \min_{y \in Y} I(y), \quad y_1(\xi^-) = y_2(\xi^+) = \eta, \tag{2.110}$$

where η is a fixed constant, but unfortunately this conjecture is false in general. To show this consider again the problem (2.96) and let $\xi = \frac{3}{4}$, $\eta = 1$. The solution $\bar{y}(x) = 1/x$, $I(\bar{y}) = 1$ of (2.96) should therefore coincide with the additive combination of

$$\bar{y}_1(x) = -1 + \frac{3}{2x}, \quad I(\bar{y}_1) = \tfrac{3}{2} \quad \text{and} \quad \bar{y}_2(x) \equiv 1, \quad I(\bar{y}_2) = 0.$$

Combining $\bar{y}_1, \bar{y}_2$ and $I(\bar{y}_1), I(\bar{y}_2)$ according to (2.108) and (2.110) yields

$$\bar{y}(x) = \left\{\begin{array}{ll} -1 + \dfrac{3}{2x}, & \frac{1}{2} \leqslant x \leqslant \frac{3}{4} \\ 1, & \frac{3}{4} \leqslant x \leqslant 1 \end{array}\right\}, \quad I(\bar{y}) = \frac{3}{2}$$

which is clearly an erroneous result.

If the formal recurrence relation (2.110) is to hold, the least value of $I(y)$ must be attained on a *unique* curve $\bar{y}(x) \in Y$, and the constant η may not be chosen arbitrarily but must be set equal to $\bar{y}(\xi)$. Furthermore, as the classical example of geodesic curves (great circles) on a sphere shows the sum $AB + BC$ of two minimal segments AB and BC does not necessarily yield the minimal segment AC. With this added restriction (2.110) constitutes essentially the optimality principle of dynamic programming [B4],[B5],[B6]. Because in practice the value $\eta = \bar{y}(\xi)$ is not known in advance, the determination of $\bar{y}(x), I(\bar{y})$ by means of (2.108), (2.109) and (2.110) requires a systematic scanning of all possible values of η. Let us note in passing, that except for the terminology, all features of dynamic programming can already be found in the work of Massé[M6],[M7],[M8],[M9].

Instead of assuming the uniqueness of $\bar{y}(x)$, and having to scan the values of η, it is possible to replace (2.110) by the much weaker relation

$$\delta I(y_1) + \delta I(y_2) = \delta I(y), \quad y_1(\xi^-) = y_2(\xi^+) = \eta, \qquad (2.111)$$

where ξ and η are two unknown constants. In particular, if $I(y)$ is of a form compatible with (2.109) and with the general expression for the first-order variation (2.95), then $\delta I(y_1)$ and $\delta I(y_2)$ can be written

$$\left.\begin{array}{l} \delta I(y_1) = \displaystyle\int_{x_0}^{\xi} \frac{\delta I}{\delta y} \delta y\, dx + F_{\dot{y}}|_{x=\xi^-}\, \delta y + (F - \dot{y}F_{\dot{y}})|_{x=\xi^-}\, \delta x, \\ \delta I(y_2) = \displaystyle\int_{\xi}^{x_1} \frac{\delta I}{\delta y} \delta y\, dx - F_{\dot{y}}|_{x=\xi^+}\, \delta y - (F - \dot{y}F_{\dot{y}})|_{x=\xi^+}\, \delta x. \end{array}\right\} \qquad (2.112)$$

If $y_1(x)$ and $y_2(x)$ are to be pieces of the extremal $y(x)$, then from (2.111), (2.112) and $\delta I(y) = 0$ it follows that

$$[F_{\dot{y}}|_{x=\xi^-} - F_{\dot{y}}|_{x=\xi^+}]\, \delta y + [(F - \dot{y}F_{\dot{y}})|_{x=\xi^-} - (F - \dot{y}F_{\dot{y}})|_{x=\xi^+}]\, \delta x = 0.$$

Taking into account the independence of the variations δx, δy yields

$$\left.\begin{array}{l} F_{\dot{y}}|_{x=\xi^-} = F_{\dot{y}}|_{x=\xi^+}, \\ (F - \dot{y}F_{\dot{y}})|_{x=\xi^-} = (F - \dot{y}F_{\dot{y}})|_{x=\xi^+}. \end{array}\right\} \qquad (2.113)$$

The two equations in (2.113) are known as the Weierstrass–Erdmann continuity conditions, and in general they furnish two algebraic equations which are sufficient to determine the two unknown constants ξ and η. The argument used in the derivation of (2.95) implies that $y(x)$ and

$\dot{y}(x)$ are usually continuous for $x = \xi$ when $F_{\dot{y}\dot{y}}(\xi, y(\xi), \dot{y}(\xi)) \neq 0$, and that at least $y(x)$ is continuous when $F_{\dot{y}\dot{y}}(\xi, y(\xi), \dot{y}(\xi)) = 0$. Since in the latter case $\dot{y}(\xi^-)$ may be different from $\dot{y}(\xi^+)$, the curve $y = \bar{y}(x)$ may have a corner at the point (ξ, η). Because of this circumstance, the equations (2.113) are also known as corner conditions.

Another application of (2.109) and (2.111) is the method used to determine 'refracted' and 'reflected' extremals, which has been evolved from problems of geometrical optics. Let the function $F(x, y, \dot{y})$ in (2.41) be discontinuous on the curve $y = \phi(x)$, where ϕ is a single-valued continuous and continuously differentiable function. It is well known that the refracted or reflected extremals $y(x)$ of (2.41) must satisfy on $y = \phi(x)$ one of the continuity conditions

$$y(\xi^-) = y(\xi^+), \left.\begin{array}{r}\\ (F - \dot{y}F_{\dot{y}} + \dot{\phi}F_{\dot{y}})|_{x=\xi^-} = (F - \dot{y}F_{\dot{y}} + \dot{\phi}F_{\dot{y}})|_{x=\xi^+},\end{array}\right\} \quad (2.114)$$

where (ξ, η) is the point of intersection of $y = y(x)$ and $y = \phi(x)$ (see for example [L 4] and [M 11]). Let us note here that the jump conditions of Pontryagin (§ 36[P 11]) are not essentially different from the continuity conditions (2.113) and (2.114). This point will be discussed with more details in chapters 3 and 4. A generalization of the problem of refracted and reflected extremals has been studied by Kerimov[K 9]–[K 14].

As an illustration of the use of local additivity consider the classical problem

$$\min_{y \in Y} I(y) = \min_{y \in Y} \int_0^2 y^2(x)(\dot{y}(x) - 1)^2 dx \quad (y(0) = 0, \quad y(2) = 1), \quad (2.115)$$

where $I(y)$ is a Riemann integral and Y is the set of continuous piecewise differentiable functions $y(x)$. The Euler equation and the extremals associated with $I(y)$ are

$$y(y\ddot{y} + \dot{y}^2 - 1) = 0, \quad y(x) \equiv 0, \quad y(x) = \pm\sqrt{[x^2 + Cx + C_1]}, \quad (2.116)$$

C and C_1 being two integration constants. It is easy to verify that neither $y(x) \equiv 0$ nor $y^2(x) = x^2 + Cx + C_1$ can be used alone to satisfy the prescribed boundary conditions. A solution of (2.115) is, however, readily obtained from (2.116) and (2.108), (2.109). In fact $I(y)$ attains its least value $I(\bar{y}) = 0$ on the curve

$$\bar{y} = \left\{\begin{array}{ll} 0, & 0 \leqslant x \leqslant 1 \\ +\sqrt{[x^2 - 2x + 1]} = x - 1, & 1 \leqslant x \leqslant 2 \end{array}\right\} \in Y, \quad (2.117)$$

which has a corner at the point $(1, 0)$. It is easily verified that the Weierstrass–Erdmann conditions (2.113) are satisfied at the point $x = 1$, $y = 0$.

14.6   Sensitivity of functionals

We have stressed on several occasions that if a nominal optimization model, for instance of form (2.1) or (2.41), is to be meaningful for control engineering design purposes, then it must be correctly set in the sense of Hadamard and it must possess low sensitivity with respect to perturbations of its structure. Let us examine first the model (2.41) from the point of view of § 4.1. To comply with the conditions (1.6), the functional $I(y)$ in (2.41) must admit a unique solution $I(x_0, x_1)$ and this solution must be continuous with respect to the data contained in (2.41). In particular, in some suitable sense, $I(y)$ must be a continuous functional of $y(x) \in Y$. Furthermore, $I(y)$ must be a continuous point-function of the parameters x_0, x_1, y_0 and y_1. Let

$$I(x_0, x_1) = I(\bar{y}) \quad (\bar{y} = \phi(x, x_0, x_1, y_0, y_1)) \tag{2.118}$$

designate the nominal solution of (2.41). To simplify the subsequent exposition, only the arguments of ϕ actually used will be written out.

The determination of the continuity of $I(y)$ with respect to $y(x)$ amounts essentially to the determination of a subspace Y_1 of Y and of a norm function N such that for $\epsilon > 0$, $\delta > 0$, $x_0 \leqslant x \leqslant x_1$,

$$N(y, \phi) < \delta \quad \text{implies} \quad \Delta I = |I(y) - I(\phi)| < \epsilon \quad (y \in Y_1). \tag{2.119}$$

Let us recall that the continuity of $I(y)$ depends not only on N and Y_1 but also on the size and location of the interval $x_0 \leqslant x \leqslant x_1$.

A quantitative estimate of ΔI can be obtained from the intermediate calculations leading to the solution (2.118). If (2.118) has been obtained by means of a minimal sequence $y_n = \phi_n(x, x_0, x_1, y_0, y_1)$, then

$$\Delta I = I(\phi_n) - I(\phi).$$

If on the contrary (2.118) has been obtained by means of the parametric imbedding (2.40), and $I(y)$ admits, subject to (2.119), the functional derivatives $I'(y, \xi)$, $I''(y, \xi_1, \xi_2)$, then according to Volterra [V 5], $I'(\phi, \xi) \equiv 0$ and

$$\Delta I = \frac{1}{2} \int_{x_0}^{x_1} \int_{x_0}^{x_1} I''(\phi, \xi_1, \xi_2)\, \eta(\xi_1)\, \eta(\xi_2)\, d\xi_1 d\xi_2 + \ldots, \tag{2.120}$$

where $\eta(x) = y(x) - \phi(x)$, $\eta(x) \in Y_1$. In the special case when $I(\phi)$ is a local minimum and the function $F(x, y, \dot{y})$ in (2.41) is twice differentiable with respect to y and $\dot{y}$, (2.120) simplifies into

$$\Delta I = \frac{1}{2} \int_{x_0}^{x_1} [F_{yy}(x, \phi, \dot{\phi})\, \eta^2(x) + 2F_{y\dot{y}}(x, \phi, \dot{\phi})\, \eta(x)\, \dot{\eta}(x)$$
$$+ F_{\dot{y}\dot{y}}(x, \phi, \dot{\phi})\, \dot{\eta}^2(x)]\, dx + \ldots. \tag{2.121}$$

A somewhat more pronounced simplification can be made when $I(y)$ is a suitably restricted line integral [G 22], because from Hilbert's invariant integral it follows that

$$\Delta I = \int_{x_0}^{x_1} E(x, \phi, \dot{\phi}, \dot{\eta})\, dx, \qquad (2.122)$$

where
$$E = F(x, \phi, \dot{\eta}) - F(x, \phi, \dot{\phi}) + (\dot{\phi} - \dot{\eta})\, F_{\dot{y}}(x, \phi, \dot{\phi})$$
$$= \tfrac{1}{2}(\dot{\phi} - \dot{\eta})^2 F_{\dot{y}\dot{y}}(x, \phi, \dot{\phi}) + \dots$$

is the Weierstrass function already encountered in (2.84).

The determination of the continuity of $I(\phi)$ with respect to the parameters x_0, x_1, y_0, y_1 is easily accomplished by the direct use of the inequalities (2.7), but such a procedure requires the knowledge of at least two minimal solutions of (2.41). The determination of the continuity of $I(\phi)$ with respect to y_0 requires, for instance, the knowledge of $\phi(x, y_0)$ and $\phi(x, y_0 + \delta)$, where δ is a sufficiently small real number. The 'close' solution $\phi(x, y_0 + \delta)$ must be known either in the form of a minimal sequence $y_n = \phi_n(x, y_0 + \delta)$ or in the form of a solution of the boundary-value problem

$$\frac{\delta I}{\delta y} = 0, \quad y(x_0) = y_0 + \delta, \quad y(x_1) = y_1. \qquad (2.123)$$

If $F(x, y, \dot{y})$ in (2.41) admits third-order partial derivatives, then an approximate solution of (2.123) can be found by the Poincaré small parameter method.

To obtain the first approximation of $\phi(x, y_0 + \delta)$ let

$$\phi(x, y_0 + \delta) = \phi(x, y_0) + \delta\psi(x) + O(\delta^2), \qquad (2.124)$$

where $\psi(x)$ is an undetermined function. Substituting (2.124) into (2.123) and collecting the terms of the first order in δ yields a linear boundary-value problem for the determination of $\psi(x)$:

$$\frac{d}{dx}(R\dot{\psi}) - P\psi = 0, \quad \psi(x_0) = 1, \quad \psi(x_1) = 0, \qquad (2.125)$$

where
$$P = \left[F_{yy}(x, \phi, \dot{\phi}) - \frac{d}{dx} F_{y\dot{y}}(x, \phi, \dot{\phi}) \right]_{\delta=0},$$
$$R = [F_{\dot{y}\dot{y}}(x, \phi, \dot{\phi})]_{\delta=0}.$$

The differential equation in (2.125) is called the Jacobi differential equation, and it is simply the Poincaré variational equation corresponding to the solution $y = \phi(x, x_0, x_1, y_0, y_1)$ of $\delta I/\delta y = 0$. The non-trivial solutions of the Jacobi equation, satisfying the boundary conditions $\psi(x_0) = 0$ or $\psi(x_1) = 0$, which vanish inside the integral $x_0 \leqslant x \leqslant x_1$, are used to determine the presence of conjugate points on the curve $y = \phi(x, x_0, x_1)$.

If the nominal solution (2.118) of (2.41) is not only continuous but also differentiable with respect to x_0, x_1, y_0, y_1, then the sensitivity of $I(\phi)$ can be tested by evaluating the partial derivatives

$$\frac{\partial I}{\partial x_0}, \quad \frac{\partial I}{\partial x_1}, \quad \frac{\partial I}{\partial y_0}, \quad \frac{\partial I}{\partial y_1}.$$

These derivatives can be found either by differentiation of the explicit solution (2.118), or by a modification of the numerical algorithm leading to a table of values of (2.118).

As a simple illustration of the sensitivity analysis of an optimization model consider the problem (2.96). In the neighbourhood of the solution $I(\phi) = 1$, $\phi = 1/x$ the functional $I(y)$ is already known to be continuous of order $k = 1$ in h. To estimate the sensitivity of $I(\phi)$ with respect to an admissible variation of ϕ let $\frac{1}{2} < a < 1$ and

$$y_a(x) = \begin{cases} 2(1-\alpha) + \dfrac{\alpha}{x}, & \frac{1}{2} \leqslant x \leqslant a \\ 0, & a \leqslant x \leqslant 1 \end{cases} \quad \left(\alpha = \frac{a}{2a-1} \right).$$

An elementary integration yields $I(y_a) = \alpha$, and hence,

$$\Delta I = I(y_a) - I(\phi) = \frac{a}{2a-1} - 1.$$

If the lower integration limit $\frac{1}{2}$ in (2.96) is replaced by x_0, $0 < x_0 < 1$, then the solution of (2.96) becomes

$$\phi(x, x_0) = \frac{1}{1-x_0} \left(1 - 2x_0 + \frac{x_0}{x} \right), \quad I(\phi) = \frac{x_0}{1-x_0}.$$

The sensitivity of $I(\phi)$ to x_0 can be estimated by means of the partial derivative $\partial I / \partial x_0 = (1-x_0)^{-2}$. An identical procedure can be followed to determine the effect on $I(\phi)$ of a variation of the other constants contained in the boundary conditions of (2.96).

As a slightly more involved illustration consider the problem (2.92), whose solution is a local lower limit attained on the straight line segment OB of Fig. 6. The conditions (1.6) of §4 imply that the boundary data in (2.92) are not only the parameters x_0, x_1, y_0, y_1, but also the shape of the curve $g(x, y) = 0$. The solution $I(\bar{y})$ of (2.92) must therefore be a continuous functional of the point-function g, i.e. the value of $I(\bar{y})$ must change little when $g(x, y) = 0$ is replaced by a 'close' curve $g_1(x, y) = 0$. The precise meaning of the closeness between g and g_1 must be specified by defining an admissible neighbourhood of g.

Suppose that near B the admissible neighbourhood of g is a neighbourhood of order zero in h, and that in the vicinity of the point B of Fig. 6 the curve $g_1(x, y) = 0$ has the shape shown in Fig. 10. Using $g_1(x, y) = 0$ to solve (2.92) yields the straight line segment OB_1 which satisfies the transversality condition at the point B_1. The local lower limit $I(\bar{y})$ of $I(y)$, attained on the segment OB, is therefore equivalent to the local minimum $I(\bar{y}_1)$ of $I(y)$, attained on the segment OB_1. From an examination of Fig. 10 it is obvious that the closeness of g and g_1 implies the closeness of the values $I(\bar{y})$ and $I(\bar{y}_1)$.

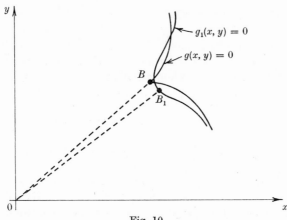

Fig. 10

The fact that a local lower limit of a functional can be changed into a local minimum by an admissible modification of the boundary conditions was used frequently by Todhunter[T 3], and many of his results are based on the implicit argument that

$$I(\bar{y}) = \lim_{g_1 \to g} I(\bar{y}_1). \tag{2.126}$$

This argument failed to receive the deserved acclaim, because Todhunter did not specify the nature of the admissible modifications of g.

Let us now adopt the point of view of §§ 4.3 and 4.4 in the examination of the sensitivity of (2.41). The solution (2.118) must then not only be interpreted as a least value of $I(y)$ when the functions $y(x)$ belong to an admissible set Y but also as a functional of the functions $F(x, y, \dot{y})$, when F belong to an admissible set Φ. Instead of (2.118) let the nominal solution of (2.41) be

$$I = I(\bar{y}, F) \quad (\bar{y} = \phi(x, x_0, x_1, y_0, y_1)). \tag{2.127}$$

If the set Φ is such that with a certain norm N the functional I admits a functional derivative $\delta I/\delta F$, then the sensitivity of I can be estimated by means of

$$\delta I_F = \int_{x_0}^{x_1} \frac{\delta I}{\delta F} \delta F \, dx,$$

where the variation δF is confined to some specified neighbourhood of the reference function F actually used in (2.41).

The direct evaluation of δI_F is generally quite laborious. A considerable simplification occurs when F can be imbedded in a parametric family F_ϵ, and δI_F is interpreted as the total differential $dI = (\partial I/\partial \epsilon)\,d\epsilon$. Another possibility of estimating δI_F consists in calculating the finite difference

$$\Delta I_F = I(F_\epsilon, \bar{y}) - I(F, \bar{y}). \tag{2.128}$$

The existence of a suitable imbedding F_ϵ implies of course that

$$\lim_{\epsilon \to 0} F_\epsilon = F_0 = F, \quad \lim_{\epsilon \to 0} \Delta I_F = \delta I_F.$$

As an illustration of the sensitivity of the nominal model (2.41) with respect to a variation of the reference function F consider the problem (2.68) and let $F = x^2 \dot{y}^2$ be an element of the parametric family

$$F_\epsilon = (x^2 + \epsilon^2)\dot{y}^2(x). \tag{2.129}$$

The Euler equation and the extremals corresponding to (2.129) are

$$\frac{d}{dx}[(x^2 + \epsilon^2)\dot{y}] = 0, \quad y(x) = \frac{C}{\epsilon} \operatorname{arc} tg \frac{x}{\epsilon} + C_1. \tag{2.130}$$

Contrary to the case $\epsilon = 0$, the boundary conditions $y(-1) = -1$, $y(1) = 1$ are now located inside the set of attainability of (2.96). The points $(-1, -1)$, $(1, 1)$ can be joined by the unique continuous extremal

$$\bar{y}_\epsilon(x) = \frac{\operatorname{arc} tg(x/\epsilon)}{\operatorname{arc} tg(1/\epsilon)}, \tag{2.131}$$

for which

$$I(\bar{y}_\epsilon) = \frac{\epsilon}{2 \operatorname{arc} tg(1/\epsilon)}. \tag{2.132}$$

Since for $\epsilon \to 0$ the expressions (2.131) and (2.132) approach the limiting expressions $I(\bar{y}) = 0$ and $\bar{y}(x) = 1$ for $x > 0$, -1 for $x < 0$, respectively, the problem (2.68) with $F = \lim_{\epsilon \to 0} F_\epsilon$ appears as a degenerate case of the family of problems

$$I_\epsilon = \min_{y \in Y} \int_{-1}^{1} (x^2 + \epsilon^2)\dot{y}^2(x)\,dx \quad (y(-1) = -1, \quad y(1) = 1). \tag{2.133}$$

Since the difference $|F_\epsilon - F|$, $-1 \leqslant x \leqslant 1$, can be made as small as desired by choosing $|\epsilon|$ sufficiently small, the problems (2.68) and (2.133) are completely equivalent from a control engineering point of view. In other words, the parametric imbedding (2.129) can be considered as a regularization algorithm in the sense of Tikhonov[T 2] for the 'pathological' optimization model (2.68). From the point of view of §§4.2 and 4.3 such a regularization algorithm is entirely analogous to the Andronov–Pontryagin theory of parasitic elements associated with nominal models of dynamical systems, and it can be used as an alternative to the generalization of the admissible functional space Y. Both alternatives were used in the study of the one time-constant multivibrator (cf. equations (1.29) and (1.30) of §4.3). Structural modifications of a nominal optimization model, used as an artifice for removing a pathological behaviour, are contained, for example, in the investigation of distributed solutions by Swinnerton-Dyer[S 7] and in the investigation of sliding regimes by Gamkrelidze[G 5]. Both authors use modifications of essentially the same type.

The application of sensitivity analysis to the study of the properties of functionals yields also a regularization algorithm for the inverse problem of the calculus of variations. Let us recall first that a necessary and sufficient condition for an ordinary differential equation

$$f(x, y(x), \dot{y}(x), \ldots, \overset{(n)}{y}(x)) = 0 \quad (x_0 \leqslant x \leqslant x_1), \tag{2.134}$$

to be the Euler equation of some functional

$$I(y) = \int_{x_0}^{x_1} F(x, y(x), \dot{y}(x), \ldots, \overset{(m)}{y}(x)) \, dx \tag{2.135}$$

is that the order n of (2.134) be *even* and that the variational equation of (2.134) be self-adjoint (see for example [D 4] and [H 13]). Differential equations of form (2.134) and of an *odd* order n are pathological in the sense that they are not the Euler equations of any functional of form (2.135). This result is in glaring disagreement with the experimentally verified equivalence of causal and teleological representations used in physics in general and in mechanics in particular (cf. §6). Since causal representations of non-conservative phenomena show no preference for even orders of n in (2.134), the corresponding modification of the Hamilton principle (1.76) required the introduction of an auxiliary 'work' function W, which is not a state variable. As can be easily seen from the trivial example

$$-(1 - y^2(x)) \dot{y}(x) + y(x) = 0, \tag{2.136}$$

the statement that W is not a state variable is equivalent to the statement that the functional

$$I(y) = \frac{1}{2} \int_{x_0}^{x_1} (L + W) \, dx \quad \left(L = y^2, \quad W = \int_0^x (1 - y^2(\xi)) \dot{y}^2(\xi) \, d\xi \right) \tag{2.137}$$

is incompatible with the so-called Lagrange transformation, permitting to explicit $\eta(x)$ or $\dot{\eta}(x)$ as a common factor in (2.75). Indirectly this means that the fundamental lemma of the calculus of variation (2.76), or the equivalent fundamental lemma

$$\int_{x_0}^{x_1} b(x) \dot{y}(x) \, dx = 0 \quad \text{for all} \quad y(x) \in Y \quad \text{implies} \quad b(x) = ct, \tag{2.138}$$

cannot be used to derive (2.136) from (2.137).

To regularize the above pathological behaviour suppose that (2.136) is a degenerate case of the parametric family of well-behaved problems

$$\epsilon \ddot{y}(x) - (1 - y^2(x))\, \dot{y}(x) + y(x) = 0, \qquad (2.139)$$

where ϵ is a real valued parameter and $|\epsilon| \ll 1$. In fact, according to a well-known theorem (see for example [B 16]), for $\epsilon \neq 0$ there exists a functional $I(y)$, of form (2.135) with $m = 1$, admitting (2.139) as its 'normal' Euler equation. Since (2.139) and (2.136) are only a particular case of the equations (1.35) and (1.34), respectively, the properties of the corresponding optimization model $I = \min\limits_{y \in Y} I(y)$ can be studied by either continuous or discontinuous extremals (cf. §4.3). In the latter case, the discrepancy in the number of admissible boundary conditions is removed automatically by the presence of the Mandelshtam conditions.

As a terminal remark let us stress the fact that the regularization of a pathological optimization problem by a means of a parametric imbedding process is not a foolproof technique, because to be successful, both the degenerated and the regularized model must be elements of the set of equivalent models discussed in §4.2.

§ 15. EXTREMA OF FUNCTIONALS IN THE PRESENCE OF CONSTRAINTS

15.1 General inequality constraints

Let us now return to the problem

$$\left.\begin{aligned} I(t_0, t_1) &= \min_{u \in U} F(y(t), u(t), t), \\[4pt] G(y(t), u(t), t) &= 0, \\[4pt] L(y(t), u(t), t) &\leqslant 0 \quad (y \in Y, \quad t_0 \leqslant t \leqslant t_1), \end{aligned}\right\} \qquad (2.1)$$

where F is a functional and L either a functional or a point-function. In problems which arise presently in control engineering L may have one of the following forms:

$$\left.\begin{aligned} L &= \int_{t_0}^{t_1} G_0(y(t), \dot{y}(t), \ldots, \overset{(m)}{y}(t), u(t), \dot{u}(t), \ldots, \overset{(k)}{u}(t), t)\, dt, \\[4pt] L &= g(y(t), u(t), t), \\[4pt] L &= g(y(t), \dot{y}(t), \ldots, \overset{(m)}{y}(t), u(t), \dot{u}(t), \ldots, \overset{(k)}{u}(t), t), \\[4pt] L &= g(y(t), \Delta y(t), \ldots, \Delta^m y(t), u(t), \Delta u, \ldots, \Delta^k u(t), t), \end{aligned}\right\} \qquad (2.140)$$

where G_0, g are point-functions, m, k positive integers, and Δ is the finite difference operator. The forms of L in (2.140) are called isoperimetric, holonomous, non-holonomous and recurrent, respectively. Recurrent constraints are rather infrequent in contemporary control problems, but they arise occasionally when the functions $y(t)$, $u(t)$ have poor differentiability properties. For example, if $y(t)$ is to be convex but is not twice differentiable, the normally used non-holonomic constraint $\ddot{y} \leqslant 0$ is replaced by the recurrent constraint $\Delta^2 y \leqslant 0$.

Similarly to the case of point-functions discussed in §11, the problem (2.1) is not essentially different from the unconstrained problem (2.41), because the conditions $y \in Y$, $G = 0$, $L \leqslant 0$ may be interpreted as a part of the complete definition of the set U. The problem (2.1) can therefore be rewritten in the equivalent form

$$I(t_0, t_1) = \min_{y,\, u \in W} F(y(t), u(t), t), \tag{2.141}$$

where W is the set of $y(t)$, $u(t)$ which satisfy the conditions $u \in U$, $y \in Y$, $G = 0$, and $L \leqslant 0$. The formulation (2.141) is sometimes more convenient than (2.1), especially when dynamic programming or a direct method of solution is to be used.

Let U_1 be the set of $u(t)$ satisfying the conditions $u \in U$, $y \in Y$, $G = 0$, $L < 0$. Since from the point of view of the definition of a field of F the set U_1 is an open region, its elements admit essentially the same neighbourhood as the elements of U. Hence, if U_1 contains a local solution of (2.1), this solution is also a local solution of the less constrained problem

$$I = \min_{u \in U} F(y(t), u(t), t), \quad G(y(t), u(t), t) = 0 \quad (y \in Y, \quad t_0 \leqslant t \leqslant t_1).$$
$$\tag{2.142}$$

Consequently, the sole effect of the constraint $L < 0$ consists in rendering inadmissible certain local solutions of the less constrained problem (2.142).

Let U_2 be the set of $u(t)$ satisfying the conditions $u \in U$, $y \in Y$, $G = 0$, $L = 0$. Since, contrary to the point-function problem (2.15), the elements $u(t)$ of U are not necessarily either elements of U_1 or elements of U_2, the problem (2.1) cannot be split up into two independent problems, one of form (2.142), and the other of form

$$I = \min_{u \in U} F(y(t), u(t), t), \quad G(y(t), u(t), t) = 0, \quad L(y(t), u(t), t) = 0$$
$$(y \in Y, \quad t_0 \leqslant t \leqslant t_1, \quad G_i = 0). \tag{2.143}$$

In fact, it is well known that if the functional F is locally additive a solution $y(t)$, $u(t)$ of (2.1) may consist of segments $y_i(t)$, $u_i(t)$, $i = 1, 2, ..., n$, some of which are extremals of (2.142) and other extremals of (2.143).

Let us now consider (2.1) and the artifice due to Valentine[V 1], which permits to transform an inequality constraint $L(y(t), u(t), t) \leqslant 0$ into an equivalent equality constraint $L_1(y(t), u(t), v(t), t) = 0$, where $v(t)$ is an additional variable. Since such a transformation is independent of the properties of the functional F to be minimized, it gives no information whatever on the nature of the minimal solutions of (2.1). Nothing is changed essentially in the problem (2.1), except the notation. In the resulting problem

$$I(t_0, t_1) = \min_{u \in U} F(y(t), u(t), t), \quad G(y(t), u(t), t) = 0, \\ L_1(y(t), u(t), v(t), t) = 0 \quad (y \in Y),$$

$$(2.144)$$

the variable $v(t)$ is *unspecified*, and because of this circumstance (2.144) must either be considered as incorrectly set in the sense of Hadamard (cf. conditions (1.6)), or its solution $I(t_0, t_1)$ must be interpreted as a functional of $v(t)$. To avoid both these complications it is necessary to fix the function $v(t)$, and the only way of achieving this purpose consists in going back to the definition of $L_1(y(t), u(t), v(t), t) = 0$ by means of

$$L(y(t), u(t), t) \leqslant 0.$$

Consequently the practical usefulness of the artifice of Valentine appears to be rather dubious. The widely used assumption that the solution of (2.144) is stationary with respect to the additional variable $v(t)$ lacks a convincing justification. This more or less implicit assumption plays an important role in the argumentation of Hestenes[H 6],[H 7] and Berkovitz [B 9],[B 10]. Transformations of extremal problems will be discussed with more details in § 15.7.

15.2   Isoperimetric inequality constraints

The simplest case of the problem (2.1) arises when $G = 0$ is absent and L is a functional of the same type as F. Let, for example,

$$I(x_0, x_1) = \min_{y \in Y} \int_{x_0}^{x_1} F(x, y(x), \dot{y}(x)) \, dx \quad (y(x_0) = y_0, \, y(x_1) = y_1), \\ L = \int_{x_0}^{x_1} G(x, y(x), \dot{y}(x)) \, dx \leqslant c,$$

$$(2.145)$$

where c is a real constant, Y is the set of continuously differentiable functions $y(x)$, and F, G are single-valued point-functions such that the above integrals exist and can be interpreted in the Riemann sense. Suppose that the solution $\bar{y}(x)$ of (2.145) is a local minimum and that it is not independent of the function F. If $\bar{y}(x)$ can be imbedded in the two-parameter family

$$y(x) = \bar{y}(x) + \lambda_1 \eta_1(x) + \lambda_2 \eta_2(x) \quad (\eta_1(x_0) = \eta_1(x_1) = \eta_2(x_0) = \eta_2(x_1) = 0),$$

$$(2.146)$$

then (2.145) is equivalent to the point-function problem

$$I(x_0, x_1) = \min_{\lambda_1, \lambda_2} f(\lambda_1, \lambda_2) = \min_{\lambda_1, \lambda_2} \int_{x_0}^{x_1} F(x, \bar{y} + \lambda_1 \eta_1 + \lambda_2 \eta_2, \dot{\bar{y}} + \lambda_1 \dot{\eta}_1 + \lambda_2 \dot{\eta}_2)\, dx, \\ g(\lambda_1, \lambda_2) = \int_{x_0}^{x_1} G(x, \bar{y} + \lambda_1 \eta_1 + \lambda_2 \eta_2, \dot{\bar{y}} + \lambda_1 \dot{\eta}_1 + \lambda_2 \dot{\eta}_2)\, dx \leqslant c. \Bigg\}$$

$$(2.147)$$

From the results of §15.1 it follows that $\bar{y}(x)$ is either a solution of the unconstrained problem $I(x_0, x_1) = \min_{\lambda_1} f(\lambda_1, 0)$, satisfying automatically the constraint $g(\lambda_1, 0) \leqslant c$, or a solution of the constrained problem $I(x_0, x_1) = \min_{\lambda_1, \lambda_2} f(\lambda_1, \lambda_2)$, $g(\lambda_1, \lambda_2) = c$. In the latter case, provided f and g are differentiable with respect to their arguments, there exists a constant Lagrange multiplier μ such that $\bar{y}(x)$ is a solution of the Euler equation

$$\frac{d}{dx} \frac{\partial (F + \mu G)}{\partial \dot{y}} - \frac{\partial (F + \mu G)}{\partial y} = 0. \qquad (2.148)$$

The general solution of (2.148) contains three parameters, two integration constants, and the Lagrange multiplier μ. In principle these three parameters are sufficient to satisfy the boundary conditions $y(x_0) = y_0$, $y(x_1) = y_1$ and the constraint

$$\int_{x_0}^{x_1} G(x, \bar{y}(x), \dot{\bar{y}}(x))\, dx = c.$$

15.3   Holonomous inequality constraints

Consider the problem

$$I(x_0, x_1) = \min_{y \in Y} I(y) = \min_{y \in Y} \int_{x_0}^{x_1} F(x, y(x), \dot{y}(x))\, dx$$

$$(L = -\phi(x, y) \leqslant 0, \quad y(x_0) = y_0, \quad y(x_1) = y_1), \quad (2.149)$$

where the integral is a line integral, Y is the set of single-valued continuous $y(x)$ admitting piecewise continuous derivatives, F is a twice differentiable point-function of its arguments, and $\phi = \phi(x, y)$ is a single-valued function admitting continuous first-order partial derivatives. Let ϕ be such that the system of equations

$$\frac{\partial \phi}{\partial x} = 0, \quad \frac{\partial \phi}{\partial y} = 0, \quad \phi = 0 \qquad (2.150)$$

admits only a finite number of real roots.

Suppose that (2.149) admits a minimizing curve $\bar{y}(x)$ which is not independent of the constraint $L \leqslant 0$. Since the functional $I(y)$ is locally additive, $\bar{y}(x)$ will consist of segments $y_i(x)$ and $y_j(x)$, satisfying the conditions $\phi(x, y) > 0$ and $\phi(x, y) = 0$, respectively. $\phi(x, y) > 0$ being an open region, the segments $y_i(x)$ admit bilateral variations and in a stationary case must therefore satisfy the Euler equation

$$F_y - \frac{d}{dx} F_{\dot{y}} = 0.$$

Contrary to the $y_i(x)$, the segments $y_j(x)$ admit only unilateral variations, and in particular if $\partial\phi/\partial y > 0$ on $y_j(x)$, then only the variations $\delta y_j \geqslant 0$ are permissible. $I(\bar{y})$ being a least value by definition,

$$\delta I(y_j) = \int_{x_j}^{x_{j+1}} \left(F_y - \frac{d}{dx} F_{\dot{y}} \right) \delta y_j \, dx \geqslant 0,$$

where x_j, x_{j+1} are the abscissae of the end-points of $y_j(x)$, and hence

$$F_y - \frac{d}{dx} F_{\dot{y}} \geqslant 0 \quad \text{on} \quad y_j(x), \quad \frac{\partial\phi}{\partial y} > 0. \tag{2.151}$$

Since $I(y)$ is locally additive,

$$\delta I(y) = \sum_i \delta I(y_i) + \sum_j \delta I(y_j) = 0.$$

Suppose that the coordinates of the junction points (x_k, y_k) of $y_i(x)$ and $y_j(x)$ are not roots of (2.150), then at the points (x_k, y_k) the following conditions are known to hold (see for example [K 17],[B 11],[B 16] and [L 4])

$$\frac{\partial\phi}{\partial y} \left[F\left(x, y(x), \left(\frac{\partial\phi}{\partial y} \right)^{-1} \frac{\partial\phi}{\partial x} \right) - F(x, y(x), \dot{y}(x)) \right]$$
$$+ \left(\frac{\partial\phi}{\partial x} + \dot{y}(x) \frac{\partial\phi}{\partial y} \right) F_{\dot{y}}(x, y(x), \dot{y}(x)) = 0. \tag{2.152}$$

When $F_{\dot{y}\dot{y}}(x, y(x_k), \dot{y}(x_k))$ exists and does not vanish for $x = x_k$, the condition (2.152) is usually equivalent to the conditions

$$\dot{y}_i(x_k^-) = \dot{y}_j(x_k^+), \quad \dot{y}_j(x_{k+1}^-) = \dot{y}_{i+1}(x_{k+1}^+), \tag{2.153}$$

i.e. to the continuity of the derivative of $\bar{y}(x)$.

The above argument can be repeated when $y(x)$ is a vector function. A two-component case has been studied by Bliss and Underhill[B 12]. In some cases the segments $y_i(x)$, $y_j(x)$ can also be joined according to the condition of refraction and reflection discussed in §14.5. If the co-ordinates of the point (x_k, y_k) happen to be a root of (2.150), then the minimal curve $\bar{y}(x)$ may have a behaviour similar to that encountered at point B in Fig. 6.

As an illustrative example consider the problem (2.149) with

$$x_0 = \tfrac{1}{2}, \quad x_1 = 1, \quad y_0 = 2, \quad y_1 = 1, \quad F = x^2 \dot{y}^2(x), \tag{2.154}$$

and $L \leqslant 0$ the trapezoidal domain bounded by the straight lines $x = \frac{1}{2}$, $x = 1$, $y = 2$ and $y = -\frac{32}{15}x + c$, where c is a fixed value in the interval $\frac{5}{2} < c < 3$ (Fig. 11).

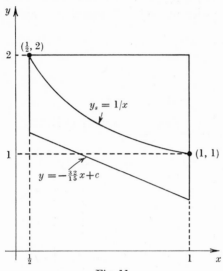

Fig. 11

When $c < \sqrt{\frac{128}{15}}$, the constraint $L \leqslant 0$ is immaterial and (2.149), (2.154) is equivalent to the problem (2.96), whose solution is

$$\bar{y}_s(x) = 1/x, \quad I(\bar{y}_s) = 1. \tag{2.155}$$

When $c = \sqrt{\frac{128}{15}}$, $\bar{y}_s(x)$ is tangent to the line $y = -\frac{32}{15}x + c$ at $x = \sqrt{\frac{15}{32}}$, (2.155) is still a solution of (2.149), (2.154), but $I(\bar{y})$ is no longer a minimum but a lower limit.

When $c = \frac{44}{15}$, the hyperbola $y_s = 1/x$ intersects the line $y = -\frac{32}{15}x + c$ at $x = \frac{5}{8}$ and $x = \frac{3}{4}$. The minimizing curve $\bar{y}(x)$ of (2.149), (2.154) consists therefore of two segments $y_1(x)$, $y_3(x)$ satisfying the Euler equation

$$-\frac{d}{dx}(x^2 \dot{y}) = 0, \tag{2.156}$$

and one segment $y_2(x)$ of the line $y = -\frac{32}{15}x + \frac{44}{15}$. Taking into account the boundary conditions $y(\frac{1}{2}) = 2$, $y(1) = 1$,

$$y_1(x) = 2(1 - C_1) + \frac{C_1}{x}, \quad y_3(x) = 1 - C_3 + \frac{C_3}{x}, \tag{2.157}$$

where C_1, C_3 are undetermined constants. Since $F_{\dot{y}\dot{y}} = 2x^2 \neq 0$ in the interval $\frac{1}{2} \leqslant x \leqslant 1$, the conditions (2.153) apply, and $y_1(x)$, $y_3(x)$ must be tangent to the line $y = -\frac{32}{15}x + \frac{44}{15}$. An elementary calculation shows that the tangency occurs at $x = \frac{1}{2} + \frac{1}{8}\sqrt{2} \approx 0\cdot677$ and $x = 1 - \frac{1}{8}\sqrt{6} \approx 0\cdot694$ and that $C_1 \approx 0\cdot977$, $C_3 \approx 1\cdot001$, $I(\bar{y}) \approx 1\cdot001$. The relative position of the minimal curves for $c \leqslant \sqrt{\frac{128}{15}}$ and $c = \frac{44}{15}$ are shown in Fig. 12. The difference between $I(\bar{y}_s)$ and $I(\bar{y})$ is very small because the curves $\bar{y}_s(x)$ and $\bar{y}(x)$ are very close with respect to a neighbourhood of order $k = 1$ in h.

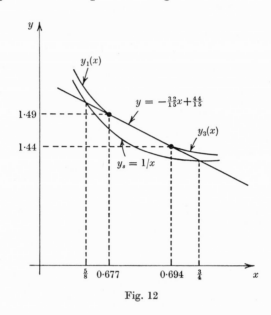

Fig. 12

The knowledge of the continuity (of order $k = 1$ in h) of $I(y)$ and of the coordinates of the points of intersection of $y_s(x)$ and $\phi = 0$ could have been used to estimate the value of $I(\bar{y})$ from the value of $I(\bar{y}_s)$.

15.4 Non-holonomous inequality constraints

Holonomous constraints constitute a limitation of the domain D in the x, y-plane where the admissible functions $y(x)$ and the minimal solution $\bar{y}(x)$ of (2.149) may be located. Except for the segments of $\bar{y}(x)$ which coincide with the boundary of D, the shapes of $y(x)$ and $\bar{y}(x)$ remain unaffected. In the case of non-holonomous constraints the admissible domain D is unchanged, but the shapes of $y(x)$ and $\bar{y}(x)$ are modified by the condition that $\max |\dot{y}(x)|$ must not exceed a prescribed bound, regardless of where in D $y(x)$ and $\bar{y}(x)$ are located.

Consider first the very simple problem

$$I(x_0, x_1) = \min_{y \in Y} I(y) = \min_{y \in Y} \int_{x_0}^{x_1} F(x, y(x), \dot{y}(x))\, dx$$

$$(a \leqslant \dot{y}(x) \leqslant b, \quad y(x_0) = y_0, \quad y(x_1) = y_1), \quad (2.158)$$

where a, b are real constants, and Y, F, $I(y)$ satisfy the same conditions as in problem (2.149). Contrary to the problem (2.149) the minimal curve $\bar{y}(x)$ of (2.158) is not restricted to a prescribed part of the x, y-plane, but its shape may coincide with the shape of extremals only if $a \leqslant \dot{\bar{y}}(x) \leqslant b$. Because of local additivity of $I(y)$, $\bar{y}(x)$ consists of segments of extremals and of segments of the lines $y = ax + C_1$, $y = bx + C_2$, where C_1, C_2 are undetermined constants. Unless $F_{\dot{y}\dot{y}}$ fails to exist or $F_{\dot{y}\dot{y}} = 0$ at the junction of two segments of $\bar{y}(x)$, the slope of $\bar{y}(x)$ is a usually continuous function of x. The details of the determination of $\bar{y}(x)$ have been worked out by Flodin[F3].

The problem (2.158) can be generalized by replacing $a \leqslant \dot{y}(x) \leqslant b$ by

$$g(x, y) \leqslant \dot{y}(x) \leqslant G(x, y). \qquad (2.159)$$

It was shown by Föllinger[F4] that the results of Flodin, described above, are basically unchanged, except that the lines $y = ax + C_1$ and $y = bx + C_2$ are replaced by the general solutions of the differential equations $\dot{y} = g(x, y)$ and $\dot{y} = G(x, y)$, respectively. With the possible appearance of multiple inequalities of form (2.159) and of singular solutions, when $\partial f / \partial \dot{y}$ fails to exist or $\partial f / \partial \dot{y} = 0$, no fundamental change occurs when (2.159) is replaced by the implicit expression

$$f(x, y(x), \dot{y}(x)) \leqslant 0.$$

15.5 Holonomous and non-holonomous equality constraints

Consider the one variable problem

$$\left. \begin{aligned} & I(x_0, x_1) = \min_{u \in U} I(y, u) = \min_{u \in U} \int_{x_0}^{x_1} F(x, y(x), u(x), \dot{y}(x), \dot{u}(x))\, dx \\ & (y(x_0) = y_0, \quad y(x_1) = y_1, \quad u(x_0) = u_0, \quad u(x_1) = u_1, \quad x_0 \leqslant x \leqslant x_1), \\ & G(x, y(x), u(x), \dot{y}(x), \dot{u}(x)) = 0 \quad (y \in Y), \end{aligned} \right\}$$

$$(2.160)$$

where $I(y, u)$ is a line integral, U, Y are the sets of the continuously differentiable functions $y(x)$, $u(x)$, respectively, and F, G are single-valued point-functions admitting second-order derivatives. The problem (2.160) has a holonomous constraint when G is a total derivative, i.e. when $G = (d/dx)\, G_0(x, y(x), u(x))$, because $G = 0$ is equivalent to $G_0(x, y, u) - C = 0$. Holonomous constraints require therefore no separate discussion.

For the study of (2.160) by means of the Euler equations the following classical result can be used:

(2.161): If the functions $\bar{y}(x)$, $\bar{u}(x)$ on which $I(y, u)$ attains a local minimum, do not satisfy the system of equations

$$\frac{d}{dx} G_{\dot{y}} - G_y = 0, \quad \frac{d}{dx} G_{\dot{u}} - G_u = 0,$$

then there exists a Lagrange multiplier $\lambda(x) \neq 0$, such that $\bar{y}(x), \bar{u}(x)$ satisfy the system of equations

$$\frac{d}{dx}\frac{\partial(F+\lambda G)}{\partial \dot{y}} - \frac{\partial(F+\lambda G)}{\partial y} = 0, \quad \frac{d}{dx}\frac{\partial(F+\lambda G)}{\partial \dot{u}} - \frac{\partial(F+\lambda G)}{\partial u} = 0.$$

Although the above result has been used by Lagrange, the first satisfactory proof of its validity was established by Hilbert[H 9],[H 10]. This proof was gradually generalized by Bliss[B 13], McShane[M 12], Graves[G 9], and more recently by Pontryagin (see for example [P 11]). The validity conditions were gradually relaxed from the existence of continuous fourth-order derivatives (see for example [B 16]) to the existence of piecewise continuous first-order derivatives.

It is well known that there exists an infinity of Lagrange multipliers $\lambda(x)$ meeting the requirement that $\bar{y}(x), \bar{u}(x)$ be extremals, but in general there exists only one $\lambda(x)$ meeting the additional requirement that $I(\bar{y}, \bar{u})$ be a local minimum (see for example[B 13]). When $I(x_0, x_1)$ is a local minimum and the existence of $\lambda(x)$ is assured, then the problem (2.160) can be transformed into several equivalent forms. For example, if $H = F + \lambda G$, (2.160) is equivalent to the unconstrained problem

$$I(x_0, x_1) = \min_{y, u, \lambda} \int_{x_0}^{x_1} H(x, y(x), u(x), \lambda(x), \dot{y}(x), \dot{u}(x))\, dx$$

$$(y(x_0) = y_0, \quad y(x_1) = y_1, \quad u(x_0) = u_0, \quad u(x_1) = u_1). \quad (2.162)$$

If desired, the point-constraints at $x = x_0$ and $x = x_1$ can be removed by means of constant Lagrange multipliers μ_i, $i = 0, 1, 2, 3$, and (2.160) becomes equivalent to the problem

$$I(x_0, x_1) = \min\left[\int_{x_0}^{x_1} H(x, y, u, \lambda, \dot{y}, \dot{u})\, dx - \mu_0(y(x_0) - y_0)\right.$$

$$\left. + \mu_1(y(x_1) - y_1) - \mu_2(\mu(x_0) - \mu_0) + \mu_3(u(x_1) - u_1)\right], \quad (2.163)$$

where the operator min applies both to the functions $y(x), u(x), \lambda(x)$ and to the constants μ_i. An equation similar to equation (2.19) of §11 can be obtained by carrying out the minimization of (2.163), while $\lambda(x)$ is held constant. Let $\mathcal{J}(\lambda)$ designate the result of this partial minimization. The solution of (2.163) can then be written

$$I(x_0, x_1) = \max_{\lambda(x)} \mathcal{J}(\lambda). \quad (2.164)$$

Equations (2.163) and (2.164) can be used to estimate the value of $I(x_0, x_1)$ both from below and from above[F 7]. For special purposes the problem (2.160) can be transformed into a variety of other forms (see for example, vol. I[C 15]).

15.6   Control problems

In control engineering there arise minimization problems of the very special form

$$I(x_0, x_1) = \min_{u \in U} I(u) = \min_{u \in U} \int_{x_0}^{x_1} F(x, y(x), u(x))\, dx$$

$$(\dot{y}(x) = f(x, y(x), u(x)), \quad |y(x)| \leqslant M_0, \quad |u(x)| \leqslant M_1, \quad\quad (2.165)$$

$$y(x_0) = y_0, \quad y(x_1) = y_1, \quad y(x) \in Y, \quad x_0 \leqslant x \leqslant x_1),$$

where M_0, M_1 are constants, the integral is a line integral or a Riemann integral, f, F are single-valued continuous or piecewise continuous point-functions of their arguments, Y is the set of piecewise differentiable functions $y(x)$, and U is the set of piecewise continuous functions $u(x)$. In the case of piecewise continuity of f some additional conditions are imposed so that a unique continuous solution of the differential equation

$$\dot{y}(x) = f(x, y(x), u(x)) \quad\quad (2.166)$$

exists in $x_0 \leqslant x \leqslant x_1$ for any fixed choice of $u(x)$.

The particularity of the problem (2.165) consists in the fact that it does not contain the derivative of $u(x)$, and that no boundary conditions are imposed on $u(x)$ at $x = x_0$ and $x = x_1$. Due to this particularity the existence of a solution of (2.165) hinges entirely on the solubility of the differential equation (2.166), which describes the control system state $y(x)$ as a function of the control input $u(x)$. The key functional in (2.165) is therefore not $I(u)$, but the non-holonomous constraint (2.166), considered as a functional of $u(x)$. The solution of (2.166) is a one-parameter family of transformations

$$y(x) = \phi(C, x, u(x)) \quad\quad (2.167)$$

of $u(x)$ into $y(x)$, $x_0 \leqslant x \leqslant x_1$. Suppose that (2.166) is such that ϕ can be expressed explicitly by means of known elementary or transcendental functions. In such a case $y(x)$ can be eliminated from the functional $I(u)$, and (2.165) reduces to the problem

$$I(x_0, x_1) = \min_{u \in U} \int_{x_0}^{x_1} F(x, u, \phi)\, dx \quad (|u(x)| \leqslant M_1,$$

$$|\phi(C, x, u(x))| \leqslant M_0, \quad \phi(C, x_0, u(x_0)) = y_0, \quad \phi(C, x_1, u(x_1)) = y_1), \quad (2.168)$$

which is continuous of order *zero* in h for all $u(x) \in U$. Consequently, subject to the solubility of (2.166), the control problem (2.165) is not essentially different from a point-function problem.

The problem (2.165) is particularly simple when the differential equation (2.166) is linear with respect to $y(x)$ and $u(x)$, because the transformation (2.167) is linear in $u(x)$, and because it admits a unique inverse transformation

$$u(x) = \psi(C, x, y(x)). \tag{2.169}$$

In practical control engineering problems inverse transformations of form (2.169) are much preferred to the direct transformations (2.167), because, if x represents time, the control input u is known as an *instantaneous* function of the system state y. The technical apparatus instrumenting (2.169) from a solution of (2.165) is often called a feedback controller. Feedback controllers, as contrasted with open-loop controllers based on (2.167), derive their popularity from the fact that simultaneously with generating the control input u, they provide a feedback compensation of various deterministic and random disturbances.

Using (2.169), $u(x)$ can be eliminated from (2.165), yielding again a problem

$$I(x_0, x_1) = \min_{y \in Y} \int_{x_0}^{x_1} F(x, y, \psi)\, dx \quad (y(x_0) = y_0, \quad y(x_1) = y_1,$$

$$|y(x)| \leqslant M_0, \quad |\psi(C, x, y(x))| \leqslant M_1), \quad (2.170)$$

which is continuous of order zero in h for all $y(x) \in Y$. In principle the problem (2.170) is even simpler to solve than the problem (2.168).

From elementary theory of differential equations it is obvious that if f in (2.166) is not linear with respect to both y and u, then in general (2.167) will not admit a unique inverse transformation (2.169). Furthermore, for some values of C the transformation (2.167) may become singular, i.e. the differential equation (2.166) may admit more than one mode of qualitatively identical behaviour (cf. §4.2). This circumstance is essentially responsible for the lack of a significant extension of the existence theorem due to Filipov[F 2]. The recent extension of Cesari[C 11] concerns only minor aspects of the problem.

15.7   Singular transformations

In §8 it was stated that a transformation of an extremal problem into another form may affect the qualitative properties of the solution. Let us now elaborate on this statement. To simplify the exposition without compromising the result it is sufficient to consider the extremal problem with a single unknown function $y(x)$ of the form

$$I(x_0, x_1) = \min_{y \in Y} I(y) = \min_{y \in Y} \int_{x_0}^{x_1} F(x, y(x), \dot{y}(x))\, dx$$

$$(y(x_0) = y_0, \quad y(x_1) = y_1), \quad (2.41)$$

with the customary limitations on $y(x)$, Y, F and $I(y)$. If (2.41) is subjected to the transformation

$$y(x) = \phi(x, z(x)), \quad z(x) = \phi^{-1}(x, y(x)) \quad (y(x) \in Y), \quad (2.171)$$

ϕ being a differentiable function of its arguments, there results an associated extremal problem

$$J(x_0, x_1) = \min_{z \in Z} \mathscr{J}(z) = \min_{z \in Z} \int_{x_0}^{x_1} G(x, z(x), \dot{z}(x)) \, dx$$

$$(z(x_0) = z_0, \quad z(x_1) = z_1), \quad (2.172)$$

where Z is the transformed admissible set defined by (2.171) and

$$G(x, z(x), \dot{z}(x)) = F\left(x, \phi(x, z(x)), \frac{\partial \phi}{\partial x} + \frac{\partial \phi}{\partial z} \dot{z}(x)\right).$$

The problem consists in determining under what conditions the solution $\bar{z}(x) \in Z$ of (2.172) is equivalent to the solution $\bar{y}(x) \in Y$ of (2.41), i.e. under what conditions $\bar{y}(x) = \phi(x, \bar{z}(x))$ and $\min_{y \in Y} I(y) = \min_{z \in Z} \mathscr{J}(z)$. Whenever this equivalence exists the transformation (2.171) is called regular; otherwise it is called singular. When (2.172) is solved independently of the restriction imposed on Z by (2.171), the singularity of the relations $\bar{y}(x) = \phi(x, \bar{z}(x)) \min_{y \in Y} I(y) = \min_{z \in Z} \mathscr{J}(z)$, may result from the properties of $G(x, z, \dot{z})$ or from the nature of $\min_{z \in Z} \mathscr{J}(z)$.

Before discussing necessary conditions insuring the regularity of $\bar{y}(x) = \phi(x, \bar{z}(x)) \min_{y \in Y} I(y) = \min_{z \in Z} \mathscr{J}(z)$, let us examine a very simple singular example. Let the original extremal problem be

$$I(x_0, x_1) = \min_{y \in Y} I(y) = \min_{y \in Y} \int_0^1 x^2 \dot{y}^2(x) \, dx \quad (y(0) = a \neq 1, \quad y(1) = 1),$$
$$(2.173)$$

where Y is the set of continuous piecewise differentiable functions and $I(y)$ is a Riemann integral. Using the transformation

$$y(x) = \frac{1}{x} z(x), \quad z(x) = xy(x), \quad (2.174)$$

the associated problem becomes

$$\mathscr{J}(x_0, x_1) = \min_{z \in Z} \mathscr{J}(z) = \int_0^1 G(x, z(x), \dot{z}(x)) \, dx \quad (z(0) = 0 \quad z(1) = 1),$$
$$(2.175)$$

where

$$G(x, z(x), \dot{z}(x)) = \left(\dot{z} - \frac{1}{x} z\right)^2$$

and $\mathscr{J}(z)$ is also a Riemann integral. The corresponding Euler equation

$$\frac{d}{dx}\left(\dot{z} - \frac{1}{x} z\right) + \frac{1}{x}\left(\dot{z} - \frac{1}{x} z\right) = 0 \quad (2.176)$$

is easily integrated to yield the two-parameter family of extremals

$$z(x) = C_1 + C_2 x. \tag{2.177}$$

The boundary conditions in (2.175) yield $C_1 = 0$, $C_2 = 1$. Taking into account the criterion (2.84) it follows that the solution

$$\bar{z}(x) = x, \quad \mathscr{J}(x_0, x_1) = \mathscr{J}(\bar{z}) = 0 \tag{2.178}$$

is a local minimum. If the transformation (2.174) were regular, the solution of (2.173) would be also a local minimum, given by

$$\bar{y}(x) = 1, \quad I(x_0, x_1) = I(\bar{y}) = 0. \tag{2.179}$$

The result (2.179) is obviously erroneous, because $\bar{y}(x)$ does not satisfy the prescribed boundary condition $y(0) = a \neq 1$. In fact, the extremal problem (2.173) does not admit a local minimum in the space Y of continuous functions. The transformation (2.174) is therefore singular in the sense given earlier.

Formally the 'solution' (2.179) of (2.173) coincides with the lower limit

$$\left. \begin{aligned} I(\bar{y}) &= \lim_{n\to\infty} I(y_n) = 0, \quad \bar{y}(x) = \lim_{n\to\infty} y_n(x), \\ y_n(x) &= \begin{cases} a + n(1-a)x, & 0 \leqslant x \leqslant (1/n) \\ 1, & (1/n) \leqslant x \leqslant 1 \end{cases} \quad (n = 2, 3, \ldots) \end{aligned} \right\} \tag{2.180}$$

unattained in the admissible space Y. The singularity of the transformation (2.174), consists in the suppression of the discontinuity of

$$\bar{y}(x) = \lim_{n\to\infty} y_n(x)$$

at the abscissa $x = 0$, and thus in the transformation of the discontinuous function

$$\bar{y}(x) = \lim_{n\to\infty} y_n(x)$$

into the continuous function $\bar{z}(x) = x$.

The above example shows that a formal use of a transformation of form (2.171) may lead to a modification of the nature of the set of functions in which minimization is carried out. This observation suggests that a properly chosen singular transformation may be used as an implementation of Hilbert's argument on the Dirichlet principle. The modalities of such an approach will be discussed in chapter 3.

Suppose now that the 'inverse part'

$$z(x) = \phi^{-1}(x, y(x)) = \psi(x, y(x)) \tag{2.181}$$

of the transformation (2.171) is unique and defines a continuous function $\psi(x, y(x))$ of its arguments in the whole interval $x_0 \leqslant x \leqslant x_1$. From the example (2.173) it is then obvious that the regularity of the transformation (2.171) implies the validity of the conditions:

(a) Let $\delta > 0, \epsilon > 0, x_0 \leqslant x \leqslant x_1, x_0 \leqslant \xi \leqslant x_1$, where ξ is a constant. For $|x - \xi| < \delta$ the inequalities $|y(x) - y(\xi)| < \epsilon$, $|z(x) - z(\xi)| < \epsilon$ hold simultaneously.

(b) Let the admissible neighbourhood in Y be defined by a norm-function N. The inequalities $N(y_1(x), y_2(x)) < \epsilon, y_1, y_2 \in Y$ and $N(z_1(x), z_2(x)) < \epsilon, z_1, z_2 \in Z$ hold simultaneously.

$$(2.182)$$

The condition (2.182a) relates the continuity of individual pairs of functions $y(z) \in Y$, $z(x) \in Z$, whereas the condition (2.182b) relates the closeness of functions in every pair of subsets of Y and Z, respectively. It is easy to verify that the transformation (2.174) fails to satisfy the conditions (2.182).

Let us now examine the artifice due to Valentine[V 1] from the point of view of regularity of a transformation. For this purpose it is sufficient to consider the one-variable case, involving the inequality

$$L(x, y(x)) \leqslant 0, \tag{2.183}$$

added as a constraint to an extremal problem of form (2.41). Let $G(x, y)$ be the open region and Γ its boundary, defined by $L < 0$ and $L = 0$, respectively. The artifice of Valentine amounts to the determination of a transformation of $y(x)$ into a new variable $z(x)$, such that (2.183) is transformed nominally into

$$L_1(x, y(x), z(x)) = 0. \tag{2.184}$$

If this objective is achieved, at least in principle (2.184) can be solved for $y(x)$, $y(x)$ can be eliminated from the functional in (2.41), and the constraint (2.183) becomes trivial. In other words, the implicit transformation (2.184) plays essentially the same role as the explicit transformation (2.171), with the added advantage that the inequality constraint (2.183) is automatically satisfied.

The simplest way to proceed is to choose L_1 in (2.184) so that the open region $G(x, y)$ is transformed into the whole x, z-plane $\bar{G}(x, z)$. In such a case the boundary Γ of G is transformed into the 'point' at infinity of $\bar{G}(x, z)$. There exist now two possibilities for the transformation of an individual curve defined by $y = y(x)$, $y(x) \in Y$. If the curve $y = y(x)$ is entirely located in G, i.e. it has no points in common with the boundary Γ of G, then with modest continuity restrictions on the function

$$L_1(x, y(x), z(x)),$$

the condition (2.182) is satisfied for the pair of functions $y = y(x)$ and $z = z(x)$. If, on the contrary, the curve $y = y(x)$ has at least one point in common with Γ, the transformation (2.184) is singular, because the condition (2.182a) is no longer satisfied. The artifice of Valentine is therefore either superfluous in principle, or it leads to unbounded functions $z(x)$. For these reasons $G(x, y)$ is usually not transformed into the whole x, z-plane, but only into an open part of it $\bar{G}(x, z)$ with a boundary $\bar{\Gamma}$, such that the points x, y located in the open region $L(x, y) > 0$, lead to imaginary or complex values of z in (2.184). The particular forms of $L_1(x, y(x), z(x))$ used among others by Leitmann[L 8], Miele[M 18] and Desoer[D 5] are of this latter type. Nothing is therefore changed intrinsically in the problem (2.41), (2.183), except that the shape of the boundary Γ of G has changed.

If a curve $y = y(x)$, $y(x) \in Y$, has a segment in common with Γ, the transformed curve $z = z(x)$ will have a segment in common with $\bar{\Gamma}$. Hence the use of unilateral variations is necessary to solve (2.41), (2.183), with or without the transformation (2.184). The artifice of Valentine is thus intrinsically irrelevant or downright useless. The only exception occurs when the equation (2.184) is in some way related to the minimizing functions of (2.41), (2.183).

The fact that explicit and implicit definitions of a boundary Γ of an open region G are essentially equivalent, and that Γ is in general topologically invariant with respect to a transformation of variables, was already known to Todhunter[T 3]. Consequently the artifice of Valentine must not be considered as a justification for the substitution of bilateral variations for unilateral variations, i.e. the use of stationarity conditions with respect to the auxiliary variable $z(x)$ is entirely heuristic. The occasional success obtained by means of the Valentine artifice and the stationarity condition is a consequence of the fact that some types of unilateral variations are automatically included in the Carathéodory formulation [C 4],[C 5], of which Pontryagin's maximum principle is a particular case. The extremal problem is thus effectively solved by means of the Carathéodory formulation, and usually by means of the explicit or implicit mechanism of Pontryagin's maximum principle. This happens in particular in the bang-bang servo-problem studied by Desoer[D 5]. If the extremal problem happens to fall outside the scope of validity of Pontryagin's maximum principle, the artifice of Valentine and the stationarity condition on the auxiliary variable or variables lead to 'pathological' solutions, an example of which was given by Lure (see §6 of [L 17]).

CHAPTER 3

EQUIVALENT FORMULATION OF EXTREMAL PROBLEMS

§ 16. TRANSFORMATION OF AN EXTREMAL PROBLEM INTO A BOUNDARY-VALUE PROBLEM ASSOCIATED WITH A PARTIAL DIFFERENTIAL EQUATION

16.1 The equivalent formulation of Carathéodory

In § 14 the extremal problem

$$I(x_0, x_1) = \min_{y \in Y} I(y) = \min_{y \in Y} \int_{x_0}^{x_1} F(x, y(x), \dot{y}(x))\, dx$$

$$(y(x_0) = y_0, \quad y(x_1) = y_1), \quad (2.41)$$

was studied by supposing that the minimizing function $\bar{y}(x)$ is imbedded in a suitably chosen parametric family $g(x, y(x), \lambda) = 0$ which in the simplest case takes the form

$$y(x) = \bar{y}(x) + \lambda \eta(x). \tag{2.73}$$

Control theory being concerned with extremal problems of limited generality, the function $F(x, y, \dot{y})$ is not completely arbitrary and some restrictions may be imposed on its properties. These restrictions must neither be so severe that a strict minimizing function $\bar{y}(x)$ of (2.41) fails to exist in the specified space Y, nor so slight that $I(\bar{y})$ fails to be a continuous point-function of x_0, x_1, y_0, y_1 and a continuous functional of F. Since the theory of functions which are only continuous is still rather incomplete, we will investigate the problem (2.41) with the somewhat stricter hypothesis that $I(\bar{y})$ is at least a piecewise differentiable function of x_0, x_1, y_0 and y_1.

At the end of §13.4 we have mentioned a particular result of Lavrentiev[L 3]: if $I(y)$ in (2.41) is a convergent Lebesgue integral, Y is the set of single-valued measurable functions whose values are bounded from below and from above, and $y = \phi(x)$ and $\dot{y} = \psi(x)$ are considered as two *independent* functions of x, then the integral

$$\int_{x_0}^{x_1} F(x, \phi(x), \psi(x))\, dx$$

attains always its smallest lower limit, but $\psi(x)$ is generally not a derivative of $\phi(x)$. Lavrentiev has shown, however[L 3], that there exist functions $y(x)$ of bounded variation such that, for any $\epsilon > 0$ given in advance,

$$\left| \int_{x_0}^{x_1} F(x, \phi(x), \psi(x))\, dx - \int_{x_0}^{x_1} F(x, y(x), \dot{y}(x))\, dx \right| \leqslant \epsilon. \tag{3.1}$$

The problem (2.41) admits therefore in general only suboptimal solutions, i.e. $I(\bar{y})$ is in general a lower limit which is not attained for any $y(x) \in Y$. As we have mentioned on several occasions, to avoid this complication it is possible either to leave $F(x, y, \dot{y})$ unchanged and generalize Y sufficiently to obtain a strict minimizing function in some weak sense, or leave Y unchanged and replace $F(x, y, \dot{y})$ by a suitable 'close' function $F_\epsilon(x, y, \dot{y})$. The details of these methods lead to a generalized method of solving (2.41). Both approaches have been used in the past with moderate success. The first method, used by many authors, is based essentially on a generalization of the operation of differentiation and the ensuing definition of a differential equation (see for example[Y 1]–[Y 3],[M 4],[M 13],[M 15], [G 2], [G 16],[W 5]–[W 8],[Z 1],[Z 2]. The second method, based on the work of Hilbert, was made popular in recent years by Sobolev[S 4], Stampacchia[S 6], Swinnerton-Dyer[S 7] and Gamkrelidze[G 5],[G 6], and it was pursued among others by Turowicz[T 8], Cesari [C 10],[C 11] and Warga[W 1]–[W 3]. An ingenious method, combining essentially both approaches, was suggested in 1905 by Carathéodory[C 1],[C 3],[C 4]. In this method the solution $y = \bar{y}(x)$, $I(\bar{y})$ of (2.41) is sought by means of an imbedding of the function

$$I(\bar{y}) = I(x_0, x_1, y_1, y_2')$$

in a family of functions

$$S(x, y) = S_1(x_0, y_0, x, y) \quad \text{or} \quad S(x, y) = S_2(x, y, x_1, y_1),$$

combined with the singular transformation

$$\dot{y} = \psi(x, y). \tag{3.2}$$

The transformation (3.2) transforms every $y \in Y$ into a unique function ψ. Let Ψ be the set of ψ corresponding to all $y \in Y$. The transformation (3.2) can therefore be interpreted as an imbedding of the space Y into the more general space Ψ. In fact for every $\psi \in \Psi$ there exists in general more than one function $y(x)$ satisfying the differential equation (3.2), and all $y(x)$ satisfying (3.2) need not be elements of Y. Consequently the operation $\min_{\psi \in \Psi}$ is certainly not less restrictive than the operation $\min_{y \in Y}$, and it is reasonable to expect that the increase in generality is in many cases sufficient to convert an unattained local lower limit of $I(y)$ into a strict local lower limit, or even into a local minimum of $S(x, y)$. This conjecture was borne out in practice[C 3], but the precise limits of its validity are still unknown.

Since the problem (2.41) does not constitute an a priori model unless it is correctly set in the sense of Hadamard, no loss of generality is incurred by supposing that S is a continuous and at least piecewise differentiable function of its arguments. In order to relate $I(y)$ to the family of functions $S(x, y)$ consider the functional

$$J(S, \psi) = \int_{x_0}^{x_1} \frac{dS}{dx} dx = \int_{x_0}^{x_1} \left(\frac{\partial S}{\partial x} + \frac{\partial S}{\partial y} \psi \right) dx. \tag{3.3}$$

The problem consists in choosing the functions ψ and S, and the boundary conditions to be imposed on S, in such a manner that

$$\min_{y \in Y} I(y) = \min_{\psi \in \Psi} J(\psi, S). \tag{3.4}$$

Taking into account the property that for every $y \in Y$ there exists at least one $\psi \in \Psi$, (3.4) can be written in the form

$$\min_{\psi \in \Psi} \int_{x_0}^{x_1} \left[F(x, y, \psi) - \left(\frac{\partial S}{\partial x} + \frac{\partial S}{\partial y} \psi \right) \right] dx = 0. \tag{3.5}$$

A sufficient but not necessary condition for (3.5) to hold is that S be a solution of the partial differential equation (cf. [R5])

$$\min_{\psi \in \Psi} \left[F(x, y, \psi) - \left(\frac{\partial S}{\partial x} + \frac{\partial S}{\partial y} \psi \right) \right] = 0, \tag{3.6}$$

and that in a sufficiently small neighbourhood of the solution

$$\left. \begin{array}{c} I(\bar{y}) = S(x_0, y_0, x_1, y_1), \\[2mm] F(x, y, \psi) - \left(\dfrac{\partial S}{\partial x} + \dfrac{\partial S}{\partial y} \psi \right) \geqslant 0. \end{array} \right\} \tag{3.7}$$

More generally, the second member of (3.6) and (3.7) can be replaced by an arbitrarily chosen but fixed function $c(x)$, but in most cases the extra freedom provided by $c(x) \not\equiv 0$ is unnecessary.

The above derivation of (3.7) was made with the assumption that the integral in (2.41) is a Riemann integral. When this integral is a line integral the same result holds. The detailed conditions of validity of (3.7) were given in the latter case by Damköhler and Hopf[D 1].

If the problem (2.41) is such that $I(\bar{y})$ is a local minimum and the function $F(x, y, \psi)$ is continuously differentiable with respect to ψ, then the expression inside the brackets of (3.6) is stationary with respect to ψ and the following equations hold:

$$F(x, y, \psi) - \left(\frac{\partial S}{\partial x} + \frac{\partial S}{\partial y} \psi \right) = 0, \quad \frac{\partial S}{\partial y} = \frac{\partial F}{\partial \psi}. \tag{3.8}$$

The two partial differential equations in (3.8) have been called by Carathéodory the fundamental equations of the calculus of variations and the family of curves $S(x, y) = \text{ct}$, S being a solution of (3.8), the family of geodesic equidistants[C 1]. Carathéodory has also shown that the curves $S(x, y) = \text{ct}$, defined by (3.8), are transversal to the family of extremals of (2.41), if the latter exist, but that the existence of the geodesic equidistants ('transversals') $S(x, y) = \text{ct}$ is *not* a consequence of the existence of extremals[C 3]. In fact a solution of (2.41) can be constructed from a solution of (3.8) without any references to the existence of the corresponding Euler equation[C 1],[C 3],[C 4]. In many cases the existence of geodesic equidistants $S(x, y) = \text{ct}$ implies the existence of a family of curves $f(x, y) = \text{ct}$, which are transversal to the family $S(x, y) = \text{ct}$, but which do not form a field of extremals in the ordinary sense[C 3]. Bliss has shown that if (2.41) admits an Euler equation, and $F_{\dot{y}\dot{y}}(x, y, \dot{y}) \neq 0$ on the extremal passing through the points (x_0, y_0), (x_1, y_1), then this Euler equation and the fundamental system (3.8) are essentially equivalent (§31[B 14]). In other words, if a field of extremals exists, then the curves $f(x, y) = \text{ct}$, deduced from $S(x, y) = \text{ct}$, coincide with this field[C 3], [H 14]. The formulation (3.8) of the variational problem (2.41) is therefore a generalization of the indirect formulation based on the Euler equation.

Consider the second equation in (3.8) as an algebraic equation in the variable ψ; x, y and $\partial S/\partial y$ playing the role of parameters, and let ψ_i, $i = 1, 2, ..., m$, $m \geqslant 1$, be its roots:

$$\psi_i = \psi_i\left(x, y, \frac{\partial S}{\partial y}\right). \tag{3.9}$$

Substituting (3.9) into the first equation of the system (3.8) yields a set of partial differential equations of the form

$$\frac{\partial S}{\partial x} + H_i\left(x, y, \frac{\partial S}{\partial y}\right) = 0, \tag{3.10}$$

where

$$H_i\left(x, y, \frac{\partial S}{\partial y}\right) = \frac{\partial S}{\partial y}\psi_i - F(x, y, \psi_i). \tag{3.11}$$

The function H_i is often called a Hamiltonian of (3.8). There exists thus one Hamiltonian for every $i = 1, 2, ..., m$ in (3.9). The case $m = 1$ is highly exceptional. It occurs, for example, when $\partial S/\partial y = \partial F/\partial \psi$ is linear in ψ. Only if $m = 1$ is it possible to speak of *the* Hamiltonian of (3.8) or of (2.41).

Partial differential equations of form (3.10) are called Hamilton–Jacobi equations. There are thus generally more than one Hamilton–Jacobi differential equations associated with an extremal problem. When all roots (3.9) are known explicitly, it is quite straightforward to replace the set of Hamilton–Jacobi equations (3.10) by a single non-linear partial differential equation

$$\Phi\left(x, y, \frac{\partial S}{\partial x}, \frac{\partial S}{\partial y}\right) = 0. \tag{3.12}$$

Since contrary to (3.10), (3.12) is always fully equivalent to the fundamental system (3.8), it will be called the Carathéodory partial differential equation.

As an illustration of the multiplicity of the roots (3.9) consider (2.41) with

$$F = \sqrt{[1 + \dot{y}^2(x)]}.$$ (3.13)

The corresponding fundamental system is

$$\frac{\partial S}{\partial x} + \frac{\partial S}{\partial y} - \sqrt{[1 + \psi'^2]} = 0, \quad \frac{\partial S}{\partial y} = \frac{\psi'}{+\sqrt{[1 + \psi'^2]}}.$$ (3.14)

Since there exist two roots

$$\psi'_{1,2} = \pm \frac{\partial S/\partial y}{1 - (\partial S/\partial y)^2},$$

there exist two Hamilton–Jacobi equations

$$\frac{\partial S}{\partial x} + \sqrt{\left[1 - \left(\frac{\partial S}{\partial y}\right)^2\right]} = 0, \quad \frac{\partial S}{\partial x} - \sqrt{\left[1 - \left(\frac{\partial S}{\partial y}\right)^2\right]} = 0.$$ (3.15)

Combining these two equations yields the Carathéodory equation

$$\left(\frac{\partial S}{\partial x}\right)^2 + \left(\frac{\partial S}{\partial y}\right)^2 - 1 = 0.$$ (3.16)

Let us now discuss briefly the relation between the Hamilton–Jacobi equations and the theory of Lagrange multipliers, as applied to extremal problems with non-holonomous equality constraints. For this purpose let us rewrite (2.41) in the equivalent form

$$I(x_0, x_1) = \min_{y \in Y} I(y) = \min_{y \in Y} \int_{x_0}^{x_1} F(x, y(x), z(x))\, dx$$

$$(y(x_0) = y_0, \quad y(x_1) = y_1, \quad \dot{y}(x) - z(x) = 0).$$ (3.17)

Let $I(x_0, x_1)$ be a local minimum of $I(y)$ and let the function $F(x, y, z)$ be twice continuously differentiable with respect to z. It is known from the works of Hilbert[H 9], Bliss[B 13] and their respective students that if $F_{zz}(x, y, z) \neq 0$ on and near the local minimum of $I(y)$, there exists a Lagrange multiplier $\lambda(x)$, such that (3.17) is equivalent to the unconstrained problem

$$I(x_0, x_1) = \min_{y \in Y} I(y) = \min_{y \in Y} \int_{x_0}^{x_1} \left[F(x, y, z) + \lambda\left(\frac{dy}{dx} - z\right)\right] dx$$

$$(y(x_0) = y_0, \quad y(x_1) = y_1), \quad (3.18)$$

where $$\lambda(x) = F_z = \frac{\partial}{\partial z} F(x, y(x), z(x)).$$

Suppose that $F(x, y, z)$ is neither linear in z nor independent of z and consider the Legendre transformation

$$\lambda = F_z(x, y, z), \quad H(x, y, \lambda) = \lambda z - F(x, y, z). \tag{3.19}$$

The function $H(x, y, \lambda)$ is called a Hamiltonian of (3.18) corresponding to the transformation (3.19). Since by definition $F_{zz}(x, y, z) \neq 0$ on and near the minimum of $I(y)$, (3.19) admits a unique inverse transformation

$$z = H_\lambda(x, y, \lambda), \quad F(x, y, z) = z\lambda - H(x, y, \lambda). \tag{3.20}$$

The problem (3.18) can therefore be written in the so-called canonical form

$$I(x_0, x_1) = \min_{y \in Y} I(y) = \min_{y \in Y} \int_{x_0}^{x_1} \left[\lambda \frac{dy}{dx} - H(x, y, \lambda) \right] dx$$

$$(y(x_0) = y_0, \quad y(x_1) = y_1). \tag{3.21}$$

Because of the relation $\dot{y}(x) = z$ in (3.17), the second-order Euler equation of (2.41) is transformed by (3.19) into two first-order equations, the so-called canonical equations

$$\frac{d\lambda}{dx} = -H_y(x, y, \lambda), \quad \frac{dy}{dx} = H_\lambda(x, y, \lambda). \tag{3.22}$$

In control theory the variable $\lambda(x)$ is called usually the adjoint variable and the first equation in (3.22) the adjoint equation.

Let us now show that the adjoint variable, i.e. the Lagrange multiplier $\lambda(x)$ can be expressed in terms of a particular solution $S(x, y)$ of (3.10). In fact, writing $z = \psi$ and applying the canonical transformation (3.19) to the fundamental system (3.8) yields the partial differential equation

$$\frac{\partial S}{\partial x} + \frac{\partial S}{\partial y} z - \lambda z + H(x, y, \lambda) = 0,$$

which reduces to a Hamilton–Jacobi equation (3.10) provided

$$\lambda = \frac{\partial S}{\partial y}. \tag{3.23}$$

This result is well known in the classical calculus of variation (see for example [C 4] and p. 207, vol. I [C 15]), but its utilization in control theory is relatively recent (see for example [R 4] and the paper of Feldbaum in [P 3]).

From an inspection of (3.8) or (3.12) it is sometimes possible to deduce that $\partial S/\partial x = 0$ or $\partial S/\partial y = 0$ on a certain curve $\phi(x, y) = 0$. This information is often a help in constructing continuous or discontinuous minimizing curves $\bar{y}(x)$ or $I(y)$. The inspection, or if necessary a short analysis of (3.8) or (3.12), replaces therefore the rather involved limiting procedure proposed by Razmadze [R 2] and more recently by Krotov [K 20]–[K 22].

16.2   The equivalent boundary-value problem

The partial differential equations (3.6), (3.8), (3.10) and (3.12) admit an infinity of solutions. In fact, it is well known that the general solution of a partial differential equation of first order contains an arbitrary function. If (3.6), (3.8), (3.12) and exceptionally (3.10) are to be equivalent to the a priori extremal problem (2.41), they must be supplemented by appropriate boundary conditions, and these boundary conditions must be so chosen that the resulting boundary-value problem is at least correctly set in the sense of Hadamard.

From the definition of $S(x, y)$ and equation (3.3) it follows that either $S(x_0, y_0)$ or $S(x_1, y_1)$ can be chosen arbitrarily or, in other words, that either $S(x_0, y_0)$ or $S(x_1, y_1)$ can be considered as a reference value for the values of $S(x, y)$. Unless the contrary is stated, we will set either

$$S(x_0, y_0) = 0, \tag{3.24}$$

or
$$S(x_1, y_1) = 0. \tag{3.25}$$

The reference value (3.25) facilitates a comparison between Carathéodory's formulation and Bellman's dynamic programming. Such a comparison will be made in chapter 4.

It is remarkable that in a few highly special cases the point-relation (3.24) or (3.25) is already sufficient to specify completely the arbitrary function contained in the general solution of a partial differential equation of order one. For example, the boundary-value problem (3.16), (3.25) admits the unique solution

$$S(x, y) = +\sqrt{[(x - x_1)^2 + (y - y_1)^2]}. \tag{3.26}$$

This fact was already known to Huyghens and it constitutes the foundation of the now classical Huyghens principle. Since $S(x, y) = ct > 0$ represents a family of concentric circles, with centre at (x_1, y_1), the family of transversals is a family of straight lines passing through the point (x_1, y_1). The extremal functions associated with (3.26) are therefore

$$y = C(x - x_1) + y_1, \tag{3.27}$$

where C is an arbitrary real constant. The complete solution

$$I(\bar{y}) = S(x_0, y_0), \quad \bar{y} = \frac{y_0 - y_1}{x_0 - x_1}(x - x_1) + y_1$$

of (2.52) is obtained by substituting simply $x = x_0$ and $y = y_0$ into (3.26) and (3.27), respectively. The relation between the geodesic equidistants (3.26) and the extremals (3.27) is therefore identical to that existing between the Huyghens and the Fermat principles. The same relation

holds more generally between the 'partial' boundary-value problem
(3.12), (3.25) and the 'ordinary' boundary-value problem

$$\frac{\delta I}{\delta y} = 0, \quad y(x_1) = y_1,$$

where $\delta I/\delta y$ represents the first-order functional derivative of $I(y)$.

When (3.12) does not have a special form, the boundary conditions
consisting of the point-relations (3.24) or (3.25) are obviously not
sufficient to define a unique solution of (3.12). As an illustrative example
consider the Weierstrass problem (2.68), where

$$F(x, y, \dot{y}) = x^2 \dot{y}^2(x) \quad (x_1 = 1, \quad y_1 = 1).$$

The corresponding fundamental equations are

$$\frac{\partial S}{\partial x} + \frac{\partial S}{\partial y} \psi - x^2 \psi^2 = 0, \quad \frac{\partial S}{\partial y} = 2x^2 \psi. \tag{3.28}$$

Since the second equation in (3.28) admits only one root ψ_i, the
Carathéodory equation (3.12) coincides in this case with the Hamilton–
Jacobi equation (3.10). Substituting

$$\psi = \frac{1}{2x^2} \frac{\partial S}{\partial y}$$

into the first equation in (3.28) yields

$$\frac{\partial S}{\partial x} + \frac{1}{4x^2} \left(\frac{\partial S}{\partial y} \right)^2 = 0. \tag{3.29}$$

To facilitate a comparison with the results one could obtain by means of
dynamic programming[G 16], suppose that $x_0 < x_1$ and that the positive
direction of x runs from x_1 to x_0. With this convention equation (3.29)
must be written in the form

$$\frac{\partial S}{\partial x} - \frac{1}{4x^2} \left(\frac{\partial S}{\partial y} \right)^2 = 0. \tag{3.30}$$

It can easily be verified by substitution that the boundary-value problem
(3.30), (3.25) admits, among others, the following solutions[G 16]:

$$
\begin{array}{ll}
(a) & S(x, y) = \dfrac{x}{1-x}(y-1)^2, \\[2mm]
(b) & S(x, y) = x(y^2 - 1)^2, \\[2mm]
(c) & S(x, y) = xy^2 - 1, \\[2mm]
(d) & S(x, y) \equiv 0.
\end{array}
\tag{3.31}
$$

The boundary-value problem (3.30), (3.25) is therefore *not* correctly set in the sense of Hadamard, and additional boundary conditions must be added to (3.25). The same conclusion holds of course for the general boundary-value problem (3.12), (3.25) or (3.12), (3.24).

The simplest, but also the least effective method to reduce the excessive latitude in (3.30), (3.25) which leads to the multiplicity of solutions (3.31), consists in imposing a 'transversality' condition on the 'extremals' resulting from (3.31). Indeed, if $I(\bar{y}) = S(x_0, y_0)$ is a non-degenerated local minimum, then the extremal $y = y(x)$ passing through the points (x_0, y_0) and (x_1, y_1) is an extremal in the ordinary sense, i.e. it is normal in the sense of Bliss[B 14] and $\bar{y}(x)$ satisfies for all $x_0 \leqslant x \leqslant x_1$ the usual transversality condition

$$[F(x, \bar{y}(x), \dot{y}(x)) - \dot{y}F_{\dot{y}}(x, \bar{y}(x), \dot{y}(x))]\frac{\partial S}{\partial y} - F_{\dot{y}}(x, \bar{y}(x), \dot{y}(x))\frac{\partial S}{\partial x} = 0. \quad (3.32)$$

The same reasoning holds of course for the boundary problem (3.12), (3.25). Unfortunately, the inverse proposition is not always true. A function $S(x, y)$ satisfying (3.12), (3.25) and (3.32) represents a stationary value of the functional $I(y)$ in (2.41), which is not necessarily a local minimum. The stationary solutions $S(x, y)$ of (3.12), (3.25) and (3.32) must still be subjected to a comparison with neighbouring solutions. This can be done, for example, by constructing admissible variations of $S(x, y)$ or of $\bar{y}(x)$ and then rejecting the unsuitable stationary solutions $S(x, y)$.

In extremal problems (2.41) which are not normal in the sense of Bliss the boundary condition (3.32) is generally of very limited usefulness, because the notion of transversality ceases to be defined by (3.32), or because (3.32) is satisfied identically. A simple illustrative example is furnished by the solution $S(x, y) \equiv 0$ in (3.31), to which corresponds for $x_0 = -1, y_0 = -1$ the generalized extremal (2.89). The extremal (2.89) is not transversal to $S(x, y) \equiv 0$ in the ordinary sense.

The above-mentioned difficulties can be avoided by an entirely different argument, which consists in the formulation of such boundary conditions that (3.10) or (3.12) become correctly set boundary-value problems in the sense of Hadamard. From the theory of partial differential equations of order one (see for example [K 4], vol. 2[C 15], or vol. 4[S 3]) it is known that (3.12) is correctly set when it is subjected to boundary conditions of the Cauchy type, i.e. when the values of $S(x, y)$ are given on a curve, defined say by $g(x, y) = 0$. Instead of the point-relations (3.24), (3.25) let us therefore consider the functional relation

$$S(x, y) \equiv 0 \quad \text{on} \quad g(x, y) = 0 \quad (3.33)$$

with
$$g(x_0, y_0) = 0 \quad \text{or} \quad g(x_1, y_1) = 0.$$

Consider (3.12), (3.33). It is well known that if Φ in (3.12) is a continuously differentiable function of its arguments, $g(x, y) = 0$ describes a rectifiable curve, and

$$\left(\frac{\partial S}{\partial x}\right)^2 + \left(\frac{\partial S}{\partial y}\right)^2 \neq 0 \quad \text{on} \quad g(x, y) = 0,$$

then the boundary-value problem (3.12), (3.33) admits a unique solution in a certain neighbourhood of the curve $g(x, y) = 0$ (see for example [M 24], vol. 2 [C 15], or [K 4]). Less restrictive versions of this theorem, as well as estimates of the domain in which $S(x, y)$ exists and is unique are available (see for example [K 2], [D 6], [H 5] and [O 1]–[O 6]).

If the problem (2.41) is such that a unique solution of (3.12), (3.33) with $g(x_1, y_1) = 0$ exists in a domain containing the point (x_0, y_0), then to every choice of $g(x, y)$ in (3.33) there corresponds one value of $S(x_0, y_0)$. Let us designate this value by $S(x_0, y_0, g)$. The solution of the extremal problem (2.41) is therefore equivalent to the solution of the boundary-value problem (3.12), (3.33) with $g(x_1, y_1) = 0$, subject to the condition that the value $S(x_0, y_0, g)$ be the least possible. This result can be written in the condensed form

$$I(x_0, x_1) = \min_{y \in Y} I(y) = \min_{g \in G} S(x_0, y_0, g), \tag{3.34}$$

where G is the set of admissible functions $g(x, y)$. The precise definition of G depends on the nature of Y, the nature of the integral in (2.41) and the validity conditions of the most general or the most convenient existence and uniqueness theorem associated with the boundary-value problem (3.12), (3.33). To perform the operation $\min_{y \in Y} I(y)$ in (3.34) it is of course sufficient to select from the set G a suitable sequence of functions $g_n(x, y)$ so that the values $S(x_0, y_0, g_n)$ form a numerical sequence converging to $I(\bar{y}) = \min_{y \in Y} I(y)$. This conclusion is simply the statement of the Dirichlet principle as formulated by Hilbert (cf. §7).

In most cases the Cauchy boundary-value problem (3.12), (3.33) can be solved by the method of characteristics (see for example [K 4], vol. 2 [C 15] or vol. 4 [S 3]). In this theory it is shown that the boundary-value problem (3.12), (3.33) with $g(x_1, y_1) = 0$ is equivalent to the solution of the simultaneous system of ordinary differential equations

$$\left. \begin{array}{l} \dfrac{dx}{\Phi_u} = \dfrac{dy}{\Phi_v} = \dfrac{dS}{u\Phi_u + v\Phi_v} = -\dfrac{du}{\Phi_x} = -\dfrac{dv}{\Phi_y}, \\[2mm] \text{with boundary conditions} \\[2mm] \quad y(x_0) = y_0, \quad y(x_1) = y_1, \quad u(x_1) = u_1, \quad v(x_1) = v_1, \\[2mm] \quad S(x_1, y_1) = 0, \quad \Phi(x_1, y_1, u_1, v_1) = 0, \end{array} \right\} \tag{3.35}$$

where $\qquad y = y(x), \quad S = S(x, y(x)), \quad u = \dfrac{\partial S}{\partial x}, \quad v = \dfrac{\partial S}{\partial y},$

$$\Phi_u = \frac{\partial \Phi}{\partial u}, \quad \Phi_v = \frac{\partial \Phi}{\partial v}, \qquad \Phi_x = \frac{\partial \Phi}{\partial x}, \quad \Phi_y = \frac{\partial \Phi}{\partial y},$$

and u_1, v_1 are real constants. The differential system in the 'ordinary' boundary-value problem (3.35) can be considered as a general condition of transversality between the curves $S(x, y) = \text{ct}$ and $y = y(x)$, which in the case of stationarity contains the ordinary conditions of transversality (3.32) as a particular case.

The 'characteristic' system (3.35) contains one free parameter, say v_1, which defines the 'initial' value of $\partial S / \partial y$ at the point (x_1, y_1). Letting $v_1 = \alpha$, the other parameter, u_1, is fixed by the equation

$$\Phi(x_1, y_1, u_1, v_1) = 0.$$

The solution of (3.35) constitutes therefore a one-parameter family of functions
$$\left.\begin{aligned} y &= y(x, \alpha), \quad S = S(x, y(x), \alpha), \\ \partial S / \partial x &= u(x, y(x), \alpha), \quad \partial S / \partial y = v(x, y(x), \alpha). \end{aligned}\right\} \tag{3.36}$$

If the equivalence between (2.41) and (3.35) exists, its solution can be obtained from the solution (3.36) be choosing the value of α so that

$$I(x_0, x_1) = S(x_0, y_0) = \min_{\alpha} S(x, y(x), \alpha). \tag{3.37}$$

Similarly to the case (3.34), it is sufficient to select from all real values of α a converging sequence α_n so that the corresponding values of $S(x, y(z), \alpha_n)$ form a convergent minimal sequence.

The characteristic system (3.35) simplifies considerably when equation (3.12) coincides with the Hamilton–Jacobi equation (3.10), and in such a case it contains the canonical system (3.22) as a particular case. It is important to note, however, that (3.22) contains less information about the extremal problem (2.41) than (3.35).

Whenever applicable, the ordinary boundary-value problem (3.35), (3.37) is quite efficient for numerical computations, because in addition to $S(x_0, y_0)$ it furnishes automatically the values of dy/dx, $\partial S / \partial x$ and $\partial S / \partial y$ in the whole interval $x_0 \leqslant x \leqslant x_1$. These derivatives can be used to investigate the sensitivity of $I(y)$ in (2.41) (cf. § 14.6).

If the solution $I(x_0, x_1) = \min_{y \in Y} I(y)$ of (2.41) cannot be represented by a function $S(x, y)$ which renders the functional equation

$$\frac{\partial S}{\partial x} + \frac{\partial S}{\partial y} \psi - F(x, y, \psi) = 0 \tag{3.38}$$

stationary with respect to ψ, then the boundary problems (3.12), (3.33), (3.34) and (3.35), (3.37) become meaningless, at least in the ordinary sense. The solution $I(x_0, x_1)$ must then be sought by solving directly the functional boundary-value problem (3.6), (3.24) or (3.6), (3.25). As will be shown later in this chapter, this solution can often be obtained by rather elementary considerations.

16.3 Solution of extremal problems by means of the Carathéodory equation

Consider first the Weierstrass problem (2.68), which led to the equations (3.28), (3.30) and (3.31) of §16.2. If $I(y)$ in (2.68) is a Riemann integral, $I(y) \geqslant 0$ for any $y(x) \in Y$, and in particular $I(y) = 0$, when $y(x) = $ ct. Under these circumstances the only appropriate solution in (3.31) of the boundary-value problem (3.30), (3.25), (3.34) is $S(x, y) \equiv 0$. The other solutions in (3.31) are negative for $x < 0$, which is inconsistent with the property $F = x^2 \dot{y}^2(x) \geqslant 0$.

To determine the shape of the discontinuous minimizing curve $y = \bar{y}(x)$ 'transversal' to $S(x, y) \equiv 0$ it is sufficient to note that $\partial S/\partial y = 0$ on the vertical line $x = 0$. This conclusion follows immediately from (3.30) after multiplication by x^2. Since $\partial S/\partial y = 0$ for $x = 0$ implies that the value of $S(x, y)$ does not vary on the vertical line $x = 0$, the minimizing curve $y = \bar{y}(x)$ consists of two horizontal segments $y(x) = -1$, $-1 \leqslant x \leqslant 0$, $\bar{y}(x) = +1$, $0 \leqslant x \leqslant +1$, respectively and of the vertical segment $x = 0$, $-1 \leqslant y \leqslant +1$. The above result coincides with that obtained previously in §13.4 and §14.2.

As a second example consider the more regular Weierstrass problem (2.96), which differs from (2.68) only by the location of the interval $x_0 \leqslant x \leqslant x_1$. If $I(y)$ in (2.96) is a Riemann integral, then the solution of the boundary-value problem (3.30), (3.25), (3.34) is

$$S(x, y) = \frac{x}{1-x}(y-1)^2, \quad x_0 = \tfrac{1}{2} \leqslant x \leqslant 1. \tag{3.31a}$$

From (3.31a), $y(x_0) = 2$ and the transversality condition (3.32) it follows that
$$I(x_0, x_1) = I(\bar{y}) = 1, \quad \bar{y}(x) = 1/x.$$

If instead of a Riemann integral, $I(y)$ in (2.96) is a line integral there exist at least two additional particular solutions of the boundary-value problem (3.30), (3.25), (3.34). It is quite straightforward to verify that these particular solutions are

$$S(x, y) = x(y-1)^2, \tag{3.31b}$$

and
$$S(x, y) \equiv 0. \tag{3.31d}$$

To the former corresponds the minimal curve C_0 as $a \to \infty$, equation (2.97), and to the latter the minimal curve C_n, as $n \to \infty$, equation (2.98), of Fig. 7. The least non-negative value of $I(y) = S(x_0, y_0)$ is given by (3.31d).

As another illustration of the increase of generality provided by the singular transformation (3.2) consider the Scheefer problem (2.65), which does not admit the existence of a two-parameter family of extremals in the ordinary sense. Substituting

$$F(x, y, \dot{y}) = (x-a)^2 \dot{y}^2 + (x-a)\dot{y}^3 \quad (x_0 < a < x_1)$$

into (3.8), and writing $\qquad \alpha = (x-a)$

for brevity, yields

$$\frac{\partial S}{\partial x} + \psi \frac{\partial S}{\partial y} - \alpha^2 \psi^2 - \alpha \psi^3 = 0, \quad \frac{\partial S}{\partial y} = 2\alpha^2 \psi + 3\alpha \psi^2. \tag{3.39}$$

The second equation in (3.39) admits two roots

$$\psi_1 = -\frac{\alpha}{3} + \sqrt{\left[\frac{\alpha^2}{9} + \frac{1}{3\alpha} \frac{\partial S}{\partial y}\right]}, \quad \psi_2 = -\frac{\alpha}{3} - \sqrt{\left[\frac{\alpha^2}{9} + \frac{1}{3\alpha} \frac{\partial S}{\partial y}\right]}, \tag{3.40}$$

which are real, provided

$$\frac{\partial S}{\partial y} \geqslant -\frac{\alpha^3}{3} = -\tfrac{1}{3}(x-a)^3. \tag{3.41}$$

The problem (2.65) admits thus two Hamilton–Jacobi equations

$$\left.\begin{aligned}
\frac{\partial S}{\partial x} - \frac{\alpha}{3}\frac{\partial S}{\partial y} - \frac{2\alpha^4}{27} + \left(\frac{2\alpha^3}{9} + \frac{2}{3}\frac{\partial S}{\partial y}\right)\sqrt{\left[\frac{\alpha^2}{9} + \frac{1}{3\alpha}\frac{\partial S}{\partial y}\right]} = 0, \\
\frac{\partial S}{\partial x} - \frac{\alpha}{3}\frac{\partial S}{\partial y} - \frac{2\alpha^4}{27} - \left(\frac{2\alpha^3}{9} + \frac{2}{3}\frac{\partial S}{\partial y}\right)\sqrt{\left[\frac{\alpha^2}{9} + \frac{1}{3\alpha}\frac{\partial S}{\partial y}\right]} = 0,
\end{aligned}\right\} \tag{3.42}$$

both subject to the constraint (3.41). The two equations (3.42) are easily combined into the single equation

$$\alpha\left(\frac{\partial S}{\partial x}\right)^2 - \tfrac{2}{3}\alpha^2 \frac{\partial S}{\partial x}\frac{\partial S}{\partial y} - \frac{4\alpha^5}{27}\frac{\partial S}{\partial x} - \frac{\alpha^3}{27}\left(\frac{\partial S}{\partial y}\right)^2 - \frac{4}{27}\left(\frac{\partial S}{\partial y}\right)^3 = 0. \tag{3.43}$$

From the inspection of (3.43) it is obvious that $S(x, y) \equiv 0$, $\bar{y}(x) \equiv 0$ is a solution of the boundary-value problem (3.43), (3.24), (3.32). This solution is valid subject to the constraint (3.41), which can be written

$$\partial S/\partial y = 2\alpha^2 \psi + 3\alpha \psi^2 \geqslant -\tfrac{1}{3}\alpha^3.$$

Rearranging the above inequality in ψ yields

$$|\psi| \leqslant \tfrac{1}{3}|\alpha|.$$

Recalling that $\dot{y} = \psi$, it follows that

$$|\ddot{y}| = \left|\frac{d\psi}{dx}\right| \leqslant \frac{1}{3}\left|\frac{d\alpha}{dx}\right| = \frac{1}{3}.$$

The solution $S(x, y) \equiv 0$, $\overline{y}(x) \equiv 0$ coincides therefore with that reported by Hadamard[H 1] (cf. §13.4).

As a second problem which is not normal in the sense of Bliss consider

$$I(x_0, x_1) = \min_{y \in Y} I(y) = \min_{y \in Y} \int_{s_0}^{s_1} e^{\alpha x} \sqrt{[\dot{x}^2 + \dot{y}^2]} \, ds$$

$$(x(s_0) = x_0, \quad x(s_1) = x_1, \quad y(s_0) = y_0, \quad y(s_1) = y_1, \quad s_0 \leqslant s \leqslant s_1), \quad (3.44)$$

where α is a real constant and the variables $x = x(s)$ and $y = y(s)$ are given parametrically. Let Y be the set of continuous piecewise differentiable functions $y(x)$, and $I(y)$ a line integral. For $\alpha = 1$ the above problem coincides with the problem (2.99), discussed in §14.4 in connection with the set of attainability, and for $\alpha = 0$ it coincides with the problem of determining the shortest distance in a Euclidean plane. The fundamental system (3.8) corresponding to (3.44) is given in the x–y representation by the two equations

$$\frac{\partial S}{\partial x} + \psi \frac{\partial S}{\partial y} - e^{\alpha x} \sqrt{[1 + \psi^2]} = 0, \quad \frac{\partial S}{\partial y} = e^{\alpha x} \frac{\psi}{\sqrt{[1 + \psi^2]}}. \quad (3.45)$$

Since ψ is a double-valued function of $\partial S / \partial y$, there exist two Hamilton–Jacobi equations

$$\frac{\partial S}{\partial x} + \sqrt{\left[e^{2\alpha x} - \left(\frac{\partial S}{\partial y} \right)^2 \right]} = 0, \quad \frac{\partial S}{\partial x} - \sqrt{\left[e^{2\alpha x} - \left(\frac{\partial S}{\partial y} \right)^2 \right]} = 0, \quad (3.46)$$

subject to the constraint $\qquad \left| \dfrac{\partial S}{\partial y} \right| \leqslant e^{\alpha x}. \qquad (3.47)$

The Carathéodory equation is therefore

$$\left(\frac{\partial S}{\partial x} \right)^2 + \left(\frac{\partial S}{\partial y} \right)^2 = e^{2\alpha x}. \quad (3.48)$$

It is easy to verify that the boundary-value problem (3.48), (3.24), (3.33) admits two distinct solutions

$$S(x \ y) = +\frac{1}{\alpha} \left[1 + e^{2\alpha(x - x_0)} - 2 \, e^{\alpha(x - x_0)} \cos \alpha(y - y_0) \right]^{\frac{1}{2}} \quad (3.49)$$

and $\qquad\qquad S(x, y) = \dfrac{1}{\alpha} \left(e^{\alpha x} - e^{\alpha x_0} \right). \qquad (3.50)$

For $\alpha \neq 0$ the first solution is related to the set of attainability of (3.44), but the second is obviously not. In fact, inside the set of attainability the transversals to $S(x, y) = ct$, defined by equation (3.32), coincide with the ordinary extremals

$$\cos \alpha y + C_1 \sin \alpha y + C_2 \, e^{-\alpha x} = 0$$

of (3.44), where C_1, C_2 are constants of integration.

The least value $S(x_1, y_1)$ of $I(y)$ in (3.44) depends on the values of x_1 and y_1, and it may be given either by (3.49) or by (3.50). The 'transversals' of (3.50) are the lines $y(x) \equiv y_0$ and $y(x) \equiv y_1$, joined by vertical segments at $x = +\infty$ for $\alpha < 0$ and at $x = -\infty$ for $\alpha > 0$. For $\alpha = 0$ the solution (3.50) does not yield the least value of $I(y)$ in (3.44) unless $y_0 = y_1$.

Let us now examine an example where the solution of an extremal problem depends strongly on the nature of the admissible minimizing functions. Consider in fact the problem

$$\min_{y \in Y} I(y) = \min_{y \in Y} \int_{-1}^{+1} y^2(x) \, [\dot{y}^2(x) - 1] \, dx \quad (y(-1) = 0, \ y(1) = 1), \quad (3.51)$$

where $I(y)$ is a line integral, and Y is in turn the set of continuous $y(x)$ admitting
    (a) continuous second-order derivatives,
    (b) continuous first-order derivatives,
    (c) piecewise continuous first-order derivatives.

For conciseness let us designate these three cases by $Y = C_2$, $Y = C_1$ and $Y = C_{10}$. The fundamental equations (3.8) associated with (3.51) are

$$\frac{\partial S}{\partial x} + \psi \frac{\partial S}{\partial y} - y^2(\psi^2 - 1) = 0, \quad \frac{\partial S}{\partial y} = 2y^2 \psi.$$

The root
$$\psi = \frac{1}{2y^2} \frac{\partial S}{\partial y}$$

being unique, the Hamilton–Jacobi equation associated with (3.51) is

$$4y^2 \frac{\partial S}{\partial x} + \left(\frac{\partial S}{\partial y}\right)^2 + 4y^4 = 0.$$

Letting $S(1, 1) = 0$ and supposing that the positive direction of x runs from $+1$ to -1 yields

$$-4y^2 \frac{\partial S}{\partial x} + \left(\frac{\partial S}{\partial y}\right)^2 + 4y^4 = 0. \quad (3.52)$$

If $Y = C_2$ the boundary conditions (3.25), (3.32) are sufficient to define the unique solution

$$S(xy) = (x-1)\left[1 + \left(\frac{x^2 + y^2 - 2}{2(x-1)}\right)^2\right] + \frac{2}{3}\left[\left(1 - \frac{x^2 + y^2 - 2}{2(x-1)}\right)^3\right.$$

$$\left. - \left(x - \frac{x^2 + y^2 - 2}{2(x-1)}\right)^3\right]. \quad (3.53)$$

The transversal to $S(x, y) = ct$ passing through the two points $(-1, 0)$, $(1, 1)$ being

$$y(x) = +\sqrt{\left[\tfrac{25}{16} - (x - \tfrac{1}{4})^2\right]} \quad (3.54)$$

(curve (a) of Fig. 13), the least value of $I(y)$ is

$$S(-1,0) = -\tfrac{37}{24}.$$

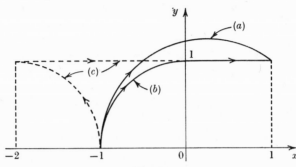

Fig. 13

If $Y = C_1$ it is necessary to use the boundary conditions (3.33), (3.34). Equation (3.52) admits then two distinct solutions, depending on whether the transversals $y = y(x)$ to $S(x,y) = ct$ must be single-valued functions of x or not. These two solutions are

$$S_1(x,y) = x-1+\tfrac{2}{3}(1-y^2)^{\tfrac{3}{2}} \tag{3.55}$$

and

$$S_2(x,y) = x-1-\tfrac{2}{3}(1-y^2)^{\tfrac{3}{2}}. \tag{3.56}$$

Provided the point (x,y) is located in the shaded area of Fig. 14, the two 'transversals', each passing through the two points $(-1,0)$, $(1,1)$ are given by

$$y_1(x) = \begin{cases} +\sqrt{[1-x^2]}, & -1 \leqslant x \leqslant 0 \\ 1, & 0 \leqslant x \leqslant 1 \end{cases} \tag{3.57}$$

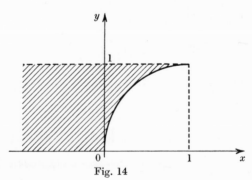

Fig. 14

(curve (b) of Fig. 13), and

$$y_2(x) = \begin{cases} +\sqrt{[1-(x+2)^2]}, & -2 \leftarrow x \leftarrow -1 \\ 1, & -2 \rightarrow x \rightarrow +1 \end{cases} \tag{3.58}$$

(curve (c) of Fig. 13), respectively. The corresponding solutions of (3.51) are therefore

$$S_1(-1, 0) = -\tfrac{4}{3}, \quad \bar{y} = y_1(x),$$

and

$$S_2(-1, 0) = -\tfrac{8}{3}, \quad \bar{y} = y_2(x).$$

Suppose now that $Y = C_{10}$. From an inspection of (3.52) it is obvious that

$$\frac{\partial S}{\partial y} = 0 \quad \text{if} \quad y(x) \equiv 0 \quad (-\infty < x < +\infty). \tag{3.59}$$

Taking into account (3.59), the transversals to $S(x, y) = ct$ have the form

$$y_\alpha(x) = \begin{cases} 0, & -1 \to x \to \alpha \\ +\sqrt{[R^2 - (x-c)^2]}, & \alpha \to x \to 1 \end{cases}, \tag{3.60}$$

where

$$c = \frac{\alpha^2 - 2}{2(\alpha - 1)}, \quad R^2 = 1 + \frac{\alpha}{2}\frac{\alpha - 2}{\alpha - 1},$$

shown in Fig. 15, where the abscissa $x = \alpha$ can be chosen freely anywhere on the x-axis. The corresponding value of $I(y)$

$$I(y_\alpha) = \frac{1}{12}\left(\frac{\alpha^2 - 2\alpha + 2}{\alpha - 1}\right)^3 + \frac{\alpha^3}{12}\left(\frac{\alpha - 2}{\alpha - 1}\right)^3 - \frac{(\alpha^2 - 2\alpha + 2)}{4(\alpha - 1)} \tag{3.61}$$

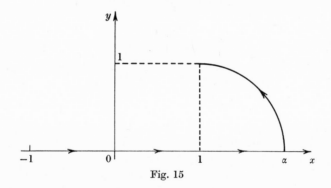

Fig. 15

can be rendered as small or as large as desired by a suitable choice of α. This conclusion is quite normal, because when $Y = C_{10}$ the functional $I(y)$ is bounded neither from below nor from above. If the problem (3.51) is to admit a meaningful solution when $Y = C_{10}$, constraints must be added on the admissible maximal values of $|x|$ or of $|y(x)|$.

Since in control theory the definition of an admissible set Y constitutes a physical and not a mathematical problem, the problem (3.51) is meaningless for control engineering purposes when the set Y is not precisely specified.

The effect of inequality constraints on the Carathéodory formulation (3.8) and (3.12) will be examined in detail in §17. It will be shown that the presence of holonomous inequality constraints is actually a help in solving an extremal problem, because the region where $S(x, y)$ needs to be known is effectively reduced. Such a reduction is especially useful in numerical computations. The presence of non-holonomous inequality constraints is a handicap from a purely theoretical point of view, because such constraints may invalidate the system of equations (3.8), and the more general equation (3.6) has to be used. From a more detailed study it follows, however, that the complications produced by the presence of non-holonomous inequality constraints can be removed by means of the results of Flodin[F 3] and Föllinger[F 4] (cf. §15.4).

As a preparatory step for the study of the influence of non-holonomous inequality constraints consider the very simple functional

$$I(y) = \int_0^{x_1} [\dot{y}^2(x) - y^2(x)]\, dx \quad (y(0) = y_0, \quad y(x_1) = y_1, \quad y \in Y), \quad (3.62)$$

where the integral is interpreted in the Riemann sense, and Y is the set of single-valued piecewise differentiable functions. The extremals associated with (3.62) are

$$y_s(x) = C \cos x + C_1 \sin x. \quad (3.63)$$

If $\sin x_1 \neq 0$ the boundary conditions in (3.62) yield

$$C = y_0, \quad C_1 = \frac{y_1 - y_0 \cos x_1}{\sin x_1}.$$

The Hamilton–Jacobi equation associated with (3.62) is

$$\frac{\partial S}{\partial x} + \frac{1}{4}\left(\frac{\partial S}{\partial y}\right)^2 + y^2 = 0 \quad (0 \to x \to x_1). \quad (3.64)$$

From an inspection of (3.63) it is obvious that not every point (x_1, y_1) of the x, y-plane is attainable from the point $(0, y_0)$, because the integration constants C, C_1 are not uniquely defined for all values of x_1 and y_1. This indetermination will be also found in the solutions $S(x, y)$ of (3.64). For example, using the boundary conditions (3.24), (3.32) it is easy to verify that the solution

$$S(x, y) = \frac{\sin 2x}{4}\left[\left(\frac{y - y_0 \cos x}{\sin x}\right)^2 - y_0^2\right] - 2y_0(y - y_0 \cos x)\sin x \quad (3.65)$$

of (3.64) is free of indeterminations only if $|x| < \pi$.

Although the functional $I(y)$ in (3.62) is unbounded both from below and from above, it admits local minima. For example, if $y_0 = 1$, $y_1 = 1$, $x_1 = \frac{1}{2}\pi$, the solution

$$y_s(x) = \cos x + \sin x, \quad S(\tfrac{1}{2}\pi, 1) = -2 \quad (3.66)$$

is a local minimum with respect to a sufficiently small neighbourhood of order $k = 0$ in h. Let us note that in the interval $0 \leqslant x \leqslant \frac{1}{2}\pi$ the extremal $y_s(x)$ in (3.66) satisfies automatically the non-holonomous constraint

$$|\dot{y}(x)| \leqslant 1.$$

16.4　Extremal problems with mobile end-points

In the study of extremal problems by means of the Euler equations the presence of mobile end-points appears as a complication, which may lead to a possible loss of physically meaningful solutions (cf. problem (2.92), §§14.3 and 14.6). No such complication arises in the study of extremal problems by means of the Carathéodory equation (3.12). Indeed, in many cases extremal problems with mobile end-points are simpler than the same problems with fixed end-points.

Consider again the a priori optimization model

$$I(x_0, x_1) = \min_{y \in Y} I(y) = \min_{y \in Y} \int_{x_0}^{x_1} F(x, y(x), \dot{y}(x))\, dx \quad (y(x_0) = y_0, \ \ y(x_1) = y_1),$$

(2.41)

and suppose first that the coordinates of one end-point, say (x_1, y_1), are not given in advance, but are known to be located on a prescribed rectifiable curve $\psi(x, y) = 0$. In order to obtain a solution of (2.41) under these circumstances it is sufficient to construct a correctly set boundary-value problem based on (2.41) and the associated Carathéodory equation (3.12).

This objective can be attained in at least two different ways. Choose first as a reference of $S(x, y)$ the value $S(x_0, y_0) = 0$, and introduce the boundary condition

$$S(x, y) = 0 \quad \text{on} \quad g(x, y) = 0, \quad g(x_0, y_0) = 0. \tag{3.67}$$

Designate the resulting solution of (3.12), (3.67) by $S(x, y, g)$. The function $S(x, y, g)$ defines a family of geodesically equidistant curves $S(x, y, g) = b$, where b is a real parameter. For the sake of definiteness suppose that g is fixed and that $b > 0$ when $S(x, y, g) \neq 0$. If the domain of existence of $S(x, y, g)$ is sufficiently large, then there exist values of b such that the curves $S(x, y, g) = b$ and $\psi(x, y) = 0$ intersect, i.e. these curves have some points in common. Let b_g be the bifurcation value of the algebraic problem

$$S(x, y, g) = b, \quad \psi(x, y) = 0 \tag{3.68}$$

(cf. §4.1 and §16.2), i.e. let b_g be such a value of b that the two curves defined by (3.68) do not intersect but have nevertheless at least one point in common. In such a case the roots $x = x_1$, $y = y_1$ of (3.68) are all degenerate, or what is the same, the Jacobian $[\partial(S, \psi)]/[\partial(x, y)]$ vanishes or fails to exist at the point (x_1, y_1). Geometrically this means that the curves $S(x, y, g) = b_g$, $\psi(x, y) = 0$ touch at the point (x_1, y_1), but are not necessarily tangent. The point (x_1, y_1) may in fact be a corner, cusp, or some other singular point of $\psi(x, y) = 0$ in the sense of differential geometry.

Choosing a sequence of functions $g(x, y) = 0$ in (3.67) produces a sequence of bifurcation values of (3.68), as well as a sequence of points (x_1, y_1). The solution of the problem (2.41), subject to the constraint $\psi(x_1, y_1) = 0$, will obviously be obtained when the sequence of bifurcation values b_g of (3.68) attains its lower limit $\bar{b}$. Let $\bar{g}(x, y)$ and $(\bar{x}_1, \bar{y}_1)$ be the corresponding limits of $g(x, y)$ and (x_1, y_1). Hence the required solution of (2.41) is given by

$$S(\bar{x}_1, \bar{y}_1, \bar{g}) = \bar{b}, \quad y = \bar{y}(x), \tag{3.69}$$

where $\bar{y}(x)$ is a transversal to the family of curves

$$S(x, y, \bar{g}) = b \quad (0 \leqslant b \leqslant \bar{b}, \quad x_0 \leqslant x \leqslant \bar{x}_1).$$

The scanning process with respect to the initial curves $g(x, y) = 0$ in (3.67) can be avoided entirely by using instead of (3.67) the unique boundary condition

$$S(x, y) = 0 \quad \text{on} \quad \psi(x, y) = 0, \quad \psi(x_1, y_1) = 0. \tag{3.70}$$

If the rectifiable initial curve $\psi(x, y) = 0$ has no corners, cusps or other similar singular points in the sense of differential geometry, then the boundary-value problem (3.12), (3.70) admits usually a unique solution $S_\psi(x, y)$ in a sufficiently small neighbourhood of the curve $\psi(x, y)$. Suppose that the domain of existence of $S_\psi(x, y)$ contains the point (x_0, y_0). The least value of the integral $I(y)$ is then given by

$$I(x_0, x_1) = \min_{y \in Y} I(y) = S_\psi(x_0, y_0).$$

The minimizing curve $y = \bar{y}(x)$ is fixed by the property that it is transversal to the family of curves $S_\psi(x, y) = \text{ct}$ in the whole interval $x_0 \leqslant x \leqslant x_1$. Since the coordinates of the point (x_0, y_0) are known, the intersection of $y = \bar{y}(x)$ with $\psi(x, y) = 0$ fixes the coordinates of the unknown points (x_1, y_1).

If the rectifiable curve $\psi(x, y) = 0$ is not free of corners and other similar singularities, then the approach based on the unique boundary condition can still be used, provided $S_\psi(x, y)$ is interpreted as a suitably defined weak solution. Weak solutions of first-order partial differential equations have been studied by many authors (see for example [D 6],[M 4], [O 1]–[O 5]). Another method of dealing with corner-type singularities of $\psi(x, y) = 0$ consists in replacing $\psi(x, y) = 0$ by a 'close' admissible curve $\psi_\epsilon(x, y) = 0$, free of singularities (cf. § 14.6).

Suppose now that both end-points of (2.41) are mobile, i.e. the points (x_0, y_0) and (x_1, y_1) are located on the rectifiable curves $\phi(x, y) = 0$ and $\psi(x, y) = 0$, respectively, but their coordinates are not known. The solution of the problem (2.41) is then quite straightforward in principle, because either $\phi(x, y) = 0$ or $\psi(x, y) = 0$ can be used to formulate a unique boundary condition of type (3.70). It is of course advantageous to use as initial curve the smoother of the curves $\phi(x, y) = 0$ and $\psi(x, y) = 0$. For the sake of definiteness suppose that it is the curve $\phi(x, y) = 0$. The function g in (3.67) is therefore known by definition, and $\bar{g} \equiv \phi$. The required bifurcation value $\bar{b}$ and the coordinates of the points (x_0, y_0), (x_1, y_1) are readily determined from (3.68) which takes the particular form

$$S(x_1, y_1, \phi) = \bar{b}, \quad \psi(x_1, y_1) = 0,$$

and the property that the minimizing curve $y = \bar{y}(x)$ is transversal to $S(x, y, \phi) = b$ for all $0 \leqslant b \leqslant \bar{b}$ and all $x_0 \leqslant x \leqslant x_1$ as well as to $\psi(x, y)$ at the point (x_1, y_1). This statement appears to be very complicated, but in practice $\bar{b}$ and (x_1, y_1) are known as soon as the solution $S(x, y, \phi) = \mathrm{ct}$ of (3.12), (3.67) has been computed sufficiently far to touch the known curve $\psi(x, y) = 0$. The point (x_0, y_0) is then readily found by determining the intersection of the minimizing curve $y = \bar{y}(x)$, which is transversal to $S(x, y, \phi) = \mathrm{ct}$ and which passes through the point (x_1, y_1) where $S(x, y, \phi) = \mathrm{ct}$ and $\psi(x, y) = 0$ touch, with the other known curve $\phi(x, y) = 0$. When neither $\phi(x, y) = 0$ nor $\psi(x, y) = 0$ have sufficient smoothness for the validity of the above argument, it is necessary to interpret the resulting boundary-value problem in a weak sense, or to replace either $\phi(x, y) = 0$ or $\psi(x, y) = 0$ by a suitable 'close' curve. When the points (x_0, y_0) and (x_1, y_1) are not unique, the minimization problem (2.41) is not correctly set in the presence of mobile end-points. Uniqueness can frequently be obtained by specifying some other desirable property on the minimizing curve $y = \bar{y}(x)$ without disrupting the relation that $I(\bar{y})$ is the least value of $I(y)$.

As an illustrative example consider the problem (2.92). The solution of the corresponding boundary-value problem, based on the Carathéodory equation, is given by the equation (3.26). The determination of $\bar{b}$, (x_1, y_1) and of the minimizing curve $y = \bar{y}(x)$ (Fig. 6), of (2.92), is then so simple that it does not require any intermediate calculations. The results are self-evident from an inspection of Fig. 6. The situation is basically unchanged in the case of a more general problem. The only difficulties which may arise are of a computational nature, when the functions $S(x, y) = \mathrm{ct}$ or $y = \bar{y}(x)$ cannot be expressed analytically.

§ 17. EXTREMAL PROBLEMS WITH CONSTRAINTS

17.1 Isoperimetric inequality constraints

Let us return to the isoperimetric problem

$$\left.\begin{aligned} I(x_0, y_1) = \min_{y \in Y} \int_{x_0}^{x_1} F(x, y(x), \dot{y}(x))\, dx, \\ \int_{x_0}^{x_1} G_0(x, y(x), \dot{y}(x))\, dx \leqslant c \quad (y(x_0) = y_0, \quad y(x_1) = y_1) \end{aligned}\right\} \quad (2.145)$$

already discussed in §15.2. Suppose that the solution of the problem (2.145) is not independent of $F(x, y, \dot{y})$. It was shown in §15.2 that two cases may arise:

(a) The value $I(x_0, x_1)$ is independent of $G_0(x, y, \dot{y})$.

(b) The value $I(x_0, x_1)$ depends only on the equality

$$\int_{x_0}^{x_1} G_0(x, y(x), \dot{y}(x)) = c.$$

In the first case the isoperimetric inequality constraint is satisfied automatically, and in the second case there exists a *constant* Lagrange multiplier μ, such that (2.145) is equivalent to the unconstrained problem

$$I(x_0, x_1) = \min_{y \in Y} \int_{x_0}^{x_1} [F(x, y, \dot{y}) + \mu G_0(x, y, \dot{y})]\, dx \quad (y(x_0) = y_0, \quad y(x_1) = y_1).$$

$$(3.71)$$

In non-degenerated cases the extremals of (3.71) contain three parameters, two integration constants, and the Lagrange multiplier μ. These three parameters are sufficient to satisfy the three boundary conditions

$$\left.\begin{aligned} y(x_0) = y_0, \ y(x_1) = y_1 \\ \int_{x_0}^{x_1} G_0(x, y(x), \dot{y}(x))\, dx = c. \end{aligned}\right\} \quad (3.72)$$

When the problem (2.145) is studied by means of the Carathéodory equation (3.12), the procedure to be followed is exactly the same as in the case of the unconstrainted problem (2.41). The only difference arises from the fact that it is possible to determine (3.12) on the basis of $F(x, y, \dot{y})$ or of $\overline{F}(x, y, \dot{y}) = F(x, y, \dot{y}) + \mu G_0(x, y, \dot{y})$. In the former case the scanning process $g \in G$ in the boundary condition (3.34) must be compatible with the functional constraint

$$\int_{x_0}^{x_1} G_0(x, y, \dot{y})\, dx \leqslant c, \quad (3.73)$$

and in the latter case with the point-constraint (3.72). In isoperimetric problems the usefulness of the Carathéodory equation is limited to a possible increase in smoothness of the curves $S(x, y) = ct$ as compared with the extremals $y = \bar{y}(x)$[M3]. The increased smoothness is convenient when the extremals $y = \bar{y}(x)$ have some anomalous behaviour.

17.2 Holonomous inequality constraints

The study of the extremal problem

$$I(x_0, x_1) = \min_{y \in Y} I(y) = \min_{y \in Y} \int_{x_0}^{x_1} F(x, y(x), \dot{y}(x))\, dx$$

$$(y(x_0) = y_0, \quad y(x_1) = y_1, \quad L = -\phi(x, y) \leqslant 0), \quad (2.149)$$

by means of the Euler equation (cf. §15.3) required the subdivision of the minimizing curve $y = \bar{y}(x)$ into segments $y_i(x)$ and $y_j(x)$, the former located in the interior of the open region $\phi(x, y) < 0$ and the latter located on the boundary curve $\phi(x, y) = 0$. This subdivision was made possible by the local additivity of the functional $I(y)$.

In spite of the admissibility of only unilateral variations of $\bar{y}(x)$ near $\phi(x, y) = 0$, the solutions of (2.149), obtained in §15.3, are not essentially different from local minima of $I(y)$. This conclusion is a consequence of the fact that in the presence of the continuity conditions (2.152) or (2.153) the sum

$$\sum_k \int_{x_k}^{x_{k+1}} F(x, y_j(x), \dot{y}_j(x))\, dx \qquad (3.74)$$

constitutes basically a constant component of $I(y)$, and exerts thus no influence on its minimal properties. The analysis of §15.3 was made with the specific provision that only such local lower limits of (2.149) are admissible which render $I(y)$ stationary. This provision is the basis for the definition of the admissible unilateral variations of $\bar{y}(x)$ near $\phi(x, y) = 0$. In other words, the segments $y_j(x)$ coinciding with the boundary $\phi(x, y) = 0$ are so chosen that their presence does not disturb the stationarity of

$$I(y) = \sum_k \int_{x_{k-1}}^{x_k} F(x, y_i(x), \dot{y}_i(x))\, dx + \sum_k \int_{x_k}^{x_{k+1}} F(x, y_j(x), \dot{y}_j(x))\, dx,$$

but affects only the terminal points of the segments $y_i(x)$. Such a procedure does not exhaust all possibilities. Non-stationary local lower limits exist for example when the extremals associated with $I(y)$ meet the curve $\phi(x, y) = 0$ at a point where the latter has a corner (cf. problem 2.92, §14.3). This theoretically unusual but practically very common behaviour was specifically excluded in §15.3 by the condition that $\phi(x, y)$ be a continuously differentiable function of its arguments, satisfying

$$\left(\frac{\partial \phi}{\partial x}\right)^2 + \left(\frac{\partial \phi}{\partial y}\right)^2 \neq 0$$

at all junction abscissae x_k (cf. equations (2.150)).

As an illustration of the appearance of a meaningful lower limit to which the above-mentioned restrictions do not apply, consider the problem, discussed already by Todhunter[T 8], of finding the shortest path between the points M, N or the points M, P running entirely in the non-shaded part of Fig. 16. The plane x, y in Fig. 16 is supposed to be

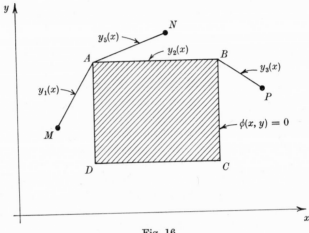

Fig. 16

Euclidean. The constraint $L = -\phi(x, y) \leqslant 0$ is such that it renders inadmissible the interior points of the rectangle $ABCD$ shaded in Fig. 16. Since $F = \sqrt{[1 + \dot{y}^2]}$, $y(x) = Cx + C_1$, and $S(x, y) = \sqrt{[(x - x_0)^2 + (y - y_0)^2]}$, the solution of the above problem can be found by inspection. The shortest path between M and N is the curve MAN, and it consists of two interior straight-line segments $y_1(x)$ and $y_5(x)$, joined at the corner point A of the boundary $\phi(x, y) = 0$. The shortest path between the points M and P of Fig. 16 is the curve $MABP$, which consists of two interior straight-line segments $y_1(x)$, $y_3(x)$ and of one boundary segment $y_2(x)$, joined respectively at the corner points A and B of the boundary curve $\phi(x, y) = 0$.

If the same problem is to be solved by the method of unilateral variations described in §15.3, the corners A and B of the boundary curve $\phi(x, y) = 0$ must be replaced by a 'close' continuously differentiable curve $\phi_\varepsilon(x, y) = 0$ free of corners. Consider, for instance, the problem of finding the shortest path MN. The corner A of $\phi(x, y) = 0$ may be rounded off by the circular segment $A_3 A_4$ of a sufficiently small radius $\rho = \varepsilon > 0$, shown in Fig. 17. The shortest path between the points M and N consists therefore of two straight-line segments $y_1(x)$, $y_5(x)$ and of the circular boundary segment $y_4(x)$ (Fig. 17). These segments meet tangentially at the points A_1 and A_5, respectively.

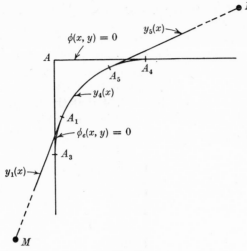

Fig. 17

The non-stationary lower limit

$$I(\bar{y}) = \sqrt{[(x_A - x_M)^2 + (y_A - y_M)^2]} + \sqrt{[(x_N - x_A)^2 + (y_N - y_A)^2]},$$

determined directly, is the limit as $\epsilon \to 0$ of the stationary lower limit

$$I(y_\epsilon) = \sqrt{[(x_{A_1} - x_M)^2 + (y_{A_1} - y_M)^2]}$$
$$+ 2\epsilon \arcsin \frac{d}{2\epsilon} + \sqrt{[(x_N - x_{A_5})^2 + (y_N - x_{A_5})^2]},$$

where
$$d = \sqrt{[(x_{A_5} - x_{A_1})^2 + (y_{A_5} - y_{A_1})^2]},$$

determined by the method of §15.3. The formal limiting process, quite trivial in the above illustrative example, is rather tedious in a more general case, and it can frequently be avoided by taking advantage of the increased smoothness of the Carathéodory formulation. In fact, equation (3.6) does not depend on any stationarity considerations, and experience has shown that limits of stationary lower limits do not usually appear as degenerate in the sense of §15.3 as far as the stationary equation (3.12) is concerned. This property is a consequence of the singular transformation (3.2). This was shown in particular in the solution of the shortest path problem of Fig. 16, where the knowledge of $S(x, y)$ made the final result self-evident. This situation is generally unchanged in more complex problems. The constraint of form $L = -\phi(x, y) \leqslant 0$ is not an obstacle but actually a help in finding the required solution, because it diminishes the region of the x, y-plane where $S(x, y)$ has to be known. The knowledge of the form of $S(x, y)$ reduces the extremal problem (2.149) to an algebraic problem of finding the junction points (x_k, y_k) of the interior and boundary segments $y_i(x)$ and $y_j(x)$.

Since the functional $I(y)$ in (2.149) is locally additive, the solution $S(x, y)$ of (3.12), (3.33) may be determined with or without the use of Lagrange multipliers. The presence of Lagrange multipliers $\lambda(x)$ is neither an advantage nor an inconvenience in solving the boundary-value problem (3.12), (3.33), (3.34), because the determination of $S(x, y)$ turns out to be completely independent of the determination of $\lambda(x)$. The reason for this property lies in the fact that the 'dimensionality' of a fixed correctly set problem is an invariant with respect to its modes of representation, i.e. it is independent of the way used to write down the problem.

To illustrate this point consider a particular case of the problem (2.154) discussed in §15.3:

$$I_0 = \min_{y \in Y} I(y) = \min_{y \in Y} \int_{\frac{1}{2}}^1 x^2 \dot{y}^2(x)\, dx$$

$$(y(x) \geqslant -\tfrac{32}{15}x + \tfrac{44}{15}, \quad y(\tfrac{1}{2}) = 2, \quad y(1) = 1). \quad (3.75)$$

Let $I(y)$ in (3.75) be a Riemann integral. Since without the inequality constraint the minimizing curve is $y = 1/x$ and it intersects the boundary $y(x) = -\tfrac{32}{15}x + \tfrac{44}{15}$ at two points, the solution of (3.75) consists of three parts: $S_1(x, y), x_B \leqslant x \leqslant 1, S_2(x, y), x_A \leqslant x \leqslant x_B$ and $S_3(x, y), \tfrac{1}{2} \leqslant x \leqslant x_A$, satisfying, respectively, the following three equations

$$\frac{\partial S_1}{\partial x} - \frac{1}{4x^2}\left(\frac{\partial S_1}{\partial y}\right)^2 = 0 \quad (S_1(1,1) = 0), \quad (3.76)$$

$$\frac{\partial S_2}{\partial x} - \frac{1}{4x^2}\left(\frac{\partial S_2}{\partial y}\right)^2 = 0 \quad (y = -\tfrac{32}{15}x + \tfrac{44}{15}, \quad S_2(x_B, y_B) = S_1(x_B, y_B)). \quad (3.77)$$

$$\frac{\partial S_3}{\partial x} - \frac{1}{4x^2}\left(\frac{\partial S_3}{\partial y}\right)^2 = 0 \quad (S_3(x_A, y_A) = S_2(x_A, y_A)), \quad (3.78)$$

where x_A, x_B, y_A and y_B are unknown constants. The equations (3.76), (3.77), (3.78) should be solved simultaneously, subject to the additional boundary condition

$$S_1(x, y) = 0 \quad \text{on} \quad g(x, y) = 0, \quad g(1, 1) = 0, \quad I_0 = \min_g S_3(\tfrac{1}{2}, 2, g). \quad (3.79)$$

The solution of the above boundary-value problem is considerably simplified by the knowledge that the unconstrained problem (3.75) admits non-degenerated extremals. A step-by-step procedure, based on the transversality condition (3.32), is therefore possible.

Using (3.76) with (3.32) and $S_1(1, 1) = 0$ yields

$$S_1(x, y) = \frac{x}{1-x}(y-1)^2 \quad (x_B \leqslant x \leqslant 1). \quad (3.80)$$

One boundary condition for the equation (3.77) becomes therefore

$$S_2(x_B, y_B) = \frac{x_B}{1-x_B}(y_B-1)^2 = \frac{x_B}{1-x_B}(-\tfrac{32}{15}x_B+\tfrac{29}{16})^2.$$

Because $S_2(x,y) = ct$ is transversal to the known boundary curve $y = -\tfrac{32}{15}x+\tfrac{44}{15}$, only x in $S_2(x,y)$ is an independent variable, and from (3.77) or directly from (3.75) it follows that

$$S_2(x,y) = \tfrac{1}{3}(\tfrac{32}{15})^2(x_B^3-x^3)+\frac{x_B}{1-x_B}(-\tfrac{32}{15}x_B+\tfrac{29}{15})^2$$

$$(y = -\tfrac{32}{15}x+\tfrac{44}{15}, \quad x_A \leqslant x \leqslant x_B), \quad (3.81)$$

Using (3.78) and (3.32) yields finally

$$S_3(x,y) = \frac{xx_A}{x_A-x}(y-y_A)^2+S_2(x_A,y_A)$$

$$(y_A = -\tfrac{32}{15}x_A+\tfrac{44}{15}, \quad \tfrac{1}{2} \leqslant x \leqslant x_A). \quad (3.82)$$

Substituting (3.81) into (3.82) after having set $x = x_A$, it becomes apparent that $S_3(x,y)$ depends only on the two unknown constants x_A, x_B, which must be real and satisfy the inequality $\tfrac{1}{2} < x_A \leqslant x_B < 1$. These constants must be so chosen that $S_3(\tfrac{1}{2}, 2)$ is a local minimum.

Since $S_3(\tfrac{1}{2}, 2)$ is a point-function of x_A and x_B, the Carathéodory formulation transforms the variational problem (3.75) into the point-problem

$$I_0 = \min_{x_A, x_B} S_3(\tfrac{1}{2}, 2).$$

After a differentiation of $S_3(\tfrac{1}{2}, 2)$ with respect to x_A and x_B an elementary analysis shows that

$$x_A = 1-\frac{\sqrt{6}}{8}, \quad x_B = \frac{1}{2}+\frac{\sqrt{2}}{8}.$$

The solution (3.82) coincides therefore with the solution found in §15.3.

Let us now solve problem (3.75) again, but this time use a Lagrange multiplier. The equations (3.76) and (3.78) are unchanged, but in order to derive an equation taking the place of equation (3.77) it is first necessary to replace

$$F = x^2\dot{y}^2(x) \quad \text{by} \quad \bar{F} = x^2\dot{y}^2(x)+\lambda(x)\,[y(x)+\tfrac{32}{15}x-\tfrac{44}{15}]$$

and interpret x, y and λ as independent variables. This procedure leads to a Hamilton–Jacobi equation with a constraint

$$\left[\frac{\partial S_2}{\partial x}-\frac{1}{yx^2}\left(\frac{\partial S_2}{\partial y}\right)^2\right]+[y+\tfrac{32}{15}x-\tfrac{44}{15}]\lambda = 0, \quad \frac{\partial S_2}{\partial \lambda} = 0 \quad (x_A \leqslant x \leqslant x_B).$$

$$(3.83)$$

From the second equation in (3.83) it follows that the point-function $S_2(x,y,\lambda)$ is idependent of λ. The same conclusion applies of course to the partial derivatives $\partial S_2/\partial x$ and $\partial S_2/\partial y$. Since by definition the first equation in (3.83) is an identity in the three variables x, y and λ, the two expressions in brackets must vanish simultaneously. Hence, (3.83) is equivalent to

$$\frac{\partial S_2}{\partial x} - \frac{1}{4x^2}\left(\frac{\partial S_2}{\partial y}\right)^2 = 0 \quad (y + \tfrac{32}{15}x - \tfrac{44}{15} = 0, \quad x_A \leqslant x \leqslant x_B)$$

and (3.83) differs from (3.77) only in notation. The method of solution is not affected by this difference.

17.3   Non-holonomous inequality constraints

Contrary to holonomous inequality constraints, non-holonomous inequality constraints affect the reduction of the fundamental system to the system of Hamilton–Jacobi equations (3.10) and especially to the Carathéodory equation (3.12), because the existence of roots

$$\psi = \psi\left(x, y, \frac{\partial S}{\partial y}\right) \quad \text{of} \quad F_\psi(x, y, \psi) = \frac{\partial S}{\partial y}$$

depends on the limitations imposed on the extremal values of $|\Psi|$. For the sake of definiteness suppose that the non-holonomous inequality constraint $\phi(x, y, \dot{y}) \leqslant 0$ added to the extremal problem (2.41), is equivalent to the inequality

$$0 \leqslant \left|\frac{\partial S}{\partial y}\right| \leqslant f(x, y), \tag{3.84}$$

where $f(x, y)$ is a known function. In order to find the solution of (2.41), subject to $\phi(x, y, \dot{y}) \leqslant 0$, it is necessary to solve the Carathéodory equation (3.12), subject to (3.84). The functional $I(y)$ in (2.41) being locally additive, the solution $S(x, y)$ of (3.12), (3.84) may be expressed as a succession of terms $S_i(x, y)$ and $S_j(x, y)$, the former satisfying the constraint

$$0 < |\partial S/\partial y| < f(x, y),$$

and the latter the constraint

$$|\partial S/\partial y| = f(x, y) \quad \text{or} \quad |\partial S/\partial y| = 0.$$

The terms $S_i(x, y)$ and $S_j(x, y)$ must be joined in such a way that

$$S(x, y) = \begin{cases} S_1(x, y) & x_0 \leqslant x \leqslant \xi_1 \\ S_2(x, y) & \xi_1 \leqslant x \leqslant \xi_2 \\ \cdots\cdots\cdots & \cdots\cdots\cdots \\ S_{n+1}(x, y) & \xi_n \leqslant x \leqslant x_1 \end{cases} \tag{3.85}$$

is a continuous and at least piecewise differentiable function of x and y.

Let $y = y_i(x)$ and $y = y_j(x)$ be segments of curves which are transversal to the curves $S_i(x,y) = ct$ and $S_j(x,y) = ct$, respectively, and consider an assembly of segments $y_i(x)$ and $y_j(x)$ which form a continuous curve $y(x)$ passing through the points (x_0, y_0) and (x_1, y_1). If (3.85) represents a local minimum or a stationary lower limit of the functional $I(y)$ in (2.41), then the segments $y_i(x)$, $y_j(x)$ meet in points (x_k, y_k), the coordinates of which are usually roots of at least one of the following algebraic equations:

$$\left.\begin{array}{ll} (a) & \dot{y}_i(x_k) = \dot{y}_j(x_k), \\[2mm] (b) & F_{\dot{y}\dot{y}}(x_k, y_i(x_k), \dot{y}_i(x_k)) = F_{\dot{y}\dot{y}}(x_k, y_j(x_k), \dot{y}_j(x_k)) = 0. \end{array}\right\} \quad (3.86)$$

Fortunately it is known from the work of Flodin and Föllinger (cf. §16.4) that the transversals $y_j(x)$ to $S_j(x,y) = ct$, are solutions of the implicit differential equation

$$\phi(x, y(x), \dot{y}(x)) = 0, \tag{3.87}$$

and this additional knowledge is generally sufficient to determine the proper number of terms in (3.85).

If (3.85) represents a non-stationary lower limit of the functional $I(y)$ in (2.41), the Carathéodory equation (3.12) is not applicable, and recourse must be taken to the more general functional differential equation (3.6). Neither the simple transversality condition (3.32) nor the continuity conditions (3.86) must then be used. In fact, (3.32) and (3.86) may be interpreted as a criterion of validity of the 'stationary' Carathéodory equation (3.12).

In control theory the segments $y_j(x)$ of $\overline{y}(x)$ and the elements $S_j(x,y)$ of $S(x,y)$ are frequently called 'bang-bang' solutions, and it is generally conjectured that the functional $I(y)$ in (2.41) attains its least value when the solution $S(x,y)$ in (3.85) is entirely 'bang-bang', i.e. when the non-boundary elements $S_i(x,y)$ are absent. This conjecture is true in many cases, but it is obviously false in general, as the following example shows.

Consider the extremal problem

$$I_0 = \min_{y \in Y} I(y) = \min_{y \in Y} \int_0^{x_1} (\dot{y}^2(x) - y^2(x))\, dx$$

$$(y(0) = y_0, \quad y(x_1) = y_1, \quad |x_1| < \pi), \tag{3.88}$$

where $I(y)$ is a Riemann integral and Y consists of continuous piecewise differentiable functions $y(x)$ satisfying the constraint

$$|\dot{y}(x)| \leqslant 1. \tag{3.89}$$

The inequality (3.89) implies that all admissible curves $y(x)$ joining the points $(0, y_0)$ and (x_1, y_1) must be confined to a quadrilateral bounded by the straight lines

$$y = \pm x + y_0 \quad \text{and} \quad y = \pm (x - x_1) + y_1. \tag{3.90}$$

If the values of y_0, x_1 and y_1 are such that the above quadrilateral does not exist, then the extremal problem (3.88), (3.89) has no solution. From $F = \dot{y}^2 - y^2$ and the second equation in (3.8) it follows that the constraint (3.89) is equivalent to the constraint

$$\left| \frac{\partial S}{\partial y} \right| \leqslant 2. \tag{3.89a}$$

From §16.3 it is known that a solution of (3.88), (3.89) exists when $y_0 = 1$, $y_1 = 1$ and $x_1 = \frac{1}{2}\pi$. This solution is given by (3.66), and with the explicit constraint (3.89) it is a strong minimum. Furthermore, this strong minimum coincides with the least value of $I(y)$, and it cannot be bettered by any bang-bang solution.

The problem (3.88), (3.89) admits of course an infinity of 'bang-bang minimizing curves' composed of straight-line segments of slope ± 1. It is easy to verify that the best bang-bang solution of (3.88), (3.89) with $y_0 = 1$, $y_1 = 1$, $x_1 = \frac{1}{2}\pi$ is given by

$$y(x) = \left\{ \begin{array}{ll} x+1, & 0 \leqslant x \leqslant \frac{1}{4}\pi \\ -(x-\frac{1}{2}\pi)+1, & \frac{1}{4}\pi \leqslant x \leqslant \frac{1}{2}\pi \end{array} \right\},$$

$$I(y) = \frac{1}{2}\pi + \frac{2}{3}[1-(1+\frac{1}{4}\pi)^3] > -1\cdot56. \tag{3.91}$$

This solution is clearly not as good as the stationary solution (3.66).

Because the constraint (3.89) bounds the functional $I(y)$ in (3.88) both from below and from above, let us replace the operator $\min_{y \in Y}$ in (3.88) by the operator $\max_{y \in Y}$, or what amounts to the same, let us change the sign of the function $F = \dot{y}^2 - y^2$ under the integral sign. The modified problem (3.88), (3.89) admits only bang-bang solutions, and for $y_0 = 1$, $y_1 = 1$, $x_1 = \frac{1}{2}\pi$ the best solution is

$$y(x) = \left\{ \begin{array}{ll} -x+1, & 0 \leqslant x \leqslant \frac{1}{4}\pi \\ x-\frac{1}{2}\pi+1, & \frac{1}{4}\pi \leqslant x \leqslant \frac{1}{2}\pi \end{array} \right\},$$

$$I(y) = \frac{2}{3}[1-(1-\frac{1}{4}\pi)^3] - \frac{1}{2}\pi \approx -0\cdot91. \tag{3.92}$$

Consider now the modified problem (3.88), (3.89) with $y_0 = \frac{1}{2}$, $y_1 = \frac{1}{2}$, $x_1 = \frac{1}{2}\pi$. The two-segment bang-bang solution

$$y(x) \equiv \left\{ \begin{array}{ll} -x+\frac{1}{2}, & 0 \leqslant x \leqslant \frac{1}{4}\pi \\ x-\frac{1}{2}\pi+\frac{1}{2}, & \frac{1}{4}\pi \leqslant x \leqslant \frac{1}{2}\pi \end{array} \right\},$$

$$I(y) = \frac{1}{2}\pi - \frac{1}{12}[1+(\frac{1}{2}\pi-1)^3] < 1\cdot44 \tag{3.93}$$

inspired from (3.92) does not, however, furnish the largest value of $I(y)$. In fact, since $I(y)$ is continuous of order $k = 1$ in h, and the minimizing curve $y = y(x)$ (3.93) crosses the x-axis at the points A ($x = \frac{1}{2}$) and B ($x = \frac{1}{2}\pi - \frac{1}{2}$) of Fig. 18, the value of $I(y)$ in (3.93) can be increased by replacing the straight-line segments AC and CB (Fig. 18), by a sawtooth curve $y = y_n(x)$, $\frac{1}{2} \leqslant x \leqslant \frac{1}{2}\pi - \frac{1}{2}$, having n triangular teeth of slope ± 1.

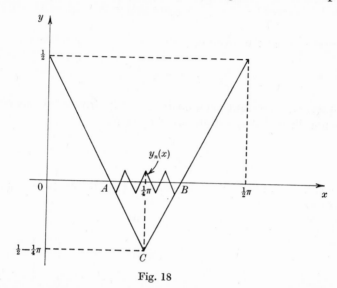

Fig. 18

If the number of teeth n is made to grow indefinitely while $\max |y_n(x)| \to 0$, the sawtooth segment $y_n(x)$ becomes a sliding regime and instead of (3.93) there results

$$y(x) = \begin{cases} -x + \frac{1}{2} & 0 \leqslant x \leqslant \frac{1}{2} \\ \lim_{n \to \infty} y_n(x) & \frac{1}{2} \leqslant x \leqslant \frac{1}{2}\pi - \frac{1}{2} \\ x - \frac{1}{2}\pi + \frac{1}{2} & \frac{1}{2}\pi - \frac{1}{2} \leqslant x \leqslant \frac{1}{2}\pi \end{cases}, \quad I(y) = \frac{1}{2}(\pi - \frac{1}{6}) > 1 \cdot 48. \quad (3.94)$$

From an examination of the various minimizing and maximizing curves obtained for the functional $I(y)$ in (3.62) we conclude that in the presence of non-holonomous inequality constraints the extremal value of a locally additive functional can be attained on functions of very diverse nature. Depending on the form of the functional $I(y)$ and on the associated boundary conditions, a continuous minimizing curve $y = y(x)$, $x_0 \leqslant x \leqslant x_1$, may consist of segments of the following curves: (a) solutions of $\delta I/\delta y = 0$, (b) solutions of $\phi(x, y, \dot{y}) = 0$, (c) sliding regimes. The main difficulty in solving extremal problems with non-holonomous inequality constraints consists in the choice of the proper segments of $y = y(x)$, or what is the same by a proper choice of 'interior' and 'boundary' elements $S_i(x, y)$ and $S_j(x, y)$ in (3.85).

Let us illustrate the last statement by determining the bang-bang solution (3.91) of (3.88), (3.89) by means of the Carathéodory formulation. Substituting $F = \psi^2 - y^2$ into (3.6) and (3.89) into (3.2) yields

$$\min_{\psi \in \Psi} \left(-\frac{\partial S}{\partial x} - \frac{\partial S}{\partial y} \psi - y^2 + \psi^2 \right) = 0 \quad (|\psi| \leqslant 1). \tag{3.95}$$

The expression in parentheses attains its lower limit when

$$\psi = + \operatorname{sgn} \frac{\partial S}{\partial y}. \tag{3.96}$$

The functional-differential equation (3.95) is reduced by means of (3.96) to the non-linear partial differential equation

$$\frac{\partial S}{\partial x} + \left| \frac{\partial S}{\partial y} \right| + y^2 - 1 = 0, \tag{3.97}$$

which can be written as a system of two linear partial differential equations

$$\left.\begin{array}{l} \dfrac{\partial S_1}{\partial x} + \dfrac{\partial S_1}{\partial y} + y^2 - 1 = 0, \\[3mm] \dfrac{\partial S_2}{\partial x} - \dfrac{\partial S_2}{\partial y} + y^2 - 1 = 0, \\[3mm] S(x, y) = \begin{cases} S_1(x, y) & \text{if } \dfrac{\partial S}{\partial y} > 0, \\[3mm] S_2(x, y) & \text{if } \dfrac{\partial S}{\partial y} < 0. \end{cases} \end{array}\right\} \tag{3.98}$$

Similarly to the case of equation (3.12), a correctly set boundary-value problem will be obtained when the boundary conditions

$$\left.\begin{array}{l} S(x_0, y_0) = 0 \quad \text{on} \quad g(x, y) = 0 \quad (g(x_0, y_0) = 0), \\[2mm] S(x_1, y_1) = \liminf_{g \in G} S(x, y, g) \end{array}\right\} \tag{3.99}$$

are added to (3.97) or to (3.98).

The solution of the boundary-value problem (3.97), (3.99) is considerably simplified by the knowledge that the two curves

$$y_1(x) = + (x - x_0) + y_0, \quad y_2(x) = - (x - x_0) + y_0, \tag{3.100}$$

resulting from the constraint (3.89) and the boundary condition $y(x_0) = y_0$, are characteristics of (3.97), (3.99). This rather self-evident result signifies that the operation $\liminf_{g \in G}$ in (3.99) involves the selection of $g(x, y)$ from a set G having at most two different elements. Hence, the presence of non-holonomous inequality constraints is not an obstacle but a help in solving an extremal problem by means of the Carathéodory formulation, no matter whether it is based on the stationary equation (3.12) or the more general equation (3.6). The situation is thus similar to that encountered in the study of extremal problems in the presence of holonomous inequality constraints.

Taking into account the known form of the characteristics passing through the point (x_0, y_0), (3.100), the first part of the boundary conditions (3.99) becomes equivalent to the point-relation (3.24). Consider now the equations (3.98). These equations are so simple that a particular solution, satisfying the condition (3.24) with $x_0 = 0$, $y_0 = 1$, can be found by inspection. In fact,

$$S_1(x,y) = x - \tfrac{1}{3}y^2 + C_1, \Big\} $$
$$S_2(x,y) = x + \tfrac{1}{3}y^2 + C_2, \Big\} \tag{3.101}$$

where $\qquad C_2 = -\tfrac{1}{3}y_0^3 = -\tfrac{1}{3}, \quad C_1 = +\tfrac{1}{3}y_0^3 = \tfrac{1}{3}.$ $\qquad$ (3.102)

In order that $S(x,y)$ be a solution of (3.88), (3.89) it is necessary to find a sequence of terms of form (3.101) so that the second part of the boundary condition (3.99) is also satisfied. Taking account of (3.100) and (3.101) it is easily verified that in the neighbourhood of the point $(0,1)$ $S(x,y)$ shows the fastest decrease when

$$S(x,y) = S_1(x,y) \quad (y = y_1(x)),$$

which yields $\qquad S(x,y) = S_1(x, y_1(x)) = x - \tfrac{1}{3}(x+1)^3 + \tfrac{1}{3}.$ $\qquad$ (3.103)

If the point $\qquad\qquad (x_1 = \tfrac{1}{2}\pi, \; y_1 = 1)$

is to be attained from the point

$$(x_0 = 0, \; y_0 = 1)$$

by a sequence of characteristic segments of form (3.100), the solution (3.103) is inappropriate for the whole interval $x_0 \leqslant x \leqslant x_1$. Suppose that it is valid in an interval $x_0 \leqslant x \leqslant \xi < x_1$, where ξ is an undetermined constant.

Since the complete solution $S(x,y)$ of (3.88), (3.89) is by definition a continuous function of x and y, the second particular solution $\bar{S}(x,y)$ of (3.98) must take at the point $(x = \xi, y = y_1(\xi))$ the value

$$S_1(\xi, y_1(\xi)) = \xi - \tfrac{1}{3}(\xi+1)^3 + \tfrac{1}{3} = \alpha(\xi).$$

Considering the above relation as analogous to (3.24), a suitable particular solution of (3.98) is found by redetermining simply the constants C_1 and C_2 in (3.101). After an examination of two possibilities it is readily found that

$$\bar{S}(x,y) = S_2(x,y) \quad (y = y_2(x)).$$

If the two particular solutions

$$S(x, y) = S_1(x, y_1(x)) \quad \text{and} \quad \bar{S}(x, y) = S_2(x, y_2(x))$$

are to be sufficient to yield a solution of the boundary-value problem (3.88), (3.89), the abscissa $x = \xi$ must be so chosen that the characteristic curve

$$y = y_2(x) = -(x - \xi) + [y_1 - (\xi - x_1)]$$

passes also through the given point $(x_1 = \tfrac{1}{2}\pi, \ y_1 = 1)$. An elementary calculation yields $\xi = \tfrac{1}{4}\pi$ and

$$\bar{S}(x, y) = S_2(x, y_2(x))$$
$$= x + \tfrac{1}{3}[(1 + \tfrac{1}{2}\pi - x)^3 - (1 + \tfrac{1}{4}\pi)^3] + \alpha(\tfrac{1}{4}\pi) - \tfrac{1}{4}\pi \quad (\tfrac{1}{4}\pi \leqslant x \leqslant \tfrac{1}{2}\pi).$$

(3.104)

The least value of $I(y)$ in (3.88), (3.89) is therefore (cf. 3.91)

$$\bar{S}(\tfrac{1}{2}\pi, 1) = 2\alpha(\tfrac{1}{4}\pi) = \tfrac{1}{2}\pi + \tfrac{2}{3}[1 - (1 + \tfrac{1}{4}\pi)^3].$$

As the example (3.88), (3.89) shows, the determination of a bang-bang solution by means of the Carathéodory formulation is very simple in principle, but rather tedious in practice. This is particularly so when the solution of

$$\lim_{\psi \in \Psi} \inf \left[F(x, y, \psi) - \left(\frac{\partial S}{\partial x} + \frac{\partial S}{\partial y} \psi \right) \right] = 0 \quad (\phi(x, y, \psi) = 0) \quad (3.105)$$

is not known explicitly, but must be determined numerically. Practical difficulties arise especially when neither the roots $\psi = \psi(x, y)$ of the algebraic equation $\phi(x, y, \psi) = 0$, nor the solution $y = y(x)$ of the ordinary differential equation $\phi(x, y(x), \dot{y}(x)) = 0$ are expressible in terms of tabulated functions. In spite of this inconvenience, which cannot be avoided anyhow, the use of the Carathéodory formulation is in general quite rewarding, because at the very least it provides additional insight into the extremal problem.

17.4   Holonomous and non-holonomous equality constraints

Extremal problems with equality constraints differ in a significant although inessential way from extremal problems without constraints or with inequality constraints. To illustrate this difference consider the holonomous problem

$$I(x_0, x_1) = \min_{y \in Y} I(y) = \min_{y \in Y} \int_{x_0}^{x_1} F(x, y(x), z(x), \dot{y}(x), \dot{z}(x))\, dx$$
$$(y(x_0) = y_0, \quad y(x_1) = y_1), \quad (3.106)$$

$$L(x, y(x), z(x)) = 0 \quad (x_0 \leqslant x \leqslant x_1), \quad (3.107)$$

where $I(y)$ is a convergent Riemann integral, F is a single-valued piecewise continuous function, Y is the set of continuous piecewise differentiable functions $y(x)$, and $L(x, y, z)$ is a piecewise continuous function of its arguments. The set Z of the functions $z(x)$ is fully defined implicitly by the conditions governing $L(x, y, z)$ and $I(y)$. For this reason it is not allowed to impose any boundary conditions, say

$$z(x_0) = z_0, \quad z(x_1) = z_1$$

on $z(x)$, unless these boundary conditions are compatible with

$$L(x, y, z) = 0 \quad (y(x_0) = y_0, \quad y(x_1) = y_1).$$

Compatible boundary conditions on $z(x)$ will be called trivial, because they have no influence on the solubility of the extremal problem (3.106), (3.107).

Appearances to the contrary, the extremal problem (3.106), (3.107) is basically one-dimensional, i.e. it involves a single unknown function, and it can thus be studied by the methods of §16. The only difference between (3.106), (3.107) and (2.41) consists in the property that the function F in $I(y)$ may have a more complex structure in (3.106), (3.107) than in (2.41). The reason for this difference lies in the implicit definition of $z(x)$ by means of (3.107), which does not necessarily constitute a unique functional relationship between $y(x)$ and $z(x)$. It is quite obvious that the equation $L(x, y(x), z(x)) = 0$ is equivalent to a set of explicit equations of the form

$$y(x) = \phi_i(x, z(x)) \quad \text{or} \quad z(x) = \psi_j(x, y(x)) \quad (i, j = 1, 2, \ldots), \quad (3.108)$$

where by definition $y \in Y$. The ϕ_i and ψ_j are thus continuous piecewise differentiable functions of their arguments. Geometrically the non-uniqueness in (3.108) means that in the Euclidean x, y, z-space the surface $L(x, y, z) = 0$ may possess folds, closed parts, and other singularities in the sense of differential geometry.

For any fixed value of j in (3.108) $z(x)$ and $\dot{z}(x)$ can be substituted into (3.106) with the result that

$$F\left(x, y, \psi_j, \dot{y}, \frac{\partial \psi_j}{\partial x} + \frac{\partial \psi_j}{\partial y} \dot{y}\right) = F_j(x, y(x), \dot{y}(x)) \quad (j = 1, 2, \ldots). \quad (3.109)$$

In general equation (3.109) represents not one but a set of single-valued functions of x, $y(x)$ and $\dot{y}(x)$. The solution $I(x_0, x_1) = I(\bar{y})$ of (3.106), (3.107) may involve more than one function F_j, with the result that the minimizing curve $y = \bar{y}(x)$ may possess angular points (corners) of a type not heretofore encountered. Even if $I(\bar{y})$ happens to be a local minimum of the functional $I(y)$, the coordinates x_k, y_k of these angular points need not necessarily satisfy the Weierstrass–Erdmann continuity conditions (2.113). In some cases, rather common in contemporary control theory (cf. §17.6), it is precisely this non-uniqueness of F_j which renders possible the existence of a continuous minimizing curve $y = \bar{y}(x)$.

The geometrical situation is more complex when $I(y)$ in (3.106) is a line integral, because in addition to angular points the minimizing curve $y = \bar{y}(x)$ may possess loops, i.e. it may be self-intersecting even inside the interval $x_0 \leqslant x \leqslant x_1$. The geometrical properties of $y = \bar{y}(x)$ may sometimes be made more transparent by interpreting

$$y = \bar{y}(x), \quad z = \psi_j(x, \bar{y}(x)), \quad x = x \quad (j = 1, 2, \ldots), \tag{3.110}$$

as a parametric representation of a continuous curve on the surface $L(x, y, z) = 0$.

When the solution of the problem (3.106), (3.107) is sought by means of the Carathéodory formulation (equation (3.6) or (3.12)), nothing is changed in principle in the method described in §16, except that the scanning process in (3.34) or (3.99) must take into account the possible non-uniqueness of F_j.

Suppose now that the holonomous constraint (3.107) is replaced by the non-holonomous constraint

$$L(x, y(x), z(x), \dot{y}(x), \dot{z}(x)) = 0. \tag{3.111}$$

An extremal problem of form (3.106), (3.111) is called a Lagrange problem. It was shown by Monge in 1784[M 25] by a geometrical argument that, provided $y(x)$ and $z(x)$ are continuously differentiable functions, an 'underdetermined' differential equation of form (3.111) defines a two-parameter family of surfaces

$$G(x, y, z, \alpha) = \beta \tag{3.112}$$

in the x, y, z-space. The parameters α and β in (3.112) take real values only. The problems (3.106), (3.107) and (3.106), (3.111) differ therefore essentially by the property that the latter possesses two free parameters. The values of these parameters have to be determined by two additional boundary conditions, either given in advance, or specified by the minimal property of $I(y)$. In problems of physical origin there exists usually enough excess information about the nominal problem (3.106), (3.111) to write the required additional boundary conditions in the form

$$z(x_0) = z_0, \quad z(x_1) = z_1. \tag{3.113}$$

The boundary conditions (3.113) are very convenient in a geometrical argument, because the two relations (3.112), (3.113), and hence also (3.111) are essentially equivalent to the single relation (3.107).

A partially degenerated case occurs when (3.111) does not contain either $\dot{y}(x)$ or $\dot{z}(x)$ explicitly. In such a case the two parameters in (3.112) are not independent, or equivalently it may be assumed that the form of G is suitably modified and that one of them is absent (cf. [M 25]). In control problems the partially degenerated constraint (3.111) takes usually the simple form

$$\dot{z}(x) = f(x, z(x), y(x)), \tag{3.114}$$

where $y(x)$ is interpreted as a control variable. In extremal problems of form (3.106), (3.114) it is therefore permissible to specify only one non-trivial boundary condition on $z(x)$. By analogy with (3.106), (3.114) the problem (3.106), (3.107) can be considered as a completely degenerated Lagrange problem of form (3.106), (3.111).

If from simplified physical considerations there arises an extremal problem of form (3.106), (3.107) with one or two non-trivial boundary conditions on $z(x)$, then this problem is incorrectly set in the sense of Hadamard, and in general it admits no solution when both $y(x)$ and $z(x)$ are required to be continuous functions. The formulation of the nominal problem (3.106), (3.107) has to be improved (cf. §2), or as an alternative, a generalized solution consistent with discontinuous $z(x)$ must be sought. Another possibility consists in introducing 'small' terms of form $\mu \dot{y}(x)$, $\epsilon \dot{z}(x)$ into (3.106) and (3.107) with $|\mu| \ll 1$ and $|\epsilon| \ll 1$. If it is attempted to carry out the limiting process $\mu \to 0$, $\epsilon \to 0$ in the solution $y(x, \mu, \epsilon)$, $z(x, \mu, \epsilon)$, $I(y)$, of the revised problem (3.106), (3.107) discontinuities will generally occur in the curves $y = \bar{y}(x, 0, 0)$, $z = \bar{z}(x, 0, 0)$. The limit process $\mu \to 0$, $\epsilon \to 0$ must therefore be supplemented by appropriate jump conditions, derived from the requirement that the revised and unrevised problems (3.106), (3.107) must be 'close' (cf. §4.3).

A similar reasoning holds when a nominal problem of form (3.106), (3.114) involves more than one non-trivial boundary condition on $z(x)$. Again there exists generally no solution when both $y(x)$ and $z(x)$ are required to be continuous functions of x. Such a nominal problem is however a correctly set problem in the sense of Hadamard when $z(x)$ is allowed to be discontinuous (cf. §17.5).

When the solution of the extremal problems (3.106), (3.107) and (3.106), (3.114) is sought by means of the Carathéodory formulation (3.8), Lagrange multipliers $\lambda_i(x)$ may be introduced whenever desired. The use of Lagrange multipliers is in general neither a hindrance nor a help, because in (3.10) this leads to the relation $\partial S/\partial \lambda_i = 0$, i.e. to the conclusion that $S(x, y)$ is independent of $\lambda_i(x)$. This property of the Carathéodory formulation (3.8) was already illustrated in §17.2.

The minimization problems (3.106), (3.107) or (3.106), (3.111) may present themselves occasionally with simultaneous holonomous or non-holonomous inequality constraints of the form

$$L_1(x, y(x), z(x)) \leqslant 0, \tag{3.115}$$

or
$$L_1(x, y(x), z(x), \dot{y}(x), \dot{z}(x)) \leqslant 0. \tag{3.116}$$

Nothing is basically changed in the method of solution, because inequality constraints can do no more than split a locally additive extremal problem into a certain number of extremal subproblems, as discussed in §§17.2 and 17.3 respectively.

17.5 Degenerate Lagrange problems

It may happen that the extremal problem of form (3.106), (3.111) does not contain explicitly the derivative $\dot{y}(x)$ of $y \in Y$. Many contemporary control problems are of such a type when $y(x)$ designates the control input and x the time. Compared to the problems discussed so far, the absence of $\dot{y}(x)$ in both (3.106) and (3.111) is a major difficulty as far as the Carathéodory equation (3.6) is concerned, because it invalidates the singular transformation (3.2). With a sacrifice of the generalization of the admissible space Y, the reduction of (3.106), (3.111) to the Carathéodory form (3.6) is still possible by assuming simply that the admissible space is the same in (3.106), (3.111) and in (3.3).

Consider first the unconstrained degenerate problem

$$I(x_0, x_1) = \min_{y \in Y} I(y) = \min_{y \in Y} \int_{x_0}^{x_1} F(x, y(x))\, dx$$
$$(y(x_0) = y_0, \quad y(x_1) = y_1), \quad (3.117)$$

which represents a degenerated version of the problem (2.41). If a meaningful comparison of $I(y)$ and the associated functional (3.3) is to be possible by means of the difference relation (3.4), $J(S, \psi)$ must be independent of ψ, or what amounts to the same, one must have

$$\frac{\partial S}{\partial y} \equiv 0. \quad (3.118)$$

Equation (3.118) implies that in the case of the problem (3.117) S must be a function of x only. Taking account of (3.118), (3.5) becomes

$$\min_{y \in Y} (I(y) - J(S)) = \min_{y \in Y} \int_{x_0}^{x_1} \left[F(x, y) - \frac{dS(x)}{dx} \right] dx = 0. \quad (3.119)$$

From (3.119) it follows that

$$S(x_1) - S(x_0) = \min_{y \in Y} \int_{x_0}^{x_1} F(x, y(x))\, dx, \quad (3.120)$$

and one is led back to the starting-point of §13.3.

Let us recall that if $I(y)$ in (3.117) is a line integral the problem (3.120) has generally no solution unless the admissible values of x and $y(x)$ are confined to a region G of the x, y-plane bounded by a curve Γ of a finite length. The minimizing curve $y = \bar{y}(x)$ of (3.120) consists in general of segments of the curves $y = y_i(x)$ defined by the algebraic equation

$$\frac{\partial F}{\partial y} = 0, \quad (3.121)$$

segments of the boundary Γ and of a certain number of vertical segments $x = ct$, connecting Γ to $y_i(x)$, and (x_0, y_0), (x_1, y_1) to either Γ or to $y(x)$.

Consider now the constrained degenerate problem

$$I(x_0, x_1) = \min_{y \in Y} I(y) = \min_{y \in Y} \int_{x_0}^{x_1} F(x, y(x), z(x), \dot{z}(x)) \, dx, \quad (3.122)$$

$$L(x, y(x), z(x), \dot{z}(x)) = 0 \quad (y(x_0) = y_0, \quad y(x_1) = y_1, \quad z(x_0) = z_0), \quad (3.123)$$

resulting from (3.106), (3.111). For simplicity suppose that $L(x, y, z, \dot{z}) = 0$ admits only one real root

$$\dot{z}(x) = f(x, y(x), z(x)). \quad (3.114)$$

By the same argument as that used in the case (3.117), $\partial S/\partial y \equiv 0$, $S = S(x, z)$ and the relation analogous to (3.5) becomes

$$\min_{y \in Y} [I(y) - J(S)] = \min_{y \in Y} \int_{x_0}^{x_1} \left[F(x, y, z, \dot{z}) - \left(\frac{\partial S}{\partial x} + \frac{\partial S}{\partial z} \dot{z} \right) \right] dx = 0. \quad (3.124)$$

Since $z(x)$ is not a free variable, but on the contrary completely specified by (3.114) for any fixed $y(x) \in Y$, substituting (3.114) into (3.124) yields

$$\min_{y \in Y} [I(y) - J(S)] = \min_{y \in Y} \int_{x_0}^{x_1} \left[F(x, y, z, f) - \left(\frac{\partial S}{\partial x} + \frac{\partial S}{\partial z} f(x, z, y) \right) \right] dx = 0. \quad (3.125)$$

As in the case of §16.1 a sufficient but not necessary condition for (3.125) to hold is

$$\min_{y \in Y} H \left(x, y, z, \frac{\partial S}{\partial x}, \frac{\partial S}{\partial z} \right) = 0, \quad (3.126)$$

where

$$H = F(x, y, z, f) - \left(\frac{\partial S}{\partial x} + \frac{\partial S}{\partial z} f \right).$$

The function H in (3.126) is often called a Hamiltonian of (3.122), (3.123) because of its rather superficial resemblance to the Hamiltonian function $H = L + W$ used in mechanics (cf. §6).

At least in principle it is quite straightforward to eliminate the minimization variable $y(x)$ from (3.126). Suppose first that the least value of $I(y)$ in (3.122) is a local minimum. The function H in (3.126) is therefore stationary with respect to y, and

$$\frac{\partial S}{\partial x} + \frac{\partial S}{\partial z} f - F = 0, \quad \frac{\partial F}{\partial y} + \left(\frac{\partial F}{\partial f} - \frac{\partial S}{\partial z} \right) \frac{\partial f}{\partial y} = 0. \quad (3.127)$$

Consider the second equation in (3.127) as an algebraic equation in y and let its real roots be

$$y_i = \phi_i \left(x, z, \frac{\partial S}{\partial z} \right) \quad (i, 1, 2, \ldots). \quad (3.128)$$

Substituting (3.128) into the first equation in (3.127) yields a set of Hamilton–Jacobi equations

$$\frac{\partial S}{\partial x} + H_i\left(x, z, \frac{\partial S}{\partial z}\right) = 0 \quad (i = 1, 2, \ldots), \tag{3.129}$$

where

$$H_i = f(x, z, \phi_i)\frac{\partial S}{\partial z} - F(x, \phi_i, z, f(x, z, \phi_i)) \tag{3.130}$$

is the 'reduced' Hamiltonian H, corresponding to the root $y_i = \phi_i$. The equations (3.124) are analogous to the equations (3.10), and they can be combined to yield a single partial differential equation

$$\Phi\left(x, z, \frac{\partial S}{\partial x}, \frac{\partial S}{\partial z}\right) = 0, \tag{3.131}$$

analogous to (3.12)

Suppose now that the least value of $I(y)$ in (3.122) is a non-trivial lower limit. In such a case the stationarity conditions (3.127) can simply be replaced by the 'minimality' condition

$$\liminf_{y \in Y} \left[F(x, y, z, f) - \left(\frac{\partial S}{\partial x} + \frac{\partial S}{\partial z}f\right)\right] = 0, \tag{3.132}$$

which also defines $y(x)$ algebraically as a function of $x, z, \partial S/\partial x$ and $\partial S/\partial z$. Let

$$y_j = h_j\left(x, z, \frac{\partial S}{\partial x}, \frac{\partial S}{\partial z}\right) \quad (j = 1, 2, \ldots) \tag{3.133}$$

be the real roots of the operator equation (3.132). Substituting (3.133) into the Hamiltonian H defined by (3.126) and taking account of (3.132) yields a system of partial differential equations

$$\liminf_{y \in Y} H_j = F(x, h_j, z, f(x, z, h_j)) - \left(\frac{\partial S}{\partial x} + \frac{\partial S}{\partial z}f(x, z, h_j)\right) = 0$$
$$(j = 1, 2, \ldots). \tag{3.134}$$

If a solution of the extremal problem (3.122), (3.123) is to be obtained by means of the system of equations (3.134) it is necessary to formulate a correctly set boundary-value problem. This objective is accomplished by combining the equations in (3.134) into a single partial differential equation of form (3.131), and then imposing boundary conditions of form (3.99). The practical use of equations (3.129) and (3.134) will be illustrated in §17.7.

17.6 Lagrange problems with more than one non-holonomous equality constraint

Consider a more general version of the extremal problem (3.106), (3.111):

$$I(x_0, x_1) = \min_{y \in Y} I(y) = \min_{y \in Y} \int_{x_0}^{x_1} F(x, z_i(x), y(x), \dot{z}_i(x), \dot{y}(x))\, dx, \quad (3.135)$$

$$L_j(x, z_i(x), y(x), \dot{z}_i(x), \dot{y}(x)) = 0, \quad (3.136)$$

$$y(x_0) = y_0, \quad y(x_1) = y_1, \quad z_i(x_0) = z_{i0}, \quad z_i(x_1) = z_{i1}$$

$$(i, j = 1, 2, ..., n), \quad (3.137)$$

which is assumed to satisfy the same smoothness conditions. The functional $I(y)$ is as before defined by a Riemann integral. The only difference between the problems (3.106), (3.111) and (3.135), (3.136), (3.137) consists in the number $n > 1$ of dependent variables $z_i(x)$ and constraints $L_j = 0$. The problem (3.135), (3.136), (3.137) contains only one minimization variable $y(x) \in Y$, and because of this property is not essentially different from the problem (2.41). The constraints (3.136) can be interpreted as a part of the definition of Y.

To study the problem (3.135), (3.136), (3.137), along similar lines as the problem (3.106), (3.111), it is necessary to know the general solution of the underdetermined differential system (3.136). Unfortunately the results of Monge[M 25] have not yet been extended to the case $n > 1$. To eliminate some complications of the general problem (3.136) let us make the simplifying assumption that (3.136) is equivalent to the single system of explicit equations

$$\dot{z}_j(x) = f_j(x, z_i(x), y(x), \dot{y}(x)). \quad (3.138)$$

A sufficient condition for (3.138) to exist is that the $L_j(x, z_i, y, \dot{z}_i, \dot{y})$ admits continuous partial derivatives with respect to $\dot{z}_i$ and that the Jacobian

$$\frac{\partial(L_1, L_2, ..., L_n)}{\partial(\dot{z}_1, \dot{z}_2, ..., \dot{z}_n)} \neq 0 \quad (3.139)$$

in the whole interval $x_0 \leqslant x \leqslant x_1$. The condition (3.139) is not necessary in general because the roots (3.138) of (3.136) need not be simple.

Let us now formulate a conjecture, which it is hoped is an extension of the results of Monge[M 25]: the differential system (3.136) is equivalent to the set of n equations

$$\phi_j(x, z_i, y, C_i, \bar{C}_i) = 0 \quad (i, j = 1, 2, ..., n), \quad (3.140)$$

where C_i and $\bar{C}_i$ are arbitrary real constants.

This conjecture is based on the argument that it holds when the L_j are linear functions of $z_i, y, \dot{z}_i$ and $\dot{y}$, and that this result can be extended to certain non-linear cases. In fact, if the system (3.136) is linear, then it can be written in the form

$$\frac{d\bar{z}}{dx} = A\bar{z} + B + y(x)\,C + \dot{y}(x)\,D, \qquad (3.141)$$

where
$$A(x) = \begin{pmatrix} a_{11}(x) & a_{12}(x) & \dots & a_{1n}(x) \\ a_{21}(x) & a_{22}(x) & \dots & a_{2n}(x) \\ \dots & \dots & \dots & \dots \\ a_{n1}(x) & a_{n2}(x) & \dots & a_{nn}(x) \end{pmatrix}$$

is a $n \times n$ matrix and

$$\bar{z}(x) = \begin{pmatrix} z_1(x) \\ z_2(x) \\ \vdots \\ z_n(x) \end{pmatrix}, \quad B(x) = \begin{pmatrix} b_1(x) \\ b_2(x) \\ \vdots \\ b_n(x) \end{pmatrix},$$

$$C(x) = \begin{pmatrix} c_1(x) \\ c_2(x) \\ \vdots \\ c_n(x) \end{pmatrix}, \quad D(x) = \begin{pmatrix} d_1(x) \\ d_2(x) \\ \vdots \\ d_n(x) \end{pmatrix},$$

are column vectors. Suppose that the elements $a_{ij}(x)$ of A are continuous functions in the interval $x_0 \leqslant x \leqslant x_1$ and consider the homogeneous matrix equation

$$\frac{d\bar{z}}{dx} = A(x)\,\bar{z}(x). \qquad (3.142)$$

Suppose, furthermore, that $y(x)$ is a fixed but otherwise arbitrary function of the admissible set Y and let

$$E(x) = y(x)\,C + \dot{y}(x)\,D. \qquad (3.143)$$

It is well known that if

$$Z(x) = \begin{pmatrix} Z_{11}(x) & Z_{12}(x) & \dots & Z_{1n}(x) \\ Z_{21}(x) & Z_{22}(x) & \dots & Z_{2n}(x) \\ \dots & \dots & \dots & \dots \\ Z_{n1}(x) & Z_{n2}(x) & \dots & Z_{nn}(x) \end{pmatrix} \qquad (3.144)$$

is a fundamental matrix of (3.142), i.e. the columns of $Z(x)$ are linearly independent particular solutions of (3.142), then the general solution of (3.141) is given by

$$\bar{z}(x) = Z(x)\,M + Z(x)\int_{x_0}^{x_1} Z^{-1}(\xi)\,[B(\xi) + E(\xi)]\,d\xi, \qquad (3.145)$$

where M is a column vector and its elements $m_1, m_2, \ldots, m_n$ are arbitrary real constants. Since $y(x)$ and $\dot{y}(x)$ occur linearly in (3.145), every equation of (3.143) can be interpreted as a non-homogeneous first-order linear differential equation in $y(x)$. Rearranging (3.145), and if necessary using (3.141) to eliminate derivatives of $z_i(x)$, each equation of (3.145) can be written in the form

$$\dot{y}(x) = \alpha(x, m_i)\, y(x) + \beta(x, m_i) + \sum_{j=1}^{n} \gamma_j(x, m_i)\, z_j(x) \quad (i = 1, 2, \ldots, n),$$

(3.146)

where the coefficients α, β and γ are functions of x and m_i only. Suppose now that the $z_i(x)$ are fixed, but otherwise arbitrary. The integration of (3.146), yields one arbitrary constant, say $\overline{m}_i$, and since there are n different equations of form (3.146), there are n independent integration constants $\overline{m}_i$. The final result is therefore equation (3.140).

The preceding argument is unchanged when the system (3.136) is non-linear, provided only the integrations with respect to $z_i(x)$ and $y(x)$ can be carried out separately. The details of such a two-step integration are too involved to be examined here.

Let us now return to the extremal problem (3.135), (3.136), (3.137). The arbitrary constants in (3.140) can be determined by means of the $2(n+1)$ boundary conditions (3.137). Substituting the values of $y(x_0)$, $z_i(x_0)$ and $y(x_1)$, $z_i(x_1)$ into (3.140) yields $2n$ equations, which in principle can be solved for the $2n$ unknown constants C_i and $\overline{C}_i$. Knowing C_i and $\overline{C}_i$ the system (3.140) can be interpreted as an algebraic system in $z_i(x)$ and $y(x)$. Solving (3.140) for $z_i(x)$, the functions $z_i(x)$ are expressed in terms of x and $y(x)$, and they can thus be eliminated from the function $F(x, z_i, y, \dot{z}_i, \dot{y})$ in (3.135). The result of this elimination is a problem of form (2.41).

Some relatively minor difficulties will arise when the system (3.140) admits more than one set of independent values of the constants C_i, $\overline{C}_i$, or when the functional relations between $z_i(x)$ and $y(x)$ are not unique. Such difficulties constitute a fundamental feature of the Lagrange problem (3.135), (3.136), (3.137) and they cannot be avoided no matter what method of solution is adopted.

The Lagrange problem (3.135), (3.136), (3.137) becomes degenerate when the system (3.140) contains less than $2n$ independent constants. This occurs usually when all equations in (3.136) are not fully independent or when some derivatives $\dot{z}_i(x)$, $\dot{y}(x)$ are missing. In such a case the number of non-trivial boundary conditions in (3.137) must be reduced accordingly. The difference p between $2n$ and the number of independent constants in (3.140) can be called the order of degeneracy of the Lagrange problem (3.135), (3.136), (3.137). If $\dot{y}(x)$ is missing in all equations of the system (3.136) $p = n$, and the degenerate 'general' Lagrange problem

(3.135), (3.136), (3.137) reduces to the non-degenerate 'special' Lagrange problem

$$
\left.
\begin{aligned}
I(x_0, x_1) = \min_{y \in Y} I(y) = \min_{y \in Y} \int_{x_0}^{x_1} F(x, z_i(x), y(x), \dot{z}_i(x), \dot{y}(x)) \, dx, \\
L_j(x, z_i(x), y(x), \dot{z}_i(x)) = 0 \quad (y(x_0) = y_0, \; y(x_1) = y_1, \\
z_i(x_0) = z_{i0}, \quad i, j = 1, 2, \ldots, n),
\end{aligned}
\right\} \quad (3.147)
$$

studied by Carathéodory[C 4],[C 5]. The case when the implicit constraints $L_j(x, z_i, y, \dot{z}_i) = 0$ are replaced by the explicit constraint

$$
\dot{z}_j(x) = f_j(x, z_i(x), y(x)) \tag{3.148}
$$

was studied by numerous authors, and in particular by Bolza, Bliss and their disciples (see for example [B 13],[B 16],[G 9],[M 12]). Following the example set by Bolza, most authors restricted their investigations to particular properties of (3.147) when a further degeneracy no long occurs, i.e. when (3.147) is 'normal' as far as the number of admissible non-trivial boundary conditions is concerned. It appears that it was Escherich[E 2] who first called attention to the now rather obvious fact that the Lagrange problem (3.147), or (3.147) with $L_j = 0$ replaced by (3.148), is not always normal. Following the work of Escherich, Carathéodory has shown[C 5] that the order of degeneracy q of (3.147) is an invariant with respect to non-singular transformations of variables, and therefore a characteristic property of that problem. Carathéodory has called q the 'class' of the Lagrange problem (3.147). When (3.147) possesses a unique Hamiltonian function H admitting continuous partial derivatives of the second order, the class q of (3.147) can be deduced from H by elementary operations (see §§ 11 and 12[C 5]).

A very particular type of degeneracy occurs when $\dot{y}(x)$ is missing completely in (3.147). This is usually the case in contemporary control problems when $y(x)$ is interpreted as a control input. Because the boundary conditions on $y(x)$ can no longer be prescribed arbitrarily, if $y(x)$ is to be a continuous function, the theory of 'normal' Lagrange problems is inappropriate for the study of such control problems. This conclusion holds a fortiori when inequality constraints are added to (3.147). Consequently the argument of Hestenes[H 7] and Berkovitz[B 9],[B 10] based on the artifice of Valentine and the theory of normal Lagrange problems is of rather limited usefulness in control theory.

The knowledge of the continuity properties of the minimizing function $\bar{y}(x) \in Y$ is very important if the Lagrange problem (3.135), (3.136), (3.137) or (3.147) has to be solved numerically, because the construction of an efficient computational algorithm depends critically on this property. The computational algorithm can be considerably simplified when

$\bar{y}(x)$ is continuous in the whole interval $x_0 \leqslant x \leqslant x_1$. In such a case the order of degeneracy q is usually zero, and the number of non-trivial boundary conditions is $2(n+1)$ in (3.137) and $n+2$ in (3.147). If $q \neq 0$, and if physical considerations require an unchanged number of non-trivial boundary conditions, $\bar{y}(x)$ may be only piecewise continuous, and there arises the problem of determining the number of continuous segments N of $\bar{y}(x)$ and the coordinates, $x_k, y_k, k = 2, \ldots,$ of the points of discontinuity. The abscissae x_k are fixed by the requirement that $\bar{y}(x)$ 'joins' the terminal points (x_0, y_0) and (x_1, y_1), and the ordinates y_k are fixed by certain jump-conditions

$$M(y(x_k^-), y(x_k^+), x_k) = 0. \tag{3.149}$$

The number N and the form of the function M in (3.149) is determined entirely by the property

$$I(\bar{y}) = \liminf_{y \in Y} I(y). \tag{3.150}$$

The jump-condition (3.149) can be deduced either directly from (3.150), by examining the continuity of $I(y)$ in the neighbourhood of $y = \bar{y}(x)$, or indirectly by imbedding the degenerate Lagrange problem in a suitably chosen family of normal Lagrange problems and then performing a limiting process. Both procedures are obviously equivalent (cf. §§ 4.3, 7 and 14.6), but the details depend on the features of the Lagrange problem in question.

17.7 Solution of degenerate Lagrange problems by means of the Carathéodory formulation

An illustrative example of the solution of a degenerate Lagrange problem by means of the Carathéodory functional equation (3.6) was given already in § 17.3. Let us now extend the method used to the more general extremal problem (3.135), (3.136), (3.137). Suppose in fact that the order of degeneracy q of the problem is already known and that the number of non-trivial boundary conditions in (3.137) has been fixed accordingly. For definiteness suppose for the moment that $q = 0$.

The first procedure is based on the observation that the lower limit $I(x_0, x_1)$ of $I(y)$ in (3.135) depends on the function $y(x)$ only. In accordance with the conjecture (3.140), which permits to express the $z_i(x)$ by means of the $y(x)$, S can be considered as a function of x and y. From this starting-point the functional equation (3.6) corresponding to (3.135), (3.136), (3.137) is

$$\min_{y \in Y} \left[F(x, z, y, \dot{z}_i, \psi) - \left(\frac{\partial S}{\partial x} + \frac{\partial S}{\partial y} \psi \right) \right] = 0, \tag{3.151}$$

where the $z_i, \dot{z}_i, i = 1, 2, \ldots, n,$ are determined implicitly by the constraint (3.136). Except for the appearance of the z_i in F, equation (3.151) coincides with the equation (3.6), and because $S = S(x, y)$, the boundary conditions (3.33) still apply. Fixing $g(x, y)$ in (3.33) makes it possible to determine the tangential derivatives $\partial S/\partial l$ along the curve $g(x, y) = 0$:

$$\frac{\partial S}{\partial l} = -\frac{\partial S}{\partial x}\frac{\partial g}{\partial y} + \frac{\partial S}{\partial y}\frac{\partial g}{\partial x}. \tag{3.152}$$

Suppose for definiteness that the curve $g(x, y) = 0$ passes through the point (x_1, y_1). Since by definition (3.151) is valid in the whole interval $x_0 \leqslant x \leqslant x_1$ it is valid at the abscissa $x = x_1$. Combining (3.151), (3.152), (3.136) and the part of (3.137) pertaining to $x = x_1$ yields $n + 2$ independent algebraic equations for the values at $x = x_1$ of the $n + 2$ unknowns

$$\alpha = \frac{\partial S}{\partial x}, \quad \beta = \frac{\partial S}{\partial y}, \quad \dot{z}_i(x).$$

The value of ψ at $x = x_1$ is fixed automatically by the equation

$$\frac{\partial S}{\partial x} + \frac{\partial S}{\partial y}\psi - F(x, z_i, y, \dot{z}_i, \psi) = 0. \tag{3.153}$$

The knowledge of the initial set of values $\alpha(x_1)$, $\beta(x_1)$, $z_i(x_1)$, $\dot{z}_i(x_1)$, $y(x_1)$ is sufficient to define at least one particular solution $z_i(x)$ of (3.136), and $y(x)$, $S(x)$, $u(x)$, $v(x)$ of the characteristic equations (3.35) associated with (3.153). In principle it is thus possible to solve (3.153) for a fixed function $g(x, y)$ in (3.33). If the corresponding solution exists in the whole interval $x_0 \leqslant x \leqslant x_1$, an element $S(x_0, y_0, g)$ of (3.34) or of (3.99) is obtained. Repeating the procedure just described for a minimal sequence of 'initial' functions $g(x, y)$ permits to satisfy the boundary condition (3.34) or (3.99) and thus to obtain a solution of (3.135), (3.136), (3.137).

When $q \neq 0$ and the number of non-trivial boundary conditions in (3.137) is chosen consistently with the number of independent equations in (3.136), the same procedure as above can be followed. Difficulties will arise only when q does not have a constant value in the whole interval $x_0 \leqslant x \leqslant x_1$. That such a possibility may occur was first pointed out by Morse (see [C 5]). If the interval $x_0 \leqslant x \leqslant x_1$ cannot be subdivided into a finite number of subintervals where q is constant, then no indirect method of solving the extremal problem (3.135), (3.136), (3.137) appears to be known.

A considerable simplification occurs in the case of the special Lagrange problem (3.147), particularly when the implicit constraints $L_j = 0$ are replaced by the explicit constraints (3.148). In such a case it is advantageous to suppose that S is not only a function of the independent variables x, y, but also of the dependent variables $z_i = z_i(x, y)$. Supposing, for simplicity, that $L_j = 0$ in (3.147) is equivalent to only one system (3.148), one obtains instead of (3.151) the functional equation

$$\left.\begin{array}{l} \min_{\psi \in \Psi} \left[F(x, z_i, y, \dot{z}_i, \psi) - \left(\dfrac{\partial S}{\partial x} + \dfrac{\partial S}{\partial y} \psi + \sum_{i=1}^{n} \dfrac{\partial S}{\partial z_i} \dot{z}_i \right) \right] = 0, \\[2mm] \dot{z}_i(x) = f_j(x, z_i(x), y(x)) \quad (i, j = 1, 2, ..., n). \end{array}\right\} \quad (3.154)$$

If neither $L_j = 0$ nor F in (3.147) contain $\dot{y}(x)$ explicitly, the above equation simplifies into

$$\left.\begin{array}{l} \min_{y \in Y} \left[F(x, z_i, y, \dot{z}_i) - \left(\dfrac{\partial S}{\partial x} + \sum_{i=1}^{n} \dfrac{\partial S}{\partial z_i} \dot{z}_i \right) \right] = 0, \\[2mm] \dot{z}_j(x) = f_j(x, z_i(x), y(x)) \quad (i, j = 1, 2, ..., n). \end{array}\right\} \quad (3.155)$$

Another type of simplification occurs when the differential equation in (3.148) can be solved independently of the extremal problem (3.147). Substituting $z_i(x)$ and $\dot{z}_i(x)$ so obtained into the functional $I(y)$, and taking account of the boundary conditions imposed on $z_i(x)$, there results an extremal problem which is continuous of the order $k = 0$ in h and which can usually be solved best by the methods of §13.3.

As an illustration of the use of the Carathéodory formulation in the study of Lagrange problems consider the well-known control problem for a double-integrator plant (see for example [F 9],[F 11],[F 12],[W 13] and §15 of [P 11]):

$$I(x_0, x_1) = \min_{y \in Y} I(y) = \min_{y \in Y} \int_{x_0}^{x_1} F(x, z_1(x), z_2(x), y(x)) \, dx, \quad (3.156)$$

$$\frac{dz_1}{dx} = z_2, \quad \frac{dz_2}{dx} = \frac{1}{a} y \quad (a > 0, \ |y(x)| \leqslant 1), \quad (3.157)$$

$$z_1(x_0) = 0, \quad z_2(x_0) = 0, \quad z_1(x_1) = z_{10}, \quad z_2(x_1) = z_{20}, \quad (3.158)$$

where x_0, x_1, z_{10}, z_{20} are real constants, $y(x)$ is the control input, x represents time, and Y is the space of piecewise continuous functions. Suppose $x_0 = 0$, $x_1 > 0$ because this can be done without loss of generality.

The problem (3.156), (3.157), (3.158) is a special Lagrange problem of form (3.147) with $n = 2$ and an additional inequality constraint. Since the function $\dot{y}(x)$ does not appear explicitly, the problem (3.156), (3.157), (3.158) is degenerate with $q = 2$. Consequently a solution with a continuous minimizing curve $y = \bar{y}(x)$ exists only when either the first or the second pair of boundary conditions in (3.158) is trivial. When both pairs of boundary conditions in (3.158) are non-trivial, the minimizing curve $y = \bar{y}(x)$ is necessarily discontinuous. With this point made clear, the method of actually finding a solution of (3.156), (3.157), (3.158) is quite straightforward.

From an inspection of (3.157) it is obvious that the differential equations contained therein can be integrated explicitly independently of (3.156):

$$z_2(x) = \phi_2(x, y(x)) = \frac{1}{a} \int_0^x y(\xi)\, d\xi + C_1,$$

$$z_1(x) = \phi_1(x, y(x)) = \frac{1}{a} \int_0^x \int_0^{\xi_1} y(\xi_2)\, d\xi_2\, d\xi_1 + C_1 x + C_2,$$

$$\tag{3.159}$$

where C_1, C_2 are constants of integration. Substituting (3.159) into (3.156) there results an extremal problem

$$I(0, x_1) = \min_{y \in Y} I(y) = \min_{y \in Y} \int_0^{x_1} F(x, \phi_1(x, y(x)), \phi_2(x, y(x)), y(x))\, dx$$

$$(|y(x)| \leqslant 1, \quad y(x_0) = y_0, \quad y(x_1) = y_1,$$

$$\phi_1(x_0, y_0) = 0, \quad \phi_2(x_0, y_0) = 0, \quad \phi_1(x_1, y_1) = z_{10}, \quad \phi_2(x_1, y_1) = z_{20}),$$

$$\tag{3.160}$$

already discussed in §15.6. The curve $y = \bar{y}(x)$ minimizing $I(y)$ in (3.160) may consist therefore of vertical segments $x = x_k$, $-1 \leqslant y \leqslant +1$, $0 \leqslant x_k \leqslant x_1$, $k = 2, \ldots$, constant horizontal segments $y(x) \equiv +1$ and $y(x) = -1$, and segments of the solution $y_s(x)$ of the implicit equation

$$(\partial/\partial y) F(x, \phi_1(x, y), \phi_2(x, y), y) = 0 \quad (-1 \leqslant y \leqslant +1). \tag{3.161}$$

As was pointed out in §17.3, segments of the stationary curve $y = y_s(x)$ cannot generally be ruled out in favour of the bang-bang segments $y(x) = \pm 1$.

For a particular form of F in (3.160) it is not always necessary to make a systematic exploration of the various possible combinations of segments. The appropriate choice is often suggested from the physical context of (3.160). Two pairs of functions $\phi_1(x, y)$, $\phi_2(x, y)$, both defined by (3.159) but having different integration constants should generally be sufficient to satisfy the boundary conditions in (3.160). From (3.157) it follows that the functions ϕ_1 and ϕ_2 are continuous and at least piecewise differentiable with respect to x.

Suppose now that the function F in (3.156) is such that the minimizing function $\bar{y}(x)$ does not contain any segments of the stationary curve $y = y_s(x)$. This is certainly the case when the equation (3.161) is an identity, which happens in particular when the function F in (3.156) does not contain $y(x)$ explicitly. Under these circumstances (cf. § 15.3) $\bar{y}(x)$ consists exclusively of the segments $y(x) \equiv +1$ and $y(x) \equiv -1$. The corresponding solutions of (3.157) are

$$z_2 = \frac{1}{a} x + C_1, \qquad z_1 = \frac{1}{2a} x^2 + C_1 x + C_2, \qquad y(x) \equiv +1, \quad (3.162)$$

$$z_2 = -\frac{1}{a} x + \bar{C}_1, \quad z_1 = -\frac{1}{2a} x^2 + \bar{C}_1 x + \bar{C}_2, \quad y(x) \equiv -1, \quad (3.163)$$

where $C_1, C_2, \bar{C}_1, \bar{C}_2$ are independent integration constants.

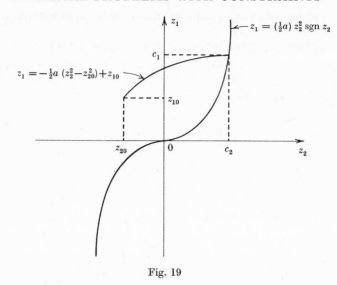

Fig. 19

The pair of boundary conditions $z_1(x_1) = z_{10}$, $z_2(x_1) = z_{20}$ is trivial when the point (z_{20}, z_{10}) is located on one of the curves, $z_1 = z_1(z_2)$ (Fig. 19), given parametrically by

$$z_2(x) = \frac{1}{a}x, \qquad z_1(x) = \frac{1}{2a}x^2, \qquad y(x) \equiv +1, \qquad (3.164)$$

$$z_2(x) = -\frac{1}{a}x, \qquad z_1(x) = -\frac{1}{2a}x^2, \qquad y(x) \equiv -1, \qquad (3.165)$$

Eliminating x in (3.164) and (3.165) yields

$$z_1 = \tfrac{1}{2}az_2^2, \qquad y(x) \equiv +1, \qquad (3.166)$$

$$z_1 = -\tfrac{1}{2}az_2^2, \qquad y(x) \equiv -1. \qquad (3.167)$$

Only the segments corresponding to $x > 0$ are significant in (3.166) and (3.167), because by definition $0 \leqslant x \leqslant x_1$.

The curve

$$z_1 = \tfrac{1}{2}az_2^2 \operatorname{sgn} z_2, \qquad (3.168)$$

resulting from the combination of (3.166), (3.167) and $x > 0$ is privileged for the control problem (3.156), (3.157), (3.158) in the sense that it represents the locus of trivial boundary points (z_{20}, z_{10}), or in other words the locus of boundary points (z_{20}, z_{10}) for which there exists a continuous minimizing curve $y = \bar{y}(x)$. This privileged minimizing curve is given either by $y(x) \equiv 1$ or $y(x) \equiv -1$. Physically this means that if (z_{20}, z_{10}) is located on the curve (3.168), the double-integrator plant can be brought from the phase-state (z_{20}, z_{10}) to the phase-state $(0, 0)$ without a change in the control input $y(x)$.

If the point (z_{20}, z_{10}) is not located on the curve (3.168), the minimizing curve $\bar{y}(x)$ must consist of two segments, one segment of (3.168) and one segment of either (3.162) or (3.163). For $z_2(x_1) = z_{20}$, $z_1(x_1) = z_{10}$ the equations (3.162), (3.163) become

$$z_2(x) = \frac{1}{a}(x - x_1) + z_{20}, \qquad z_1(x) = \frac{1}{2a}(x - x_1)^2 + z_{20}(x - x_1) + z_{10},$$

$$y(x) \equiv +1, \qquad (3.169)$$

$$z_2(x) = -\frac{1}{a}(x - x_1) + z_{20}, \quad z_1(x) = -\frac{1}{2a}(x - x_1)^2 + z_{20}(x - x_1) + z_{10},$$

$$y(x) \equiv -1, \qquad (3.170)$$

and after the elimination of $(x - x_1)$

$$z_1 = \tfrac{1}{2}a(z_2^2 - z_{20}^2) + z_{10}, \qquad y(x) \equiv +1, \qquad (3.171)$$

$$z_1 = -\tfrac{1}{2}a(z_2^2 - z_{20}^2) + z_{10}, \quad y(x) \equiv -1. \qquad (3.172)$$

Suppose for definiteness that the phase-point (z_{20}, z_{10}) corresponds to a curve $z_1 = z_1(z_2)$ with $y(x) \equiv -1$, i.e. to the curve defined by (3.172). In such a case it is located on the left side of the curve (3.168) shown in Fig. 19. Let (c_2, c_1) be the point of intersection of the curves defined by (3.168) and (3.172), and let γ be the corresponding value of x. An elementary calculation yields

$$c_2 = +\frac{1}{\sqrt{a}}\sqrt{[\tfrac{1}{2}az_{20}^2 + z_{10}]}, \quad c_1 = \tfrac{1}{4}az_{20}^2 + \tfrac{1}{2}z_{10}, \quad \gamma = x_1 + \sqrt{[\tfrac{1}{2}a^2z_{20}^2 - az_{10}]}.$$

$$(3.173)$$

If the point (z_{20}, z_{10}) were located on the right side of the curve (3.168), then instead of (3.173) one would have

$$c_2 = \frac{1}{\sqrt{a}}\sqrt{[\tfrac{1}{2}az_{20}^2 - z_{10}]}, \quad c_1 = -\tfrac{1}{4}az_{20}^2 + \tfrac{1}{2}z_{10}, \quad \gamma = x_1 - \sqrt{[\tfrac{1}{2}a^2z_{20}^2 - az_{10}]}.$$

$$(3.174)$$

The time x_1 required to move the plant from the state (z_{20}, z_{10}) to the state $(0, 0)$ is obviously the time required to move it from the state (z_{20}, z_{10}) to the state (c_2, c_1) and from the state (c_2, c_1) to the state $(0, 0)$. Hence,

$$x_1 = ac_2 - a(z_{20} - c_2) = az_{20} + 2\sqrt{[\tfrac{1}{2}a^2z_{20}^2 - az_{10}]}, \qquad (3.175)$$

when (z_{20}, z_{10}) is on the left side of the curve (3.168), and

$$x_1 = -az_{20} + 2\sqrt{[\tfrac{1}{2}a^2z_{20}^2 - az_{10}]}, \qquad (3.176)$$

when it is on the right side.

From the preceding elementary results it is possible to solve the problem (3.156), (3.157), (3.158) for a variety of forms of F in (3.156):

Example 1: $F \equiv 1$ (minimum time problem).

Since

$$\min_{y \in Y} I(y) = \int_0^{x_1} dx = x_1,$$

the solution of this problem is given by either (3.175) or (3.176).

Example 2: $F = |z_1(x)|$ (integral modulus-error criterion).

The value of $\min_{y \in Y} I(y)$ is found by a substitution of $z_1(x)$ into $I(y)$ either from (3.164) and (3.172) or (3.165) and (3.171), depending on whether (z_{20}, z_{10}) is on the left or on the right of the curve (3.168) (Fig. 19). In both cases the integration is performed separately over the intervals $0 \leqslant x \leqslant \gamma$ and $\gamma \leqslant x \leqslant x_1$. A tedious but elementary integration yields

$$\min_{y \in Y} I(y) = (-az_{10}z_{20} + \tfrac{1}{3}a^2z_{20}^3)\,\mathrm{sgn}\, z_{10}$$
$$+ \tfrac{1}{3}\sqrt{a}(\sqrt{[2+\sqrt{5}]} + \mathrm{sgn}\, z_{10})\,(-2z_{10} + az_{20}^2)^{\frac{3}{2}}, \quad (3.177)$$

and

$$\min_{y \in Y} I(y) = (az_{10}z_{20} + \tfrac{1}{3}a^2z_{20}^3)\,\mathrm{sgn}\, z_{10}$$
$$+ \tfrac{1}{3}\sqrt{a}(\sqrt{[2+\sqrt{5}]} - \mathrm{sgn}\, z_{10})\,(2z_{10} + az_{20}^2)^{\frac{3}{2}}, \quad (3.178)$$

respectively.

Example 3: $F = z_1^2(x)$ (integral square-error criterion).

A substitution of $z_1(x)$ into $I(y)$, and an integration carried out in the same way as in Example 2, yields

$$\min_{y \in Y} I(y) = -az_{10}^2z_{20} + \tfrac{2}{3}a^2z_{10}z_{20}^3 - \tfrac{2}{15}a^3z_{20}^5 + \frac{\sqrt{a}}{20}(222 + 2\sqrt{33})\,(-z_{10} + \tfrac{1}{2}az_{20}^2)^{\frac{5}{2}}$$

$$(3.179)$$

for (z_{20}, z_{10}) located on the left side of the curve (3.168), and

$$\min_{y \in Y} I(y) = az_{10}^2z_{20} + \tfrac{2}{3}a^2z_{10}z_{20}^3 + \tfrac{2}{15}a^3z_{20}^5 + \frac{\sqrt{a}}{20}(222 + 2\sqrt{33})\,(z_{10} + \tfrac{1}{2}az_{20}^2)^{\frac{5}{2}}$$

$$(3.180)$$

for (z_{20}, z_{10}) located on the right side.

As it is the custom in dynamic programming[B 4],[B 5],[B 6] suppose now that the positive direction of x runs from $x = x_1$ to $x = 0$. The Carathéodory equation (3.155), corresponding to the extremal problem (3.156), (3.157), (3.158) and the special forms of F used in Examples 1 to 3, is

$$\min_{y \in Y}\left[1 + z_2\frac{\partial S}{\partial z_1} + \frac{y}{a}\frac{\partial S}{\partial z_2}\right] = 0, \quad (3.181)$$

$$\min_{y \in Y}\left[|z_1| + z_2\frac{\partial S}{\partial z_1} + \frac{y}{a}\frac{\partial S}{\partial z_2}\right] = 0, \quad (3.182)$$

and

$$\min_{y \in Y}\left[z_1^2 + z_2\frac{\partial S}{\partial z_1} + \frac{y}{a}\frac{\partial S}{\partial z_2}\right] = 0. \quad (3.183)$$

In the examples discussed the least value of $I(y)$ turned out to be a non-trivial lower limit (bang-bang solution). Replacing in (3.181), (3.182), (3.183) the symbolic operator min by the specific operator lim inf and $y \in Y$ $y \in Y$
carrying out the limit process yields

$$y(x) = -\operatorname{sgn} \frac{\partial S}{\partial z_2} \tag{3.184}$$

and

$$1 + z_2 \frac{\partial S}{\partial z_1} - \frac{1}{a}\left|\frac{\partial S}{\partial z_2}\right| = 0, \tag{3.185}$$

$$|z_1| + z_2 \frac{\partial S}{\partial z_2} - \frac{1}{a}\left|\frac{\partial S}{\partial z_2}\right| = 0, \tag{3.186}$$

$$z_1^2 + z_2 \frac{\partial S}{\partial z_1} - \frac{1}{a}\left|\frac{\partial S}{\partial z_2}\right| = 0. \tag{3.187}$$

The preceding non-homogeneous partial differential equations are rather tedious to solve directly subject to the boundary conditions (3.99). Their solution is considerably simplified by the use of the rather easily found particular solutions

$$S_0 = az_2, \tag{3.188}$$

$$S_0 = az_2|z_1| + \tfrac{1}{3}a^2 z_2^3 \operatorname{sgn} z_1, \tag{3.189}$$

$$S_0 = \pm az_1 z_2^2 + \tfrac{2}{3}a^2 z_1 z_2^3 \pm \tfrac{2}{15}a^3 z_2^5. \tag{3.190}$$

Replacing S by $\bar{S} + S_0$ yields in all three cases the same homogeneous equation

$$z_2 \frac{\partial \bar{S}}{\partial z_1} - \frac{1}{a}\left|\frac{\partial \bar{S}}{\partial z_2}\right| = 0. \tag{3.191}$$

The piecewise linear equation (3.191) is readily solved by means of the characteristic equations (3.35). From the first pair of these equations

$$\frac{dz_1}{z_2} = \pm a\, dz_2 \tag{3.192}$$

there results

$$z_1 = \pm \tfrac{1}{2}az_2^2 + C. \tag{3.193}$$

The integration constant C is readily determined by means of the boundary conditions $z_1 = 0$ when $z_2 = 0$ and $z_1 = z_{10}$ when $z_2 = z_{20}$. The constant C may therefore have two different values

$$C = 0 \quad \text{and} \quad C = z_{10} \mp \tfrac{1}{2}az_{20}^2. \tag{3.194}$$

The signs $+$ or $-$ in (3.190), (3.193), (3.194) are fixed by the sign of $\partial S/\partial z_2$, i.e. according to (3.184), by the sign of the minimizing segment $y(x) = \pm 1$. The segments of characteristics, defined by (3.193), (3.194) are so chosen that the continuous curve $z_1 = z_1(z_2)$ joins the points (z_{20}, z_{10}) and $(0, 0)$. Knowing the minimizing curve $z_1 = z_1(z_2)$, the required particular solutions of (3.185), (3.186), (3.187) are found by integrating the characteristic equations of (3.35) omitted in (3.192), together with the equations (3.185), (3.186), (3.187). Since the details of this integration coincide with the integration with respect to x carried out in the Examples 1 to 3, they will not be repeated here. The junction points of the two segments of $z_1 = z_1(z_2)$ are given as before by the equations (3.173), (3.174), and the result of the integration over one segment of $z_1 = z_1(z_2)$ is given by the particular solutions (3.188) to (3.190). With the knowledge of the particular solutions (3.188) to (3.190) the amount of calculation required by the use of the Carathéodory equations (3.185) to (3.187) is about the same as in the direct method used in the examples. Without this knowledge the amount of necessary calculations is considerably larger (cf. [F 12]).

17.8 Lagrange problems with more than one minimizing function

Up to now we have examined exclusively extremal problems with a single minimizing function. This was done not only to avoid inessential generality, but also to leave aside certain specific properties of functionals of more than one variable. The theory of such functionals is very complex, and its systematic study would go beyond our objective of stimulating an exchange of ideas between theoreticians and designers. Because optimization problems with more than one control input arise rather frequently in contemporary control practice, we will examine some elementary features of these problems. Consider for this purpose the two-variable Lagrange problem

$$I(x_0, x_1) = \min_{y_1 \in Y_1,\, y_2 \in Y_2} I(y_1, y_2)$$

$$= \min_{y_1 \in Y_1,\, y_2 \in Y_2} \int_{x_0}^{x_1} F(x, z_i(x), y_1(x), y_2(x), \dot{z}_i(x), \dot{y}_1(x), \dot{y}_2(x))\, dx,$$

$$\tag{3.195}$$

$$L_j(x, z_i(x), y_1(x), y_2(x), \dot{z}_i(x), \dot{y}_1(x), \dot{y}_2(x)) = 0, \tag{3.196}$$

$$z_i(x_0) = z_{i0}, \quad z_i(x_1) = z_{i1}, \quad y_1(x_0) = y_{10}, \quad y_1(x_1) = y_{11},$$

$$y_2(x_0) = y_{20}, \quad y_2(x_1) = y_{21} \quad (i,j = 1, 2, ..., n), \tag{3.197}$$

where x_0, x_1, z_{i0}, z_{i1}, y_{10}, y_{11}, y_{20}, y_{21} are real constants, $I(y_1, y_2)$ is a converging Riemann integral and Y_1, Y_2 are the specified sets of admissible functions $y_1(x)$ and $y_2(x)$. To render the notation a bit more concise we will denote the symbolic 'least value' operator

$$\min_{y_1 \in Y_1,\, y_2 \in Y_2} \quad \text{by} \quad \min_{y_1,\, y_2}.$$

Before discussing any properties of the solution

$$y_1 = \bar{y}_1(x), \quad y_2 = \bar{y}_2(x), \quad I(\bar{y}_1, \bar{y}_2) = \min_{y_1,\, y_2} I(y_1, y_2)$$

of the problem (3.195), (3.196), (3.197) it is necessary to ascertain whether this problem is correctly set in the sense of Hadamard. This preliminary study requires the precise definition of the sets Y_1 and Y_2 and some knowledge of the smoothness properties of the functions F and L_j. Suppose for definiteness that Y_1 and Y_2 are the sets of piecewise continuous and piecewise differentiable functions, and that the functions F and L_j are continuous with respect to their arguments. With this hypothesis it is likely that in every common interval of continuity of y_1 and y_2, the extension of the conjecture (3.140) is given by the system of algebraic equations

$$\phi_j(x, z_i, y_1, y_2, C_i, \bar{C}_i) = 0 \quad (i, j = 1, 2, ..., n), \tag{3.198}$$

where $C_i, \bar{C}_i$ are arbitrary real constants.

Using (3.198), or some other more reliable means, the order q of degeneracy of (3.195), (3.196), (3.197) is determined, and thus also a necessary condition for the existence of continuous minimizing functions $\bar{y}_1(x), \bar{y}_2(x)$. If the number of non-trivial boundary conditions in (3.197) exceeds $2(n+2) - q$, at least one of the minimizing functions $\bar{y}_1(x), \bar{y}_2(x)$ must be discontinuous. The functions $z_i(x)$ are of course presumed continuous and at least piecewise differentiable in the whole interval $x_0 \leqslant x \leqslant x_1$.

In extremal problems with a single minimizing function the least value $I(x_0, x_1)$ was either a local minimum or a non-trivial local lower limit. This property holds only exceptionally for extremal problems with more than one minimizing function, because $I(x_0, x_1)$ may well be a local minimum with respect to one minimizing function and a local lower limit with respect to the others. When the Carathéodory formulation is used, it is necessary to make the proper distinction at the beginning, but as the single-variable case shows, this distinction is made naturally as one of the steps necessary to solve the corresponding boundary-value problem. This is not necessarily so when a less general method is used, and special precautions have to be taken in the latter case.

Let us note finally that the functional $I(y_1, y_2)$ in (3.195) need not possess the same type of continuity with respect to both $y_1(x)$ and $y_2(x)$. This conclusion is true a fortiori when the functional I depends on more than two functions.

When (3.196) is equivalent to the explicit system

$$\dot{z}_j(x) = f_j(x, z_i(x), y_1(x), y_2(x), \dot{y}_1(x), \dot{y}_2(x)) \tag{3.199}$$

the Carathéodory equation of (3.195), (3.196), (3.197) becomes

$$\min_{\psi_1 \in \Psi_1, \psi_2 \in \Psi_2} \left[F(x, z_i, y_1, y_2, f_j, \psi_1, \psi_2) - \left(\frac{\partial S}{\partial x} + \frac{\partial S}{\partial y_1} \psi_1 + \frac{\partial S}{\partial y_2} \psi_2 \right) \right.$$
$$\left. - \sum_{k=1}^{n} \frac{\partial S}{\partial z_k} f_k(x, z_i, y_1, y_2, \psi_1, \psi_2) \right] = 0 \quad (i = 1, 2, ..., n), \tag{3.200}$$

where ψ_1, ψ_2 are singular transforms of y_1 and y_2, respectively. If at least one of the derivatives $\dot{y}_1(x), \dot{y}_2(x)$ does not occur explicitly in both (3.195) and (3.196), then the corresponding function ψ in the operator $\min_{\psi_1, \psi_2}$ should be replaced by y, and the term $\partial S/\partial y$ in (3.200) should be omitted.

The boundary conditions to be associated with (3.200) are identical with those described in §16.2, except that the scanning in (3.33) must be performed over surfaces $g(x, y_1, y_2) = 0$ instead of curves $g(x, y) = 0$.

CHAPTER 4

THE RELATIONSHIP BETWEEN THE CALCULUS OF VARIATIONS, DYNAMIC PROGRAMMING AND THE MAXIMUM PRINCIPLE

§ 18. THE CALCULUS OF VARIATIONS AND DYNAMIC PROGRAMMING

18.1 Dynamic programming in the case of a continuous extremal problem

Dynamic programming grew out of methods developed to deal with economic problems of an essentially discrete nature. The basic recurrence approach appears already in the works of Massé[M6]–[M9], but the present terminology, the term dynamic programming included, is due to Bellman and his collaborators[B4]–[B6]. The argument of dynamic programming can be described as follows.

Consider an economic process described by a state vector

$$\overline{W} = (W_1, W_2, \ldots, W_n),$$

n finite, depending on time, a parameter vector

$$\overline{\alpha} = (\alpha_1, \alpha_2, \ldots, \alpha_m),$$

m finite, and the initial state $\overline{W}_0$. Let the state $\overline{W}_0$ correspond to the instant t_0. By definition the evolution of $\overline{W}$ is judged on the basis of a certain performance criterion leading to a specific economic goal, for example to a maximization of profits or a minimization of tax to be paid. The possibility to make a change in $\overline{W}$ is called a decision, and the way the decisions are implemented is called a policy. Since in practice an economic process is never isolated, but on the contrary tied to other economic processes, the possible decisions and policies pertaining to W must satisfy a certain set of constraints. A policy consistent with the imposed constraints is called an admissible policy, and an admissible policy which attains the specified economic goal is called an optimal policy.

The principal feature of dynamic programming is a simple consequence of the fact that, at the time $t_1 = t_0 + \Delta t_0$, the state $\overline{W}$ is written in the form

$$\overline{W}_1 = \overline{W}(t_1) = f_1(t_0, \overline{\alpha}, \overline{W}_0), \tag{4.1}$$

where f_1 is a known function, describing the transition from $\overline{W}_0$ to $\overline{W}_1$. Since the values t_0 and Δt_0 of t are not essentially different from the components α_i, $i = 1, 2, \ldots, m$ of $\overline{\alpha}$, the above equation is equivalent to

$$\overline{W}_1 = f_1(\overline{\alpha}, \overline{W}_0),$$

where $\overline{\alpha} = (\alpha_1, \alpha_2, \ldots, \alpha_m, t_0, \Delta t_0)$. If, after another lapse of time Δt_1, the state $\overline{W}$ is changed again, there results $t_2 = t_1 + \Delta t_1$ and

$$\overline{W}_2 = \overline{W}(t_2) = f_2(\overline{\alpha}, \overline{W}_1),$$

where, by definition, f_2 is also a known function. After k changes it is thus possible to describe the state of $\overline{W}$ in the recurrent form

$$t_k = t_{k-1} + \Delta t_{k-1}, \tag{4.2}$$

$$\overline{W}_k = \overline{W}(t_k) = f_k(\overline{\alpha}, \overline{W}_{k-1}). \tag{4.3}$$

Since the solution of (4.2) is quite obvious

$$t_k = t_0 + \sum_{i=0}^{k-1} \Delta t_i,$$

the value t_k plays in (4.3) a role which is not essentially different from that of the subscript k, and need thus not be written explicitly. Using (4.3) the evolution of the economic process is represented by means of its initial state $\overline{W}_0$ and the results f_i, $i = 1, 2, \ldots, k$, of the decisions already accomplished. When $\overline{W}_k$ is considered in relation to a specific terminal state $\overline{W}_K$, where K is a fixed value of k, the recurrence (4.3) is said to constitute an invariant imbedding of $\overline{W}_K$, with respect to $\overline{W}_0$, and the argument leading to (4.3) is called the principle of invariant imbedding.

After these preliminaries, let us recall that control problems in general, and optimization problems in particular, can be studied by either continuous or discrete models, provided both representations are elements of an admissible set of 'close' models (cf. §§4.2 and 4.5). In the study of continuous dynamical systems, which are exclusively considered here, a discrete representation is admissible only if the continuous model is inert with respect to discretization (cf. §4.5). This restriction defines implicitly the scope of validity of dynamic programming when the latter is applied to the study of continuous systems. We will suppose for the moment that the validity of dynamic programming is taken for granted.

Consider now the continuous variational problem

$$I(x_0, x_1) = \min_{y \in Y} I(y) = \min_{y \in Y} \int_{x_0}^{x_1} F(x, y(x), \dot{y}(x)) \, dx$$
$$(y(x_0) = y_0, \; y(x_1) = y_1), \quad (2.41)$$

where $I(y)$ is a convergent Riemann integral and Y is the set of continuously differentiable functions. In order to apply dynamic programming to the problem (2.41), it is first necessary to note that the functional $I(y)$ is locally additive (cf. § 14.5), i.e. that

$$I(y) = \int_{x_0}^{x_1} F(x, y, \dot{y})\,dx = \int_{x_0}^{\xi} F(x, y, \dot{y})\,dx + \int_{\xi}^{x_1} F(x, y, \dot{y})\,dx \quad (4.4)$$

for any ξ in the interval $x_0 \leqslant \xi \leqslant x_1$. A function $y(x) \in Y$ passing through the points (x_0, y_0), (x_1, y_1) is then an admissible policy of the problem (2.41) and its solution $\bar{y}(x)$ an optimal policy. Let us assume that the minimizing function $\bar{y}(x)$ is unique, single-valued and continuous in the whole interval $x_0 \leqslant x \leqslant x_1$. Taking account of the local additivity of $I(y)$, the uniqueness of $\bar{y}(x)$ can now be expressed in the following way:

> An optimal policy has the property that, whatever the initial decision is, the remaining decisions must constitute an optimal policy with regard to the state resulting from the initial decision.   (4.5)

The above statement is known in dynamic programming as the principle of optimality.

Before applying the principle of optimality to (2.41) it is necessary to have recourse to the principle of invariant imbedding. If $\bar{y}(x)$ is known, then $I(\bar{y}) = \min\limits_{y \in Y} I(y)$ is obviously only a point-function of the four 'boundary' variables x_0, x_1, y_0, y_1. Fixing say the variables x_1, y_1, $I(\bar{y})$ becomes only a function of the two variables x_0, y_0. Consider a point (ξ, η) located on the minimizing curve $y = \bar{y}(x)$, and let

$$V(\xi, \eta) = \int_{\xi}^{x_1} F(x, y(x), \dot{y}(x))\,dx \quad (\eta = \bar{y}(\xi)). \quad (4.6)$$

The function $V(\xi, \eta)$ satisfies by definition the boundary conditions

$$V(x_1, y_1) = 0, \quad (4.7)$$

$$I(\bar{y}) = V(x_0, y_0), \quad (4.8)$$

and constitutes therefore an invariant imbedding of $I(\bar{y})$ with respect to the initial state $V(x_1, y_1) = 0$, when $x = x_1$, $y = y_1$.

In order to apply the principle of optimality suppose that the function $V(\xi, \eta)$ and $\eta = \bar{y}(\xi)$ are already known in the interval

$$x_0 + \Delta \leqslant \xi \leqslant x_1 \quad (0 < \Delta < x_1 - x_0).$$

Taking into account (4.4), the problem (2.41) can be written in the form (cf. § 14.5)

$$\min_{y \in Y} I(y) = \min_{y \in Y} \left[\int_{x_0}^{x_0 + \Delta} F(x, y(x), \dot{y}(x))\,dx + \int_{x_0 + \Delta}^{x_1} F(x, y(x), \dot{y}(x))\,dx \right], \quad (4.9)$$

and a substitution of (4.6) into (4.9) yields the required recurrence equation

$$V(x_0, \bar{y}(x_0)) = \min_{y \in Y} \left[\int_{x_0}^{x_0 + \Delta} F(x, y(x), \dot{y}(x)) \, dx + V(x_0 + \Delta, y(x_0 + \Delta)) \right]$$

$$(0 < \Delta < x_1 - x_0). \quad (4.10)$$

Suppose now that $0 \leqslant \Delta \ll 1$ and that $V(x_0, y(x_0))$ is a continuously differentiable function of x_0, i.e. $V(x_0, y_0)$ has low sensitivity with respect to x_0 (cf. §14.6). Recalling that all admissible policies $y(x)$ are also continuously differentiable, it follows that

$$y(x_0 + \Delta) = y(x_0) + \dot{y}(x_0) \cdot \Delta + O(\Delta^{1+\epsilon_1}) \quad (\epsilon_1 > 0),$$
$$\int_{x_0}^{x_0 + \Delta} F(x, y(x), \dot{y}(x)) \, dx = F(x_0, y(x_0), \dot{y}(x_0)) \cdot \Delta + O(\Delta^{1+\epsilon_2}) \quad (\epsilon_2 > 0).$$

$$(4.11)$$

Recalling that the value $y(x_0)$ is fixed by the boundary condition $y(x_0) = y_0$ and taking into account the fact that the interval

$$x_0 \leqslant x \leqslant x_0 + \Delta$$

is by definition extremely small, the functional operator $\min\limits_{y \in Y}$ is 'practically' equivalent to the 'algebraic' operator $\min\limits_{\dot{y}(x_0)}$, where $\dot{y}(x_0)$ is a real parameter. Formally this conclusion can be written

$$\min_{y(x) \in Y} = \min_{\dot{y}(x_0)} + O(\Delta^{1+\epsilon_3}) \quad (x_0 \leqslant x \leqslant x_0 + \Delta, \quad \epsilon_3 > 0). \quad (4.12)$$

Substituting (4.11) and (4.12) into (4.10) yields

$$V(x_0, y_0) = \min_{\dot{y}(x_0)} [F(x_0, y_0, \dot{y}(x_0)) \cdot \Delta + V(x_0 + \Delta, y_0 + \dot{y}(x_0) \cdot \Delta)]$$
$$+ O(\Delta^{1+\epsilon_4}) \quad (\epsilon_4 > 0). \quad (4.13)$$

Since

$$V(x_0 + \Delta, y_0 + \dot{y}(x_0) \cdot \Delta) = V(x_0, y_0) + \frac{\partial V}{\partial(x_0 + \Delta)} \Delta + \dot{y}(x_0) \frac{\partial V}{\partial(y_0 + \dot{y}(x_0)\Delta)} \Delta$$
$$+ O(\Delta^{1+\epsilon_5}) \quad (\epsilon_5 > 0), \quad (4.14)$$

equation (4.13) is equivalent to

$$\min_{\dot{y}(x_0)} \left\{ \left[F(x_0, y_0, \dot{y}(x_0)) + \frac{\partial V}{\partial(x_0 + \Delta)} + \dot{y}(x_0) \frac{\partial V}{\partial(y_0 + \dot{y}(x_0)\Delta)} \right] \cdot \Delta \right\}$$
$$+ O(\Delta^{1+\epsilon_6}) = 0 \quad (\epsilon_6 > 0). \quad (4.15)$$

Dividing (4.15) by Δ and letting $\Delta \to 0$ yields

$$\min_{\dot{y}(x_0)} \left[F(x_0, y_0, \dot{y}(x_0)) + \frac{\partial V}{\partial x_0} + \dot{y}(x_0) \frac{\partial V}{\partial y_0} \right] = 0. \quad (4.16)$$

Since by definition the point (ξ, η) is located on the minimal curve $y = \bar{y}(x)$, (4.16) remains valid when the terminal point (x_0, y_0) is replaced by the mobile point (ξ, η). Introducing the new variable

$$u = \dot{\eta}(\xi, \eta(\xi)), \tag{4.17}$$

the 'algebraic' equation (4.16) can be transformed back into a functional equation:

$$\min_{u \in U} \left[F(\xi, \eta, u) + \frac{\partial V}{\partial \xi} + u \frac{\partial V}{\partial \eta} \right] = 0 \quad (x_0 \leqslant \xi \leqslant x_1), \tag{4.18}$$

where U is the set of admissible functions $u(\xi)$, defined implicitly by the singular transformation (4.17). Equation (4.18) is known as Bellman's partial differential equation, or as the functional equation of dynamic programming.

When the limiting process leading from (4.10) to (4.18) is legitimate, the solutions of the corresponding equations are 'close', i.e. they are qualitatively the same and their quantitative difference is small (cf. §4.5).

In principle the recurrence equation (4.10) can be solved quite easily. In fact, it is sufficient to subdivide the interval $x_0 \leqslant x \leqslant x_1$ into n sufficiently small subintervals, and then calculate

$$y(x_0 + \Delta) \quad \text{and} \quad V(x_0 + \Delta, y(x_0 + \Delta))$$

over every such subinterval, using for example the approximation formulae (4.11) and (4.13) and rejecting at every stage the curves $y = y(x_0 + \Delta)$ which do not appear to lead to the point (x_0, y_0)[B6]. When the terminal point (x_0, y_0) is reached, the value $V(x_0, y_0)$ is known, and the shape of the curve $y = \bar{y}(x)$ can be found by back-tracing through the table of values of $y(x_0 + \Delta)$. The method just described amounts of course to a systematic scanning of all admissible shapes of the curve $y = y(x)$ (passing through the points (x_0, y_0) and (x_1, y_1)), and it leads always to a solution. Unfortunately, if the number of subintervals n is not very small, the amount of intermediary data, pertaining to both $y = \bar{y}(x)$ and $V = V(x, y)$, is so large that it exceeds the memory capacity of even the largest available digital computers. As was already pointed out by Bellman[B6], the apparently foolproof scanning method described above, is defeated in practice by the 'curse of dimensionality', i.e. by the prohibitively rapid growth of the amount of intermediate data to be stored as n increases.

In addition to systematic scanning, it is possible to attempt to solve equation (4.10) by a method of successive iterations. Two different methods of iteration have been suggested by Bellman (see for example [B6]). In the first method the iterations are performed directly on the function $V(\xi, \eta)$, $x_0 \leqslant \xi \leqslant x_1$, and the shape of the corresponding curve $\eta = \eta(\xi)$, $\eta(x_0) = y_0$, $\eta(x_1) = y_1$, is deduced from $V(\xi, \eta)$. In the second method the iterations are performed on the policy function $\eta(\xi)$, $x_0 \leqslant \xi \leqslant x_1$, and the corresponding shape of $V(\xi, \eta)$ is deduced from $\eta(\xi)$. These two types of iterations are referred to as iterations in the performance criterion space and iterations in the policy space, respectively.

Using the variables ξ, η and taking into account the boundary condition (4.7) in the first case the equation (4.10) becomes approximately

$$V_{n+1}(\xi, \eta) = \min_{\dot{\eta}(\xi)} [F(\xi, \eta(\xi), \dot{\eta}(\xi)) \cdot \Delta + V_n(\xi + \Delta, \eta(\xi + \Delta)), V_{n+1}(x_1, y_1) = 0$$

$$(x_0 \leqslant \xi \leqslant x_1, \quad 0 < \Delta < x_1 - x_0, \quad n = 0, 1, 2, \dots). \quad (4.19)$$

The initial shape $V_0(\xi, \eta)$ necessary to start the iterations defined by (4.19), is deduced from some excess information, or in the absence of the latter, it is simply guessed. If $V_0(\xi, \eta)$ has been so chosen that $V_n(\xi, \eta)$ converges as $n \to \infty$, then $V_n(\xi, \eta)$ will usually converge to a solution of (4.10). Neither necessary nor sufficient conditions for the validity of the relation

$$\lim_{n \to \infty} V_n(\xi, \eta) = V(\xi, \eta)$$

are yet known.

In the second type of iterations the initial shape of $\dot{\eta}(x)$, $x_0 \leqslant x \leqslant x_1$, is chosen by some consideration and the shape of $V_0(\xi, \eta)$ is deduced by means of the approximate implicit equation

$$\left.\begin{array}{l} V_0(\xi, \eta) = F(\xi, \eta_0(\xi), \dot{\eta}_0(\xi)) \cdot \Delta + V_0(\xi + \Delta, \eta_0(\xi + \Delta)), \\ V_0(x_1, y_1) = 0 \quad (0 < \Delta < x_1 - x_0). \end{array}\right\} \quad (4.20)$$

The function $V_0(\xi, \eta)$ can either be used in (4.19), or an improved shape of $\dot{\eta}(\xi)$ can be deduced from $\dot{\eta}_0(\xi)$ be means of the inequality

$$F(\xi, \eta(\xi), \dot{\eta}(\xi)) \Delta + V_0(\xi + \Delta, \eta(\xi + \Delta))$$

$$< F(\xi, \eta_0(\xi), \dot{\eta}_0(\xi)) \Delta + V_0(\xi + \Delta, \eta_0(\xi + \Delta)). \quad (4.21)$$

Once $\dot{\eta}(x)$ is known, the corresponding improved shape of $V(\xi, \eta)$ is determined from (4.20). If $\dot{\eta}_0(x)$ is well chosen and the iterations converge, in all likelihood they will converge to a solution of (4.10), but the conditions of validity of this statement are not yet known.

18.2 Dynamic programming, Euler's equation and the Carathéodory formulation

In order to deduce some properties of the solution of the Bellman equation (4.18) suppose that the solution (4.8) of the extremal problem (2.41) represents a non-degenerated local minimum. In such a case equation (4.18) is stationary with respect to u, as defined by (4.17), and it reduces to the system of two first-order equations

$$F(\xi, \eta, u) + \frac{\partial V}{\partial \xi} + \frac{\partial V}{\partial \eta} u = 0, \quad \frac{\partial F}{\partial u} + \frac{\partial V}{\partial \eta} = 0. \quad (4.22)$$

Differentiating the first equation in (4.22) partially with respect to η, the second totally with respect to ξ, and supposing that $V(\xi, \eta)$ has sufficient smoothness to allow the validity of the identity

$$\frac{\partial^2 V}{\partial \xi\, \partial \eta} = \frac{\partial^2 V}{\partial \eta\, \partial \xi},$$

yields the equation

$$\frac{d}{d\xi}\left(\frac{\partial F}{\partial u}\right) + \frac{\partial^2 V}{\partial \eta\, \partial \xi} + \frac{\partial^2 V}{\partial \eta^2}\, u - \frac{\partial F}{\partial \eta} + \frac{\partial V}{\partial \eta}\, \frac{\partial u}{\partial \eta} - \frac{\partial^2 V}{\partial \xi\, \partial \eta}$$

$$- \frac{\partial^2 V}{\partial \eta^2}\, u - \frac{\partial V}{\partial \eta}\, \frac{\partial u}{\partial \eta} = 0.$$

After simplifying there results the Euler equation

$$\frac{d}{d\xi}\left(\frac{\partial F}{\partial u}\right) - \frac{\partial F}{\partial \eta} = 0 \quad (u = \dot{\eta}). \tag{4.23}$$

The optimal policy $\eta = \bar{y}(\xi)$ is therefore an extremal of the problem (2.41).

Consider now the second equation in (4.22) as an algebraic equation in u, and let

$$u_i = u_i\left(\xi, \eta, \frac{\partial V}{\partial \eta}\right) \quad (i = 1, 2, \ldots)$$

be its real roots. For each u_i the first equation in (4.22) takes the form

$$\frac{\partial V}{\partial \xi} + H_i\left(\xi, \eta, \frac{\partial V}{\partial \eta}\right) = 0, \tag{4.24}$$

where $\qquad H_i\left(\xi, \eta, \dfrac{\partial V}{\partial \eta}\right) = \dfrac{\partial V}{\partial \eta}\, u_i + F(\xi, \eta, u_i) \quad (i = 1, 2, \ldots).$

The equations (4.24) can of course be combined into a single first-order partial differential equation

$$\Psi\left(\xi, \eta, \frac{\partial V}{\partial \xi}, \frac{\partial V}{\partial \eta}\right) = 0. \tag{4.25}$$

Comparing the equations (4.18), (4.22), (4.25) to the Carathéodory equations (3.6), (3.8), (3.12) and taking into account the difference in notation, due to the reference value $V(x_1, y_1) = 0$ instead of $V(x_0, y_0) = 0$, it is immediately apparent that these two sets of equations are identical. Dynamic programming is therefore a particular case of the Carathéodory formulation, and the particularity of dynamic programming consists only in the rather involved and circuitous method of deriving equation (4.18). Both the power and the limitations of dynamic programming become quite clear, when dynamic programming is examined in the context of the Carathéodory formulation, as summarized in chapter 3.

For example, the problem of equivalence between the discrete equation (4.10) and the continuous equation (4.18) is solved immediately by noting that the approximations (4.11), (4.12), (4.13) are of first order in Δ. (4.10) is thus equivalent to the direct discretization of the extremal problem (2.41) by means of the classical Euler method (see for example vol. I [C 15]), to be discussed in chapter 5. The discrete version (4.10) of (4.18) will therefore possess all inherent weaknesses of the Euler method. Another weakness of a straightforward application of dynamic programming consists in the fact that no systematic distinction is made between local minima and non-trivial local lower limits. As is well known, the failure to distinguish between local minima and local lower limits leads easily to erroneous results, affecting especially the notion of transversality and the notion of an admissible neighbourhood. Various misunderstandings concerning the precise relation between dynamic programming and the classical calculus of variations can be traced to this source, because the latter is often arbitrarily limited to the material centred on the Euler equation (see for example [B 4]–[B 6]). The direct methods and the fundamental contributions of Hilbert, Hadamard, Wiener, Carathéodory and Tonelli are excluded by such an arbitrary point of view.

§ 19. THE CALCULUS OF VARIATIONS AND THE MAXIMUM PRINCIPLE

19.1　The Mayer problem

It was stressed on several occasions that an extremal problem can be transformed into several more or less equivalent forms. A form introduced by Mayer[H 1] plays a key role in the theory of the maximum principle. This form is characterized by the property that the extremal problem in question does not contain explicitly a functional which is defined by an integral. The transformation of extremal problems discussed so far into the Mayer form[H 1] is quite straightforward. Consider for example the single-variable problem

$$I(x_0, x_1) = \min_{y \in Y} I(y) = \min_{y \in Y} \int_{x_0}^{x_1} F(x, y(x), \dot{y}(x)) \, dx$$

$$(y(x_0) = y_0, \quad y(x_1) = y_1). \qquad (2.41)$$

Introducing two new variables

$$\left.\begin{array}{l} \dot{y}(x) = u(x), \\ \dot{v}(x) = F(x, y(x), u(x)), \end{array}\right\} \qquad (4.26)$$

yields the Mayer problem

$$I(x_0, x_1) = \min_{y \in Y} I(y) = \min_{y \in Y} v(x_1),$$

which is to be solved subject to the constraints (4.26) and the boundary conditions

$$v(x_0) = 0, \quad y(x_0) = y_0, \quad y(x_1) = y_1.$$

By a similar procedure the Lagrange problem

$$
\left.
\begin{aligned}
I(x_0, x_1) = \min_{y \in Y} I(y) &= \min_{y \in Y} \int_{x_0}^{x_1} F(x, y(x), z(x), \dot{y}(x), \dot{z}(x))\, dx, \\
f(x, y(x), z(x), \dot{y}(x), \dot{z}(x)) &= 0, \\
y(x_0) = y_0, \quad y(x_1) = y_1, \quad z(x_0) &= z_0, \quad z(x_1) = z_1
\end{aligned}
\right\}
\qquad (4.27)
$$

is transformed into the Mayer problem

$$
\left.
\begin{aligned}
I(x_0, x_1) = \min_{y \in Y} I(y) &= \min_{y \in Y} v(x_1), \\
\dot{v}(x) &= F(x, y(x), z(x), \dot{y}(x), \dot{z}(x)), \\
f(x, y(x), z(x), \dot{y}(x), \dot{z}(x)) &= 0, \\
y(x_0) = y_0, \quad y(x_1) &= y_1, \\
z(x_0) = z_0, \quad z(x_1) = z_1, \quad v(x_0) &= 0.
\end{aligned}
\right\}
\qquad (4.28)
$$

If desired the two implicit differential equations in (4.28) can be replaced by three explicit differential equations and one holonomous constraint

$$
\left.
\begin{aligned}
\dot{y}(x) &= u_1(x), \\
\dot{z}(x) &= u_2(x), \\
\dot{v}(x) &= F(x, y(x), z(x), u_1(x), u_2(x)), \\
0 &= f(x, y(x), z(x), u_1(x), u_2(x)).
\end{aligned}
\right\}
\qquad (4.29)
$$

Like Lagrange problems, Mayer's problems can be classified into classes according to their order of degeneracy q. Carathéodory appears to be the first to have shown that the properties of the solutions of a Mayer problem depend critically on the value of q, and that the distinction between the cases $q = 0$ and $q > 0$ is particularly important[C 5],[C 6]. Carathéodory has also shown that a Mayer problem with $q > 0$ can sometimes be transformed into another Mayer problem with $q = 0$ but with a smaller number of dependent variables. Contrary to a conjecture of Bliss (§ 68[B 14]), when the set of minimizing functions Y is kept fixed, such a transformation is not always possible. In fact we have already presented several examples when a solution of a Lagrange problem, with $q > 0$ and Y the set of continuous functions, fails to admit a minimizing function $\bar{y}(x)$ in the set Y. The lack of existence of a solution in a specified set Y can obviously not be remedied by means of a non-singular transformation.

Keeping the preceding remarks in mind let us now trace the evolution of the argument which led to the contemporary version of the maximum principle.

19.2   The maximum principle and degenerated Mayer problems

The maximum principle, expressed as a conjecture, originated in the study of the Mayer problem

$$
\left.\begin{aligned}
I(x_0, x_1) &= \min_{y \in Y} I(y) = \min_{y \in Y} (x_1 - x_0), \\
\dot{z}_i(x) &= f_i(z_j(x), y(x)) \quad (i, j = 1, 2, \ldots, n, \quad x_0 \leqslant x \leqslant x_1), \\
z_i(x_0) &= z_{i0}, \quad z_i(x_1) = z_{i1},
\end{aligned}\right\} \quad (4.30)
$$

where Y is the set of continuous functions and x_0, z_{i0}, z_{i1} are fixed real constants. The value of the constant x_1 is of course unknown. The functions f_i in (4.30) are supposed to be sufficiently smooth so that for any fixed $y(x) \in Y$ the initial-value problem

$$
\dot{z}_i(x) = f_i(z_j(x), y(x)) \quad (z_i(x_0) = z_{i0}, \quad i, j = 1, 2, \ldots, n) \quad (4.31)
$$

admits a unique and continuous solution $z_i(x)$ in the whole interval $x_0 \leqslant x \leqslant x_1$. The extremal problem (4.30) is degenerated, because it does not contain explicitly the derivative $\dot{y}(x)$ of the minimization variable $y(x) \in Y$. This conclusion is immediate when the functional $I(y)$ in (4.30) is thought of as resulting from a Lagrange problem with the functional

$$
\left.\begin{aligned}
I(y) &= \int_{x_0}^{x_1} F(x, z_i(x), y(x), \dot{z}_i(x), \dot{y}(x)) \, dx \quad (i = 1, 2, \ldots, n), \\
F(x, z_i(x), y(x), \dot{z}_i(x), \dot{y}(x)) &\equiv +1.
\end{aligned}\right\} \quad (4.32)
$$

Unless the boundary conditions $z_i(x_1) = z_{i1}$ are trivial, the degenerated Mayer problem has therefore no solution in the space of continuous functions $y(x)$. A meaningful solution of (4.30) becomes, however, possible when the functions $y(x) \in Y$ are allowed to be piecewise continuous and bounded.

If x represents time, the Mayer problem (4.30) represents physically a minimum time problem. In control theory it may be thought of as the problem of transferring most rapidly the control system from the state $z_i = z_{i0}$ to the state $z_i = z_{i1}$.

Let us now suppose that a solution $I(\bar{y}) = \bar{x}_1 - x_0$, $y = \bar{y}(x)$ of the degenerated Mayer problem (4.30) exists in the set Y of continuous functions. The corresponding functions $z_i = \bar{z}_i(x)$ are then uniquely determined by the initial-value problem (4.31). Following the original method of Pontryagin[P 8], let us determine a set of necessary conditions to be satisfied by $\bar{y}(x)$, $\bar{z}_i(x)$ and $I(\bar{y})$.

Since by definition Y is the set of continuous functions $y(x)$, and $I(\bar{y})$ is a local minimum, there exists a bilateral neighbourhood of order $k \geqslant 0$ in h of $\bar{y}(x)$ and $\bar{z}_i(x)$. It is therefore permissible to use the method of (bilateral) variations based on a linear parametric imbedding

$$y(x) = \bar{y}(x) + \delta y(x), \qquad (4.33)$$

$$z_i(x) = \bar{z}_i(x) + \delta z_i(x) \quad (i = 1, 2, ..., n). \qquad (4.34)$$

Substituting (4.33) and (4.34) into (4.31) yields

$$\frac{d}{dx}(\delta z_i) = \sum_{\alpha=1}^{n} \frac{\partial f_i}{\partial z_\alpha} \partial z_\alpha + \frac{\partial f_i}{\partial y} \delta y + \dots \quad (\delta(z_i(x_0)) = 0). \qquad (4.35)$$

Supposing that the local minimum $I(\bar{y})$ is simple, the non-linear terms in (4.35) are qualitatively unimportant. The linear approximations $\delta(z_i^0(x))$ of $\delta(z_i(x))$ are solutions of the linear differential system

$$\frac{d}{dx}(\delta z_i^0) = \sum_{\alpha=1}^{n} \frac{\partial f_i}{\partial z_\alpha} \delta z_\alpha^0 + \frac{\partial f_i}{\partial y} \delta y \quad (i = 1, 2, ..., n) \qquad (4.36)$$

and the vector $\delta(z_i^0(x_1))$ is a linear functional of δy. The locus of the phase points $(x_1, \delta(\bar{z}_i(x_1)) + \delta(z_i^0(x_1)))$ is therefore a linear subspace $S(x_1)$ of the Euclidean space R_n of the variables $z_i(x)$. $S(x_1)$ contains the point $(x_1, z_i(x_1))$. Pontryagin has shown[P 8] that for $y = \bar{y}(x)$ the dimensionality m of $S(x_1)$ is necessarily smaller than the dimensionality of R_n. Suppose that $m = n - 1$. In such a case $S(x_1)$ contains a hyperplane R_{n-1} described by

$$\sum_{\alpha=1}^{n} \psi_\alpha(x_1) [\zeta_\alpha - z_\alpha(x_1)] = 0, \qquad (4.37)$$

where ζ_α are current variables and $\psi_\alpha(x_1)$ are the covariant coordinates of R_{n-1} at $x = x_1$. If the ψ_α were normalized, they would be the direction cosines of the normal to the hyperplane R_{n-1}. Because of uniqueness of $\bar{y}(x)$ and $\bar{z}_i(x)$ the result (4.37) applies not only to the abscissa $x = x_1$, but to any abscissa in the interval $x_0 \leqslant x \leqslant x_1$. Pontryagin has pointed out[P 8] that an equivalent definition of the vector $\psi(x) = (\psi_1(x), \psi_2(x), ..., \psi_n(x))$ is given by the system of differential equations

$$\frac{d\psi_i}{dx} = -\sum_{\alpha=1}^{n} \frac{\partial f_i}{\partial z_\alpha} \psi_\alpha \quad (i = 1, 2, ..., n). \qquad (4.38)$$

Consider the set of $2n$ differential equations (4.31) and (4.38). This set is not 'complete', because it contains $2n + 1$ unknowns, and furthermore it is incorrectly set in the sense of Hadamard, because the boundary conditions for the functions $\psi_i(x)$ are not specified.

To remove the lack of determinacy in the set of equations (4.31), (4.38) consider the auxiliary function

$$H(z_1, ..., z_n, \psi_1, ..., \psi_n, y) = \sum_{\alpha=1}^{n} \psi_\alpha(x) . f_\alpha(z_1(x), ..., z_n(x), y(x)). \quad (4.39)$$

Interpreting H as a Hamiltonian, the set (4.31), (4.38) can be written in the canonical form

$$\frac{dz_i}{dx} = \frac{\partial H}{\partial \psi_i}, \quad \frac{d\psi_i}{dx} = -\frac{\partial H}{\partial z_i} \quad (i = 1, 2, ..., n). \quad (4.40)$$

$I(\bar{y})$ being by definition a non-degenerated local minimum of the extremal problem (4.30), the Hamiltonian (4.39) is stationary for $y = \bar{y}(x)$, i.e. a necessary condition for a local minimum of $I(y)$ is

$$\frac{\partial H}{\partial y} = 0. \quad (4.41)$$

Equation (4.41) can be considered as an algebraic equation defining $y(x)$ implicitly as a function of $z_i(x)$ and $\psi_i(x)$. Except for the boundary conditions on the $\psi_i(x)$, the 'completed' system (4.40), (4.41) is fully determinate, i.e. each real root

$$y = y_j(z_i(x), \psi_i(x)) \quad (j = 1, 2, ..., n)$$

of (4.41) defines a unique and complete system (4.40).

From (4.40) and (4.41) it follows that along $y = \bar{y}(x)$ and $z_i = \bar{z}_i(x)$

$$H(z_i, \psi_i, y) \equiv \mathrm{ct} \quad (i = 1, 2, ..., n),$$

and since the functions $\psi_i(x)$ still contain undetermined constants, it is possible to choose these constants so that

$$H(\bar{z}_i(x), \psi_i(x), \bar{y}(x)) \geqslant 0 \quad (x_0 \leqslant x \leqslant x_1). \quad (4.42)$$

It is well known that the study of the second-order approximation of $\delta(z_i(x))$ is equivalent to the study of the second-order derivatives of H. From an examination of (4.41) it is obvious that (4.42) holds provided

$$\frac{\partial^2 H}{\partial y^2} < 0 \quad (x_0 \leqslant x \leqslant x_1), \quad (4.43)$$

or, in other words, provided $H(\bar{z}_i, \psi_i, \bar{y})$ is a local maximum.

The result of the preceding argument can be summarized as follows. If the continuous functions $\bar{y}(x)$, $\bar{z}_i(x)$ define a non-degenerated local minimum of the functional $I(y)$ in (4.30), then there exists a continuous and a non-identically vanishing vector function

$$\psi(x) = (\psi_1(x), \psi_2(x), ..., \psi_n(x))$$

such that $\bar{y}(x)$, $\bar{z}_i(x)$ and $\psi_i(x)$ satisfy the system of equations

$$\frac{dz_i}{dx} = \frac{\partial H}{\partial \psi_i}, \quad \frac{d\psi_i}{dx} = -\frac{\partial H}{\partial z_i}, \\ \bar{H}(\bar{z}_i, \psi_i) = \text{local } \underset{y \in Y}{\text{maximum}} \, H(z_i, \psi_i, y) = 0. \quad \Bigg\} \tag{4.44}$$

Equations (4.44) represent the stationary version of the Pontryagin maximum principle.

Let us note again that (4.44) holds in the case of the problem (4.30) only if the boundary conditions $z_i(x_1) = z_{i1}$ are trivial. In fact, if (4.41) possesses a unique root $y = y(z_i(x), \psi_i(x)) = \bar{y}(x)$ then the minimal value of

$$I(x_0, x_1) = x_1 - x_0$$

must be given simultaneously by each of the n equations

$$x_1 - x_0 = \int_{z_{i0}}^{z_{i1}} \frac{dz_i}{f_i(\bar{z}_j, \bar{y})} \quad (i,j = 1, 2, ..., n). \tag{4.45}$$

For fixed f_i, the validity of (4.45) is clearly impossible for arbitrary values of z_{i1}.

The existence of a meaningful solution of the problem (4.30) requires therefore the existence of more than one real root of the equation (4.41). These distinct roots lead to distinct solutions $z_i(x)$, $\psi_i(x)$ of the canonical equations. In order to construct a continuous vector function

$$z(x) = (z_1(x), ..., z_n(x))$$

joining two arbitrary points (x_0, z_{i0}) and (x_1, z_{i1}) in the phase-space R_n, it is necessary to combine suitably the segments of the distinct solutions $z_i(x)$, corresponding to the distinct forms of $\bar{y}(x)$. In general the resulting combination of the segments of $\bar{y}(x)$ will not form a continuous function in the interval $x_0 \leqslant x \leqslant x_1$. The admissible set Y in (4.30) must therefore contain piecewise continuous functions $y(x)$. This conclusion contradicts unfortunately the validity conditions for the bilateral variations (4.33), (4.34), because a discontinuous minimizing function $\bar{y}(x)$ does not admit a neighbourhood of order $k \geqslant 0$ in h.

To remove the above-mentioned difficulty and to allow for degenerated local minima and stationary local lower limits Pontryagin has conjectured[P 8] that the relations (4.44) remain valid, provided the operation local maximum H is replaced by the operation $\lim \sup H$. This conjecture $\underset{y \in Y}{} \qquad \qquad \underset{y \in Y}{}$ turned out to be correct[P 10],[P 11], and its proof was given by an extension of the results of Bliss[B 11],[B 13], Underhill[B 12] and McShane[M 12] on Lagrange multipliers and on unilateral variations (cf. §15.3). The generalized version of the Pontryagin maximum principle applies to degenerated Mayer problems of the form

$$
\left.
\begin{aligned}
I(x_0, x_1) &= \min_{y \in Y} I(y) = \min_{y \in Y} v(x_1) \quad (x_0 \leqslant x \leqslant x_1), \\
\dot{v}(x) &= F(x, z_i(x), y(x)) \quad (v(x_0) = 0), \\
\dot{z}_j(x) &= f_j(x, z_i(x), y(x)) \quad (z_i(x_0) = z_{i0}, \quad z_i(x_1) = z_{i1}), \\
&\quad (i, j = 1, 2, \dots, n),
\end{aligned}
\right\}
\tag{4.46}
$$

where Y is the set of piecewise continuous functions, and it states that a necessary condition for $I(y)$ to be a local minimum or a stationary local lower limit is the existence of a non-identically vanishing continuous vector function $\psi(x)$ satisfying in $x_0 \leqslant x \leqslant x_1$ the system of equations

$$
\left.
\begin{aligned}
\frac{dz_i}{dx} &= \frac{\partial H}{\partial \psi_i}, \quad \frac{d\psi_i}{dx} = -\frac{\partial H}{\partial z_i}, \quad \psi_{n+1}(x_1) = 0 \quad (x_0 \leqslant x \leqslant x_1), \\
H(x, z_i, \psi_0, \psi_i, y) &= \sum_{\alpha=1}^{n+1} \psi_\alpha(x) \cdot f_\alpha(x, z_i(x), y(x)) + \psi_0(x) F(x, z_i(x), y(x)), \\
\bar{H}(x, \bar{z}_i, \psi_0, \psi_i) &= \limsup_{y \in Y} H(x, z_i, \psi_0, \psi_i, y(x)) = 0 \quad (i = 1, 2, \dots, n+1),
\end{aligned}
\right\}
\tag{4.47}
$$

where $\qquad \dfrac{dz_{n+1}}{dx} = f_{n+1}(x, z_i(x), y(x)) \equiv 1, \quad z_{n+1}(x_0) = x_0.$

A further validity condition of (4.47) consists in a restriction of the admissible values of the function $y(x) \in Y$. In fact, if (4.47) is to hold, the region $G(x, y)$ in which the $y(x)$ take their values must be both bounded and closed. The necessity of the boundedness of $y(x)$ is obvious from the Mayer problem (4.30) when the functions f_i are linear in $z_i(x)$ and $y(x)$.

Let Γ be the boundary of $G(x, y)$. The roots $y = \bar{y}(x)$ of the algebraic equation

$$
\bar{H}(x, z_i, \psi_i) = \limsup_{y \in Y, (x,y) \in G} H(x, z_i, \psi_0, \psi_i, y) = 0
\tag{4.48}
$$

are thus either roots of (4.41) or pieces of Γ. The conclusion drawn from the maximum principle is thus essentially the same as that resulting from the work of Tonelli[T6] (cf. §13.2). The degeneracy of the Mayer problem (4.46) manifests itself also through the property that at least in the neighbourhood of $y = \bar{y}(x)$ the functional $I(y)$ in (4.46) is continuous of order zero in h (cf. §15.6).

The order of degeneracy of the Mayer problem (4.46) increases when some of the differential equations

$$
\dot{z}_j(x) = f_j(x, z_i(x), y(x))
$$

form total derivatives, because these equations are then equivalent to holonomous equality constraints

$$p_j(x, z_i(x), y(x)) = 0$$

$$(i = 1, 2, ..., n, \quad j = 1, 2, ..., m, \quad m < n). \quad (4.49)$$

To deal with such additional degeneracy it is either possible to eliminate the m excess variables z_i defined by (4.49) or to modify the statement of the maximum principle (4.47) accordingly. The latter possibility was used by Pallu de la Barrière[P 1], who has shown that the equations

$$\frac{d\psi_i}{dx} = -\frac{\partial H}{\partial z_i}$$

in (4.47) need only be replaced by the equations

$$\frac{d\psi_i}{dx} = -\frac{\partial H}{\partial z_i} + \sum_{\alpha=1}^{m} \lambda_\alpha(x) \frac{\partial p_\alpha}{\partial x} \quad (i = 1, 2, ..., n), \quad (4.50)$$

where $\lambda_\alpha(x)$ are Lagrange multipliers. In using (4.50) it is best to suppose that n represents the number of differential equations in (4.46) which are not equivalent to total derivation.

Consider now the Mayer problem (4.46) with the holonomous inequality constraint

$$p(z_1(x), z_2(x), ..., z_n(x)) \leqslant 0, \quad (4.51)$$

where p is a twice-differentiable function of its arguments. Since the functional $I(y)$ in (4.46) is locally additive (cf. §14.5), the minimizing curves $y = \bar{y}(x)$, $z_i = \bar{z}_i(x)$ can be split into segments, some located in the interior and others on the boundary C of the closed region $B(z_i)$ defined by (4.51). Let

$$\sum_{i=1}^{n} \left(\frac{\partial p}{\partial z_i}\right)^2 \neq 0 \quad (x_0 \leqslant x \leqslant x_1), \quad (4.52)$$

that is let the boundary C of $B(z_i)$ be free of singular points in the sense of differential geometry. For the interior segments the stationary version (4.44) of the maximum principle applies, whereas the boundary segments require the generalized version (4.47) subject to the holonomous constraint

$$p(z_1(x), z_2(x), ..., z_n(x)) = 0. \quad (4.53)$$

As was pointed out by Gamkrelidze[G 4], a segment of the boundary C of $B(z_i)$ can only be a segment of $z = \bar{z}_i(x)$ if

$$\sum_{i=1}^{n} \frac{\partial p}{\partial z_i} \dot{z}_i = \sum_{i=1}^{n} \frac{\partial p}{\partial z_i} f_i \equiv 0. \quad (4.54)$$

Equation (4.54) means geometrically that the boundary C of $B(z_i)$ must be a trajectory $z_i = z_i(x)$ of the differential system in (4.46).

If a solution of (4.46), (4.51), satisfying (4.54) and the maximum principle does exist, the interior and boundary segments of the minimizing curve $z_i = \bar{z}_i(x)$ must satisfy some rather stringent conditions at their junction points. Let $x = \xi$ be the abscissa of a junction point and suppose that the abscissa $x = \xi - \epsilon$, $\epsilon > 0$, corresponds to an internal segment of $z_i = \bar{z}_i(x)$. According to Gamkrelidze [G4],[P11] the corresponding adjoint variable vector $\psi(x)$ must verify either

$$\psi(\xi^-) = \psi(\xi^+) + \mu \operatorname{grad} p \neq 0, \tag{4.55}$$

or $\qquad \psi(\xi^-) = 0, \quad \psi(\xi)^+ + \mu \operatorname{grad} p = 0 \quad (\mu \neq 0), \tag{4.56}$

where μ is a real constant and

$$\operatorname{grad} p = \left(\frac{\partial p}{\partial z_1}, \frac{\partial p}{\partial z_2}, \dots, \frac{\partial p}{\partial z_n} \right)$$

is the normal to the boundary surface (4.53). Gamkrelidze has called (4.55) and (4.56) the jump conditions.

Let us note that since the functions $f_i(x, z_i(x), y(x))$ in (4.46) and the function $p(z_i)$ in (4.51) are formulated by different considerations, the condition (4.54) is almost never satisfied in practical cases, and the extremal problem (4.46), (4.51) has generally no solution satisfying the maximum principle. Since this situation occurs also in the unconstrained problem (4.46), Gamkrelidze has made use of the classical notion of a 'limiting solution', defined by a minimal sequence, and has called it a sliding regime [G5] (cf. §14.4).

Various possibilities are open to derive a suitable minimal sequence and the required limiting solution. Gamkrelidze has considered the problem (4.30) together with the 'close' problem

$$\left. \begin{aligned} I(x_0, x_1) &= \min_{y \in Y} I(y) = \min_{y \in Y} (x_1 - x_0), \\ \dot{z}_i(x) &= g_i(z_j(x), y(x)) = \sum_{\alpha=1}^{n} \lambda_\alpha(x) f_i(z_j(x), y_\alpha(x)), \\ z_i(x_0) &= z_{i0}, \quad z_i(x_1) = z_{i1} \quad (x_0 \leqslant x \leqslant x_1, \quad i, j = 1, \dots, n), \end{aligned} \right\} \tag{4.57}$$

where $y_\alpha(x)$ are suitably chosen elements of Y and the $\lambda_\alpha(x)$ are weighting functions satisfying the relations

$$\lambda_\alpha(x) \geqslant 0, \quad \sum_{\alpha=1}^{n} \lambda_\alpha(x) = 1 \quad (x_0 \leqslant x \leqslant x_1). \tag{4.58}$$

If the set of functions $y_\alpha(x), \lambda_\alpha(x), \alpha = 1, 2, \dots, n$, is properly chosen, the limiting solution of (4.30) will be a solution in the ordinary sense of (4.57), and can be found by means of the maximum principle. Gamkrelidze has shown that the maximization of the Hamiltonian

$$H = \sum_{\alpha=1}^{n} \lambda_\alpha(x) \sum_{\beta=1}^{n} \psi_\beta(x) f_\beta(z_i(x), y_\alpha(x)) \quad (i \dots 1, 2, \dots, n) \tag{4.59}$$

must then be carried out with respect to both $y_\alpha(x)$ and $\lambda_\alpha(x)$. A necessary condition for a limiting solution of (4.30) to exist is the existence of at least two distinct real roots $y = y(z_i, \psi_i)$ of the algebraic equation

$$\limsup_{y \in Y} \sum_{\alpha=1}^{n} \psi_\alpha(x) f_\alpha(z_i(x), y(x)) = 0. \tag{4.60}$$

19.3   Relation to the calculus of variations

From the summary given in § 19.2 it is immediately obvious that the maximum principle applies directly only to degenerate Mayer problems, i.e. to problems in which the derivative of the minimizing function $y(x)$ does not appear explicitly. A modification of the maximum principle is needed whenever $\dot{y}(x)$ does appear explicitly. Consider, for example, the extremal problem (2.41) which leads to the Mayer problem (4.26), (4.27). If the latter is to be studied by means of the maximum principle it is first necessary to carry out the singular transformation

$$\dot{y}(x) = u(x, y(x)) \quad (y \in Y). \tag{4.61}$$

Using (4.61) the Mayer problem (4.26), (4.27) is transformed into the Mayer problem

$$I(x_0, x_1) = \min_{y \in Y} I(y) = \min_{u \in U} v(x_1),$$
$$\left. \frac{dv}{d\xi} = F(x(\xi), y(\xi), u(\xi)), \quad \frac{dy}{d\xi} = u(\xi), \quad \frac{dx}{d\xi} = 1 \quad (x_0 \leqslant \xi \leqslant x_1), \right\} \tag{4.62}$$
$$v(x_0) = 0, \quad y(x_0) = y_0, \quad y(x_1) = y_1,$$

where U is the enlarged set corresponding to Y.

Forming the Hamiltonian

$$H = \psi_0(\xi) F(x(\xi), y(\xi), u(\xi)) + \psi_1(\xi) u(\xi) + \psi_2(\xi)$$

and applying the stationary version (4.44) of the maximum principle yields

$$\frac{d\psi_0}{d\xi} = 0, \quad \frac{d\psi_1}{d\xi} = -\frac{\partial F}{\partial y} \psi_0, \quad \frac{d\psi_1}{d\xi} = -\frac{\partial F}{\partial x} \psi_0, \tag{4.63}$$

$$\psi_0 F(x, y, u) + \psi_1 u + \psi_2 = 0, \quad \psi_0 \frac{\partial F}{\partial u} + \psi_1 = 0, \tag{4.64}$$

and

$$\frac{\partial^2 F}{\partial u^2} \psi_0 < 0. \tag{4.65}$$

Since $\psi_0(\xi)$ is a constant and the ψ_i are only determined within a constant factor, it is possible to choose $\psi_0 = -1$ without loss of generality. Differentiating the second equation in (4.64) totally with respect to ξ, and substituting into the result the second equation in (4.63), yields

$$-\frac{d}{d\xi}\left(\frac{\partial F}{\partial u}\right) + \frac{\partial F}{\partial y} = 0 \quad (u = \dot{y}), \tag{4.66}$$

which coincides with the Euler equation of the problem (2.41). Equation (4.65) represents the strong Legendre condition $F_{\dot{y}\dot{y}} > 0$. In the stationary case the optimum trajectories of the maximum principle coincide therefore with the extremals.

Let us now compare the equations (4.64) to the stationary version (3.8) of the Carathéodory equations. These two sets of equations are identical provided

$$\psi_1 = \frac{\partial S}{\partial y}, \quad \psi_2 = \frac{\partial S}{\partial x}. \tag{4.67}$$

The above result has a very simple geometrical meaning: the adjoint variables $\psi_1(x)$, $\psi_2(x)$ represent components of the normal to the family of surfaces $S(x, y) = $ ct. The relations (4.67) furnish also a very simple explanation of one classical property of Lagrange multipliers. In fact, according to Bliss[B 13] and McShane[M 12] a problem of form (4.62) admits an infinity of pairs of Lagrange multipliers $\lambda_1(x)$, $\lambda_2(x)$ to render $v(x_1)$ stationary, but generally only a single pair to render $v(x_1)$ a minimum. Since according to equation (3.23) of §16.1 and equation (4.67) of the present paragraph the adjoint variables ψ_i, $i = 1, 2, ..., n$, can also be interpreted as Lagrange multipliers, the result of Bliss and McShane follows immediately from the properties of the solution $S(x, y)$ of the partial differential equations (4.64). The equations (4.64) admit an infinity of boundary conditions of the Cauchy type, capable of specifying a unique particular solution $S(x, y)$, but in general only a single boundary condition of the Cauchy type specifying a solution making $v(x_1) = S(x_1, y_1)$ a local minimum. The requirement in the maximum principle of the existence of a non-identically vanishing adjoint vector $\psi(x)$ is thus equivalent to the requirement of the existence of a solution $S(x, y)$ of (3.8) making $S(x_1, y_1)$ a local minimum. Since more is known about the existence of solutions of partial differential equations of order one[K 4],[O 1]–[O 6], than about the existence of non-identically vanishing adjoint vectors $\psi(x)$, the Carathéodory formulation furnishes more information about the extremal problem than the maximum principle. This conclusion is strengthened by noting that the canonical equations of the maximum principle represent only a part of the characteristic equations (3.35) of the Carathéodory formulation.

If the Mayer problem, for example (4.46), to be studied by the maximum principle does not contain the derivative of the minimization variable $y(x)$, the maximization of the Hamiltonian

$$H(x, z_i(x), \psi_i(x), y(x)) = \psi_0(x) F(x, z_i, y) + \sum_{\alpha=1}^{n+1} \psi_\alpha(x) f_\alpha(x, z_i, y) \tag{4.68}$$

must be carried out in a manner consistent with the method of unilateral variations (cf. §§15.3 and 17.2), because the latter are used in the proof of the maximum principle. For comparison with the Carathéodory formulation it is therefore necessary to write the maximum principle in the form

$$
\left.
\begin{aligned}
&\frac{dz_i}{dx} = \frac{\partial H}{\partial \psi_i}, \quad \frac{d\psi_0}{dx} = 0, \quad \frac{d\psi_i}{dx} = -\frac{\partial H}{\partial z_i}, \\
&\bar{H}(x, \bar{z}_i, \psi_i) = \text{stat.}\limsup_{y \in Y} \left[\psi_0 F(x, z_i, y) + \sum_{\alpha=1}^{n+1} \psi_\alpha f_\alpha(x, z_i, y) \right] = 0 \\
&\hspace{7cm} (i = 1, 2, \ldots, n+1),
\end{aligned}
\right\} \quad (4.69)
$$

where the symbol 'stat.' signifies that the nature of the upper limit is subject to the validity of unilateral variations.

Again without loss of generality it is possible to assume that $\psi_0 = -1$. Comparing the equation

$$
\text{stat.}\limsup_{y \in Y} \left[-F(x, z_i, y) + \sum_{\alpha=1}^{n+1} \psi_\alpha f_\alpha(x, z_i, y) \right] = 0 \quad (4.70)
$$

to the corresponding Carathéodory equation (cf. §17.7)

$$
\min_{y \in Y} \left[F(x, z_i, y) - \frac{\partial S}{\partial x} - \sum_{\alpha=1}^{n} \frac{\partial S}{\partial z_\alpha} f_\alpha(x, z_i, y) \right] = 0, \quad (4.71)
$$

it is immediately apparent that the former is a particular case of the latter, provided of course

$$
\psi_{n+1} = \frac{\partial S}{\partial x}, \quad \psi_i = \frac{\partial S}{\partial z_i} \quad (i = 1, 2, \ldots, n). \quad (4.72)
$$

Equation (4.70) is less general than the equation (4.71), because the operator $\min_{y \in Y}$ in (4.71) is less restrictive, and in particular it is not tied to the theory of unilateral variations.

The difference in the scope of validity of (4.70) and (4.71) is easily demonstrated by means of the non-degenerated problem (2.92) studied in §§14.3 and 14.6. In the problem (2.92) the boundary of the admissible values of the state variables x, y is given by a curve $g(x, y) = 0$ which is only piecewise differentiable, and the least value of $I(y)$ corresponds to a minimizing curve $y = \bar{y}(x)$ which attains the boundary $g(x, y) = 0$ at a corner point. Such a situation is clearly excluded in the proof of the maximum principle (cf. chapter 6[P 11]), where $g(x, y) = 0$ must be twice differentiable and free of corners. Except in the case when the functions $f_\alpha(x, z_i, y)$ are linear in z_i and y, the scope of validity of (4.70) seems to coincide with the scope of validity of the transversality condition (3.32).

The Carathéodory formulation permits to assign a simple geometric meaning to the Gamkrelidze jump-conditions (4.55), (4.56). Since equation (4.72) implies that the vector $\psi(x) = (\psi_1(x), \psi_2(x), ..., \psi_n(x))$ represents the exterior normal

$$\mathbf{N}(x) = \left(\frac{\partial S}{\partial z_1}, \frac{\partial S}{\partial z_2}, ..., \frac{\partial S}{\partial z_n}\right) \qquad (4.73)$$

to the family $S(z_1, z_2, ..., z_n) = \mathrm{ct}$, $x_0 \leqslant x \leqslant x_1$, and the vector

$$\mathrm{grad}\, p = \left(\frac{\partial p}{\partial z_1}, \frac{\partial p}{\partial z_2}, ..., \frac{\partial p}{\partial z_n}\right) = \mathbf{n}$$

represents by definition the exterior normal of the boundary (4.53), the jump condition (4.55), (4.56) can be written

$$\mathbf{N}(\xi^-) = \mathbf{N}(\xi^+) + \mu\mathbf{n}(\xi^+) \neq 0, \qquad (4.74)$$

$$\mathbf{N}(\xi^-) = 0, \quad \mathbf{N}(\xi^+) + \mu\mathbf{n}(\xi^+) = 0 \quad (\mu \neq 0). \qquad (4.75)$$

Interpreting equation (4.53) as a boundary between two regions where the solutions $S(z_1, z_2, ..., z_n)$ of the Carathéodory equation have different qualitative properties, when non-trivial, the jump conditions (4.74), (4.75) are equivalent either to the Weierstrass–Erdmann conditions (cf. §14.5) or to the continuity conditions of the variational theory of reflected and refracted waves (cf. §14.5).

The main advantage of the maximum principle consists in the use of a uniform notation in the study of a variety of extremal problems, but like dynamic programming, the maximum principle is less general than the classical calculus of variations. In fact, both dynamic programming and the maximum principle are less general than the theory of the Carathéodory formulation.

CHAPTER 5

APPROXIMATE SOLUTION OF OPTIMIZATION PROBLEMS

§ 20. DIRECT METHODS

20.1 Close extremal problems

Optimization problems which can be solved explicitly by means of elementary functions are highly exceptional. This situation is not changed appreciably when tabulated transcendental functions are taken into consideration. The study of a concrete optimization problem leads therefore invariably to the study of approximate methods.

There exists of course a multitude of approximate methods and the main problem, when looking for an approximate solution of an optimization problem, consists in selecting the most efficient one. Since no approximate method is universal, the efficiency of the approximating method depends on the compatibility of the salient properties of the exact solution with the salient properties of the approximating functions. For example, there is generally very little sense in approximating a continuous monotonic function $\bar{y}(x)$, $x_0 \leqslant x \leqslant x_1$, by means of a sum

$$y_n(x) = \sum_{i=0}^{n} a_i \phi_i(x)$$

of rapidly oscillating continuous or discontinuous functions $\phi_i(x)$. The maximal absolute error of the approximate solution will not diminish uniformly in the whole interval $x_0 \leqslant x \leqslant x_1$ as n increases.

In order to select an appropriate approximation method for the study of a given optimization problem it is necessary to know some properties of the latter (cf. [H 11]). When no excess information about the properties of the optimization problem is available, beyond the mathematical formulation of this problem, a natural starting-point for the selection of an approximate method is the determination of the continuity properties of the functional contained in the mathematical formulation. By the continuity properties of the functional are meant not only the type of continuity with respect to close admissible functions (cf. §§ 12.2–12.4) but also the continuity with respect to the data contained in the boundary conditions (cf. § 14.6). In other words, the approximate method should not constitute a structural perturbation which destroys the inertness of the nominal optimization model (cf. §§ 4.2–4.6).

The process of applying successfully an approximate method to a given optimization problem can be envisaged as a process of constructing a sequence of increasingly accurate deductive a priori models (cf. §2) converging to the corresponding nominal optimization model. The degree of approximation is therefore a measure of the 'inessential' properties of the optimization problem, and the system of approximating equations constitutes a close formulation of the original optimization problem, i.e. a close model of the nominal optimization model. If the approximation decided upon is to be useful for design purposes, the close model must be an element of the set of physically equivalent models (cf. §4.6).

As an illustration consider the extremal problem (2.41), written with a slightly different notation

$$I(\bar{y}) = \min_{y \in Y} I(y) = \min_{y \in Y} \int_a^b F(x, y(x), \dot{y}(x))\, dx \quad (y(a) = \alpha, \quad y(b) = \beta), \quad (5.1)$$

where $I(y)$ is a Riemann integral and Y is the set of continuous piecewise differentiable functions. The problem (5.1) can be approximated by means of four distinct types of modifications:

(a) Replacing the nominal point-function $F(x, y, \dot{y})$ by a close point-function $F_\epsilon(x, y, \dot{y})$,

(b) Replacing the boundary data $y(a) = \alpha$, $y(b) = \beta$ by some 'close' boundary data $f_1(a, b, \alpha, \beta) = 0$, $f_2(a, b, \alpha, \beta) = 0$.   (5.2)

(c) Replacing the nominal admissible set Y by a 'close' set Y_ϵ.

(d) Replacing the point-function $I(\bar{y}) = S(a, b, \alpha, \beta)$ by a close point-function $S_\epsilon(a, b, \alpha, \beta)$.

The approximation of type (5.2a) is used extensively in the solution of partial differential equations by means of variational methods, and its implementation is known as the method of Hilbert[C 15] or the method of Sobolev[S 4]. Although different in appearance this type of approximation is essentially equivalent to the replacement of the exact Euler equations $\delta I / \delta y = 0$ by a set of 'close' differential or difference equations $L_\epsilon y = 0$, with $\lim_{\epsilon \to 0} L_\epsilon y = \delta I / \delta y$. The approximation of type (5.2b) was used already by Todhunter[T 3] and some of its implications were studied by Hadamard[H 3] (cf. §§14.3 and 14.6). The third type of approximation leads to a diversity of procedures, based for example on minimal sequences, on iterations using admissible variations and on Bellman's iterations in the policy space (cf. §18.1). The approximations of type (5.2d) are relatively unexplored. The various possibilities enumerated in (5.2) can be combined in various ways with each other or with Hadamard's method of descent (cf. §7). For example, the Carathéodory equation can be solved approximately by interpreting it as a degenerate case of a parabolic equation of second order. A fair amount of information is available about numerical properties of boundary-value problems associated with second-order parabolic equations (see for example [B 7], [F 5] and [F 6]). Keeping in mind the preceding remarks let us now examine some often-used approximations from the point of view of their compatibility with the properties of the functional $I(y)$ in (5.1).

20.2 The method of an enumerable set of independent variables

The set Y of continuous and continuously differentiable functions which occurs in (5.1) is non-enumerable and non-ordered, and because of the latter property there exists no natural sequence of testing $I(y)$ for its least value in Y. The set Y can of course be ordered by defining a norm-function $N(\bar{y}(x), y(x))$, $\bar{y}(x), y(x) \in Y$, but this can be done in more than one way. To find an appropriate form of N it is necessary to know the type of continuity of $I(y)$. Let us recall, however, that the dimensionality of Y in (5.1) is enumerable (cf. §12.1), and because of this property it might be possible to deduce the solution of (5.1) by converting (5.1) to an 'algebraic' problem with an enumerable number of independent variables, and then solving this algebraic problem by the method of successive approximations.

The possibly simplest procedure consists in subdividing the interval $a \leqslant x \leqslant b$ into m subintervals, say of equal width Δx_m, and interpret the ordinates $y_i = y(x_i)$, $x_i = a + i\Delta x_m$, $i = 0, 1, \ldots, m$, as the required independent variables. Supposing that the minimal curve $y = \bar{y}(x)$ of (5.1) is approximated by a polynomial curve $y = \bar{y}_m(x)$ passing through the $m+1$ points (x_i, y_i), $i = 0, 1, \ldots, m$, and that at least approximately the integration contained in $I(\bar{y}_m(x))$ can be carried out in closed form, the functional $I(y)$ is converted into a point-function, say of the form

$$I_m = \Delta x_m \sum_{i=0}^{m} \alpha_i F(x_i, y_i, \dot{y}_i^0) + O(1/m), \qquad (5.3)$$

where α_i are known real constants and $\dot{y}_i^0$ is a suitable finite difference approximation of $\dot{y}(x_i)$.

Dropping the error term $O(1/m)$ in (5.3) and supposing that the least value $I(\bar{y})$ of $I(y)$ is a non-degenerated local minimum, it follows that the unknown constants y_i, $i = 1, 2, \ldots, m-1$, should verify the system of algebraic equations

$$\frac{\partial I_m}{\partial y_i} = f_i(y_1, y_2, \ldots, y_{m-1}) = 0 \quad (i = 1, 2, \ldots, m-1). \qquad (5.4)$$

The conclusion expressed by (5.4) is equivalent to the statement that the functional problem (5.1) and the point-function problem

$$I(\bar{y}_m) = \min_{y_i} I_m \qquad (5.5)$$

are stationary simultaneously. Since in the system (5.4) there are as many unknowns as there are equations, this system is completely determinate, and its roots can be found, for example, by successive iterations. Let $y_i^{(m)}, i = 1, 2, ..., m-1$, be one root of (5.4), which in general is not unique. Substituting $y_i^{(m)}$ into (5.3) yields the unknown value I_m. If $y_i^{(m)}$ is not unique, then a set of values of I_m is obtained, only the smallest of which should be retained.

In order to estimate the accuracy of the approximation it is necessary to repeat the calculations leading to (5.3) and (5.4) for a larger value of m, say $m+1$, and to compare the closeness of $\bar{y}_m(x)$, $\bar{y}_{m+1}(x)$ and $I(\bar{y}_m)$, $I(\bar{y}_{m+1})$, respectively. If the difference $I(\bar{y}_m) - I(\bar{y}_{m+1}) > 0$ is less than the required precision, it may be assumed that a satisfactory approximation of $\bar{y}(x)$ has been obtained.

In the above statement we have made the implicit assumption that the polynomial curves $y = y_m(x)$ converge to the minimal curve $\bar{y} = y(x)$ and the approximate values $I(\bar{y}_m)$ converge to the value $I(\bar{y})$ as the integer m increases indefinitely. In other words we have assumed that the required root of the infinite system of equations

$$f_i(y_1, y_2, ..., y_m, ...) = 0 \quad (i = 1, 2, ..., m, ...) \tag{5.6}$$

can be found by the method of reduction (see for example §2[K6]). Unfortunately the roots of an infinite system of algebraic equations are not always the limits of the roots of a truncated system of order $m < \infty$, because the former may admit more roots than the latter. As an illustration consider the infinite linear system

$$\left[1 - \frac{1}{(i+1)^2}\right] x_{i+1} - x_i + \frac{1}{(i+1)^2} = 0. \tag{5.7}$$

It can be easily verified by substitution that (5.7) has two distinct roots:

$$x_i = 1 \quad \text{and} \quad x_i = 1/(i+1).$$

This fact should be contrasted with the fact that a non-degenerated linear system of finite order admits only one root.

Let us now recall that as a rule the closeness of the functions $y_m(x)$ and $y_{m+1}(x)$, with some given norm, does not guarantee the closeness of the constants $I(y_m)$ and $I(y_{m+1})$. Hence it may happen that the closeness of $\bar{y}_m(x)$ and $\bar{y}_{m+1}(x)$ appears to be satisfactory while the closeness of $I(\bar{y}_m)$ and $I(\bar{y}_{m+1})$ leaves much to be desired. The reason for this behaviour lies in the fact that the continuity of the functional $I(y)$ does not necessarily coincide with the specified neighbourhood of $\bar{y}_m(x)$. As an illustration suppose that the functional $I(y)$ is continuous of order one in h and that the curves $y = \bar{y}_m(x)$ are piecewise linear. Such curves consist of straight segments joining the points (a, α), (x_i, y_i), $i = 1, 2, ..., m-1$, and (b, β).

The curves $y = y_m(x)$ are piecewise linear when the derivatives $\dot{y}(x_i)$, $i = 0, 1, \ldots, m$, are approximated by finite differences of the first order. In such a case the approximation method is called the method of Euler (see for example [H 1] and vol. I[C 15]). Since two piecewise linear curves $y = \bar{y}_m(x)$ and $y = \bar{y}_{m+1}(x)$ which are close with respect to a neighbourhood of order zero in h are not necessarily close with respect to a neighbourhood of order one in h, the closeness properties of $\bar{y}_m(x)$, $\bar{y}_{m+1}(x)$ are not matched to the continuity properties of $I(y)$ and the differences

$$|I(\bar{y}_{m+1}) - I(\bar{y}_m)| \quad \text{and} \quad |\bar{y}_{m+1}(x) - \bar{y}_m(x)| \quad (a \leqslant x \leqslant b)$$

are not necessarily small simultaneously (cf. § 13.4). The choice of a suitable finite difference representation of $\dot{y}(x_i)$, $i = 0, 1, \ldots, m$, is thus intimately linked with the continuity properties of the functional $I(y)$.

Another method of reducing the extremal problem (5.1) to a point-function problem consists in representing the functions $y(x) \in Y$ by means of the formal series

$$y(x) = \sum_{i=0}^{\infty} a_i \phi_i(x), \tag{5.8}$$

where the $\phi_i(x)$ are linearly independent functions and the a_i are undetermined real constants. If the set $\phi_i(x), i = 0, 1, \ldots$, is complete in the interval $a \leqslant x \leqslant b$ the series (5.8) can be expected to converge to the exact minimizing function $\bar{y}(x)$. The functions $\phi_i(x)$ are called coordinate functions associated with the extremal problem (5.1). In order to satisfy the boundary conditions in (5.1) the $\phi_i(x)$ are usually so chosen that one of the following conditions is satisfied:

$$\left.\begin{array}{l} \phi_0(a) = \alpha, \quad \phi_0(b) = \beta, \quad \phi_i(a) = \phi_i(b) = 0 \quad (i > 0), \\[2mm] \text{or} \quad \sum_{i=0}^{m} a_i \phi_i(a) = \alpha, \quad \sum_{i=0}^{m} a_i \phi_i(b) = \beta \quad (m = 1, 2, \ldots). \end{array}\right\} \tag{5.9}$$

Provided the series (5.8) can be differentiated term by term, the substitution of (5.8) into (5.1) results in a point-function problem for the coefficients a_i:

$$I(\bar{y}) = \min_{a_i} I(y) = \min_{a_i} \int_a^b F\left(x, \sum_{i=0}^{\infty} a_i \phi_i(x), \sum_{i=0}^{\infty} a_i \dot{\phi}_i(x)\right) dx. \tag{5.10}$$

If $I(y)$ is a non-degenerated local minimum of (5.1), the unknown constants a_i are determined by the enumerable system of equations

$$\frac{\partial I(y)}{\partial a_j} = 0 \quad (j = 0, 1, \ldots). \tag{5.11}$$

When this system is solved by the method of reduction, it is said that the solution of (5.1) is established by the method of Ritz.

Early experience with the method of Ritz has shown that if one starts with the truncated development

$$y_m(x) = \sum_{i=0}^{m} a_i^{(m)} \phi_i(x), \qquad (5.12)$$

then the constants $a_i^{(m)}$, called usually Ritz coefficients, are not necessarily identical with the constants a_i in (5.8). It was also observed that numerically the method of Ritz is satisfactory for low precision problems, requiring a relatively small value of m. When the required precision is high, say more than three significant digits in the value of $I(y)$, the value of m becomes larger than about five, and the method of Ritz develops computational instabilities.

Let N be the norm function defining a neighbourhood in the set of coordinate functions $\phi_i(x)$, Mikhlin has shown[M 19] that numerical instabilities usually do *not* occur when the set of the $\phi_i(x)$ is minimal, i.e. when it is impossible to find two integers j and M, and M real constants $\alpha_0, \alpha_1, ..., \alpha_{j-1}, \alpha_{j+1}, ..., \alpha_M$ such that the value of

$$N\left(\phi_j(x), \sum_{\substack{i=0 \\ i \neq j}}^{M} \alpha_i \phi_i(x)\right) \qquad (5.13)$$

can be rendered as small as desired. A set of coordinate functions is minimal when it is, for example, orthogonal and normalized. The set of linearly independent functions

$$x, \quad \sin x, \quad x^2, \quad \sin 2x, \quad ..., \quad \sin mx, \quad ...$$

is obviously non-minimal in the interval $-\pi \leqslant x \leqslant \pi$, because in this interval x and x^2 can be expressed by means of convergent Fourier series. Mikhlin has shown that the set of integer powers of x

$$1, \quad x, \quad x^2, \quad ..., \quad x^m, \quad ...,$$

used in the McLaurin series, is also non-minimal[M 19].

To illustrate the practical use of the method of Ritz consider the problem of finding a non-degenerated local minimum of the functional

$$I(y) = \int_0^1 (\dot{y}^2(x) - y^2(x) - 2xy(x))\,dx \quad (y(0) = y(1) = 0, \quad y \in Y), \quad (5.14)$$

where $I(y)$ is a Riemann integral and Y is the set of continuously differentiable functions. Consider the truncated series

$$y_m(x) = x(1-x) \sum_{i=1}^{m} a_i^{(m)} x^{i-1} \qquad (5.15)$$

satisfying the second set of conditions in (5.9). Letting $m = 1$ and substituting (5.15) into (5.14) and (5.11) yields $a_1^{(1)} = \frac{5}{8}$ and

$$\bar{y}_1(x) = \tfrac{5}{8}x(1-x). \qquad (5.16)$$

If $m = 2$, (5.16) is replaced by

$$\bar{y}_2(x) = x(1-x)\left(\tfrac{71}{369} + \tfrac{7}{41}x\right). \tag{5.17}$$

Since $a_1^{(1)} = 0\cdot625$ and $a_1^{(2)} = 0\cdot193$ there is a considerable difference between the values of $a_1^{(1)}$ and $a_1^{(2)}$.

The exact solution of the problem (5.14) is

$$\bar{y}(x) = \frac{\sin x}{\sin 1} - x, \quad I(\bar{y}) = \frac{1}{2}\frac{\sin 2}{\sin 1} - \frac{2}{3}. \tag{5.18}$$

In order to compare the precision obtained consider the abscissae $x = (\tfrac{1}{4}, \tfrac{1}{2}, \tfrac{3}{4})$. The corresponding values of $\bar{y}(x)$, $\bar{y}_1(x)$ and $\bar{y}_2(x)$ are $(0\cdot044, 0\cdot070, 0\cdot060)$, $(0\cdot052, 0\cdot069, 0\cdot052)$ and $(0\cdot044, 0\cdot069, 0\cdot060)$ respectively. In spite of the considerable difference between the values of $a_1^{(1)}$ and $a_1^{(2)}$ the precision of the second approximation is fully adequate for engineering purposes. This is however not true for the approximation of $I(y)$, where the minimal values are

$$I(\bar{y}) \approx -0\cdot1263, \quad I(\bar{y}_1) = 0\cdot01302, \quad I(\bar{y}_2) = -0\cdot02457.$$

To improve the precision more than two terms are needed in (5.15).

In order to illustrate the appearance of computational instabilities consider the determination of a non-degenerated local minimum of the functional

$$I(y) = \int_0^1 \left(\dot{y}^2(x) - \frac{2}{x+1}y(x)\right) dx, \quad y(0) = \dot{y}(1) = 0 \quad (y \in Y), \tag{5.19}$$

where as before $I(y)$ is a Riemann integral and Y is the set of continuously differentiable functions. The exact solution of the problem (5.19) is

$$\left.\begin{aligned}\bar{y}(x) &= (1 + \log 2)x - (x+1)\log(x+1),\\ I(\bar{y}) &= (\log 2)^2 - 2\log 2 - 2.\end{aligned}\right\} \tag{5.20}$$

Let the Ritz system (5.12) be given by

$$y_m(x) = \sum_{i=1}^m a_i^{(m)} x^i. \tag{5.21}$$

Consider the values $m = 1$, $m = 2$ and $m = 3$. The Ritz coefficients calculated with twelve decimals and then rounded off to four decimals are

$$a_1^{(1)} = 0\cdot3069, \quad a_1^{(2)} = 0\cdot6480, \quad a_1^{(3)} = 0\cdot6869,$$

$$a_2^{(2)} = -0\cdot3411, \quad a_2^{(3)} = -0\cdot4579, \quad a_3^{(3)} = 0\cdot0788.$$

If the same coefficients are calculated with only four decimals the result is:

$$a_1^{(1)} = 0\cdot3069, \quad a_1^{(2)} = 0\cdot6483, \quad a_1^{(3)} = 0\cdot6883,$$

$$a_2^{(2)} = -0\cdot3411, \quad a_2^{(3)} = -0\cdot4614, \quad a_3^{(3)} = 0\cdot0800.$$

The precision has not suffered much yet. If m is made larger, say $m = 8$, then the results of the twelve and four decimals calculations are

$$a_1^{(8)} = 0\cdot6931, \quad a_2^{(8)} = -0\cdot5000, \quad a_3^{(8)} = 0\cdot1664,$$

and $\qquad a_1^{(8)} = 1\cdot6266, \quad a_2^{(8)} = -24\cdot5448, \quad a_3^{(8)} = 204\cdot8340,$

respectively. For $m = 8$, the computation with only four decimals shows a severe computational instability.

In spite of the large variation of the Ritz coefficients $a_i^{(m)}$, the precision of the approximate minimizing curve $y = \bar{y}_m(x)$ is sufficient for engineering purposes, if m is not too large. In fact, at the abscissa $x = 1$ the exact and approximate values of $y(x)$ computed by means of the more precise $a_i^{(m)}$ are:

$$\bar{y}(1) = 1 - \log 2 \approx 0\cdot3069, \quad \bar{y}_1(1) = 0\cdot3069,$$

$$\bar{y}_2(1) = 0\cdot3069, \quad \bar{y}_3(1) = 0\cdot3071.$$

There is more dispersion in the minimal value of the functional $I(y)$:

$$I(\bar{y}) \approx -0\cdot1332, \quad I(\bar{y}_1) = 0\cdot01302,$$

$$I(\bar{y}_2) = -0\cdot1009, \quad I(\bar{y}_3) = -0\cdot1332.$$

No improvement occurs by making m larger. For example, using the Ritz coefficients $a_i^{(8)}$ computed with twelve decimals and then rounded off to four decimals yields

$$a_4^{(8)} = -0\cdot0817, \quad a_5^{(8)} = 0\cdot0446, \quad a_6^{(8)} = -0\cdot0218,$$

$$a_7^{(8)} = 0\cdot0075, \quad a_8^{(8)} = -0\cdot0013$$

and $\qquad\qquad\qquad I(\bar{y}_8) = -0\cdot1334.$

There is thus again some evidence of computational instability. Whenever computational instability is noticed, it is necessary to repeat all numerical computations with a larger number of digits. The highest accuracy is obtained in the presence of computational instability when the differences

$$\int_a^b |\bar{y}_m(x) - \bar{y}_{m+1}(x)| \, dx \quad \text{and} \quad |I(\bar{y}_m) - I(\bar{y}_{m+1})|$$

pass through a minimum.

The computational difficulties encountered with the Ritz developments (5.15) and (5.21) are due to the fact that the sets of coordinate functions $\phi_i(x) = (1-x)x^i, \phi_i(x) = x^i, i = 1, 2, \ldots$, are non-minimal. It appears thus that a natural way to avoid these difficulties consists in using orthogonal sets $\phi_i(x)$, i.e. sets satisfying the condition

$$\int_a^b \phi_i(x)\,\phi_j(x)\,p(x)\,dx = \delta_i^j, \tag{5.22}$$

where δ_i^j is the Kronecker symbol and $p(x)$, $a \leqslant x \leqslant b$, is a weighting function. It is well known that with such a choice of $\phi_i(x)$ the mean-square error

$$\epsilon_0 = N_0(\bar{y}(x), \bar{y}_m(x)) = \int_a^b [\bar{y}(x) - \bar{y}_m(x)]^2 \, dx \tag{5.23}$$

is a minimum. Unfortunately (5.23) does not imply anything about the smallness of the errors

$$\epsilon_i = N_0 \left(\frac{d^i\bar{y}}{dx^i}, \frac{d^i\bar{y}_m}{dx^i} \right) \quad (i = 1, 2, \ldots).$$

On the contrary it is known that a small value of ϵ_0 occurs in general simultaneously with a rather large value of ϵ_1, because the main feature of the mechanism of successive approximation of $\bar{y}(x)$ by means of a sum of orthogonal functions

$$\bar{y}_m(x) = \sum_{i=0}^m a_i^{(m)} \phi_i(x) \tag{5.12}$$

is an increasingly rapid oscillation of the values of $\bar{y}_m(x)$ around the corresponding values of $\bar{y}(x)$. The approximation by means of orthogonal functions is therefore efficient only when the functional $I(y)$ in (5.1) is continuous of order zero in h.

If $I(y)$ is continuous of order $k > 0$ in h, the norm function N, analogous to N_0 in (5.23), must be so chosen that the errors $\epsilon_0, \epsilon_1, \ldots, \epsilon_k$ are small simultaneously. A way to achieve this purpose was suggested by Lewis[L 14], who proposed that instead of (5.23) the following expression

$$N_k(\bar{y}(x), \bar{y}_m(x)) = \sum_{i=0}^k \int_a^b \left(\frac{d^i\bar{y}}{dx^i} - \frac{d^i\bar{y}_m}{dx^i} \right)^2 p_i(x) \, dx \tag{5.24}$$

be made a minimum, where $p_i(x)$, $i = 0, 1, \ldots, k$, is a set of weighting functions. The condition to be satisfied by the coordinate functions $\phi_i(x)$ is then

$$\sum_{\alpha=0}^k \int_a^b \frac{d^\alpha \phi_i(x)}{dx^\alpha} \frac{d^\alpha \phi_j(x)}{dx^\alpha} p_\alpha(x) \, dx = \delta_i^j. \tag{5.25}$$

Equation (5.25) can be interpreted as a generalization of the 'ordinary' orthogonality condition (5.22). Lewis has shown[L 14] that if $p_\alpha(x) \geqslant 0$ are continuous functions in the interval $a \leqslant x \leqslant b$ and the $\phi_i(x)$ are polynomials, then (5.25) defines a unique enumerable set of these polynomials. Unfortunately very little is known about the properties of polynomial solutions of (5.25) when $k > 1$. The case $k = 1$, $p_0(x) = 1$, $p_1(x) = \lambda = \mathrm{ct}$, $a = -1$, $b = 1$ has been studied by Althammer[A 1] and the case $k = 1$, $p_0(x) = 1$, $p_1(x) = \lambda = \mathrm{ct}$, $a = 0$, $b = 1$ by Unterkircher[U 1]. The condition (5.25) simplifies then into

$$\int_a^b [\phi_i(x) \phi_j(x) + \lambda \dot{\phi}_i(x) \dot{\phi}_j(x)] \, dx = \delta_i^j. \tag{5.26}$$

For $a = 0, b = 1$ the set of polynomial solutions of (5.26) is given by [U 1]

$$
\left.\begin{aligned}
\phi_n(x) &= C_n \sum_{j=0}^{[\frac{1}{2}n]} \lambda^j \sum_{i=0}^{n} (-1)^i \left[\binom{n}{i} \frac{(2n-i)!}{(n-i-2j)!} x^{n-i-2j} \right. \\
&\qquad\qquad\qquad \left. - \alpha_n \binom{n-1}{i} \frac{(2n-i-2)!}{(n-i-2j-2)!} x^{n-i-2j-2} \right], \\
C_n &= \frac{1}{n!} \sqrt{\frac{n(2n+1)}{n+2(2n+1)\alpha_n}}, \\
\alpha_n &= \frac{\displaystyle\sum_{i=0}^{[\frac{1}{2}(n-1)]} \lambda^{i+1}(n+2i+1)! \binom{n}{2i+1}}{\displaystyle\sum_{i=0}^{[\frac{1}{2}(n-1)]} \lambda^{i}(n+2i-1)! \binom{n-1}{2i}} \quad (n = 0, 1, \ldots),
\end{aligned}\right\} \quad (5.27)
$$

where the brackets in the limits of the summations stand for 'integer part of...' and

$$
\binom{m}{k} = \frac{m!}{(m-k)!\,k!}
$$

are binomial coefficients. For $\lambda = 1$ the first five polynomials of (5.27) are

$$
\left.\begin{aligned}
\phi_0(x) &= 1, \quad \phi_1(x) = \sqrt{\tfrac{3}{13}}(2x-1), \\
\phi_2(x) &= \sqrt{\tfrac{5}{61}}(6x^2 - 6x + 1), \\
\phi_3(x) &= \frac{7}{\sqrt{143 \cdot 1538}}(20x^3 - 30x^2 + \tfrac{132}{13}x - \tfrac{1}{13}), \\
\phi_4(x) &= \frac{3}{\sqrt{254 \cdot 7705}}(70x^4 - 140x^3 + \tfrac{2052}{61}x^2 - \tfrac{860}{61}x + \tfrac{1}{61}).
\end{aligned}\right\} \quad (5.28)
$$

In spite of the rather large coefficients, $|\phi_n(x)| \leqslant 1$ for all x in $0 \leqslant x \leqslant 1$.

For $a = -1, b = +1$ the polynomial solutions of (5.26) are very similar to (5.28) [A 1]:

$$
\left.\begin{aligned}
\phi_n(x) &= -\sum_{i=1}^{[\frac{1}{2}n]} \lambda^i \left[A_n \frac{d^{2i+1}L_{n+1}}{dx^{2i+1}} + B_n \frac{d^{2i+1}L_{n-1}}{dx^{2i+1}} \right], \\
A_n &= -\frac{k_n}{n+1}, \\
B_n &= \frac{2\lambda(2n+1)-1}{2(2n+1)} nk_n - \frac{l_n}{n-1}, \\
k_n^2 &= \frac{(2n+1)\,2^{2n}}{2} \left[\frac{(2n-1)!!}{(2n)!!} \right]^2 \frac{\displaystyle\sum_{i=0}^{[\frac{1}{2}(n-1)]} \lambda^i \frac{(n+2i-1)!}{(n-2i-1)!} \frac{1}{(2i)!\,2^{2i}}}{\displaystyle\sum_{i=0}^{[\frac{1}{2}(n+1)]} \lambda^i \frac{(n+2i+1)!}{(n-2i-1)!} \frac{1}{(2i)!\,2^{2i}}} \quad (k_n > 0), \\
l_n &= \frac{n(n-1)k_n}{2(4n^2-1)} \frac{[2\lambda(2n+1)-1](2n-1)\alpha_{n-1} - 2\alpha_{n+1}}{\alpha_{n-1}}, \\
\alpha_n &= \sum_{i=0}^{[\frac{1}{2}n]} \lambda^i \frac{(n+2i)!}{(n-2i)!} \frac{1}{(2i)!\,2^{2i}} \quad (n = 0, 1, \ldots),
\end{aligned}\right\}
$$

$$(5.29)$$

where L_n is the Legendre polynomial with the coefficient of x^n equal to $+1$:

$$L_n(x) = \frac{n!}{(2n)!} \sum_{i=0}^{n} (-1)^{i-1} \binom{n}{i} \frac{(2i)!}{(2i-n)!} x^{2i-n}.$$

The negative powers of x in $L(x)$ are of course to be omitted. For $\lambda = 1$ the first five polynomials $\phi_i(x)$ of (5.29) are

$$\left. \begin{aligned} \phi_0(x) &= \tfrac{1}{2}\sqrt{2}, \quad \phi_1(x) = \tfrac{1}{4}\sqrt{6}\,x, \\ \phi_2(x) &= \tfrac{3}{16}\sqrt{10}\,(x^2 - \tfrac{1}{3}), \\ \phi_3(x) &= 5\sqrt{\tfrac{7}{302}}\,(x^3 - \tfrac{9}{10}x), \\ \phi_4(x) &= \frac{105}{2\sqrt{2102}}\,(x^4 - \tfrac{33}{28}x^2 + \tfrac{27}{140}). \end{aligned} \right\} \tag{5.30}$$

If $I(\bar{y})$ is not a non-degenerated local minimum of $I(y)$, the reduction of the extremal problem (5.1) to a point-function problem (5.3) or (5.10) can still be used, provided the corresponding truncated system is replaced by the condition that the points $(y_1, y_2, ..., y_{m-1})$ or $(a_1^{(m)}, a_2^{(m)}, ..., a_m^{(m)})$ define a local lower limit of $I(y)$. Since there exists no simple analytic expression similar to (5.4) or (5.11) which characterizes a local lower limit, the unknown constants in (5.3) or in (5.10) must be determined by an appropriate algorithm. Such an algorithm can be based on the method of steepest descent or on some form of successive iterations. Whether a given type of successive iterations is efficient or not depends of course on the specific features of the problem (5.1), and particularly on the continuity properties of $I(y)$ in (5.1). At present very little is known about the computational aspects of the determination of a local lower limit. The same situation exists when the problem (5.1) is replaced by the more complex problem (3.135), (3.136), (3.137).

Let us recall that a finite system of algebraic equations

$$f_j(a_i, a_2, ..., a_n) = 0 \quad (j = 1, 2, ..., n) \tag{5.31}$$

can be solved by the so-called method of differentiation with respect to a parameter (see for example § 72 [M 19]). Consider an auxiliary set of functions

$$F_j(a_1, a_2, ..., a_n, \lambda) = 0 \quad (0 \leqslant \lambda \leqslant 1, \quad j = 1, 2, ..., n), \tag{5.32}$$

where λ is a parameter. Suppose that the F_j are continuously differentiable with respect to the parameter λ and with respect to the unknowns a_i, $i = 1, 2, ..., n$. The functions F_j are chosen in such a manner that

$$F_j(a_1, a_2, ..., a_n, 1) = f_j(a_1, a_2, ..., a_n), \tag{5.33}$$

and that the system $\quad F_j(a_1, a_2, ..., a_n, 0) = 0 \tag{5.34}$

is easier to solve than the system (5.31). Let a_{i0} be a solution of (5.34). Since the roots a_i of (5.32) are continuous functions of λ, i.e. $a_i = a_i(\lambda)$, it is sufficient to solve (5.32) for the single value $\lambda = 1$. By differentiating (5.32) with respect to λ it is obvious that (5.32) is equivalent to the initial-value problem

$$\sum_{i=1}^{n} \frac{\partial F_j}{\partial a_i}\frac{da_i}{d\lambda} + \frac{\partial F_j}{\partial \lambda} = 0 \quad (a_i(0) = a_{i0}, \quad 0 \leqslant \lambda \leqslant 1, \quad i,j = 1,2,...,n). \quad (5.35)$$

The problem (5.35) is of a type which can be easily solved on an analog or digital computer, and the terminal values $a_i(1)$, $i = 1, 2, ..., n$, represent a solution of (5.31). Experience has shown that the calculation of the roots of (5.32) by means of (5.35) leads generally to results of a rather low accuracy.

§ 21. THE METHOD OF AN ASSOCIATED ORDINARY BOUNDARY-VALUE PROBLEM

21.1   The nature of approximate solutions

In chapter 2 it was shown that with suitable smoothness conditions the solution of the extremal problem (5.1) can be expressed by means of the solution of the boundary-value problem

$$\delta I/\delta y = 0 \quad (y(0) = \alpha, \quad y(b) = \beta). \quad (5.36)$$

If $I(\bar{y})$ is a non-degenerated local minimum of $I(y)$, and the minimal curve $y = \bar{y}(x)$ can be imbedded in the linear parametric family (2.73), then

$$\frac{\delta I}{\delta y} = \frac{\partial F}{\partial y} - \frac{d}{dx}\left(\frac{\partial F}{\partial \dot{y}}\right).$$

The problem (5.36) is a particular case of the more general boundary-value problem

$$G(x, y(x), \dot{y}(x), \ddot{y}(x)) = 0 \quad (a \leqslant x \leqslant b), \quad (5.37)$$

$$g_1(y(a), y(b), \dot{y}(a), \dot{y}(b)) = 0, \quad g_2(y(a), y(b), \dot{y}(a), \dot{y}(b)) = 0. \quad (5.38)$$

Unless the contrary is stated we will suppose that there exists at least one solution $y = \bar{y}(x)$ of (5.37), (5.38). With the existence of $\bar{y}(x)$ assured, there are many methods to compute it. Consider in fact a twice-differentiable function

$$y_m(x) = \phi(x, C_1, C_2, ..., C_m) \quad (5.39)$$

containing m undetermined parameters C_i, $i = 1, 2, ..., m$. The function $y_m(x)$ is so chosen that it satisfies the boundary conditions (5.38) for arbitrary values of the parameters C_i. Since in general $y_m(x)$ is not a solution of (5.37), substituting (5.39) into (5.37) yields the error

$$R(x, C_1, C_2, ..., C_m) = G(x, \phi(x), \dot{\phi}(x), \ddot{\phi}(x)) \quad (a \leqslant x \leqslant b). \quad (5.40)$$

To make $y_m(x)$ an approximate solution of (5.37), (5.38) it is sufficient to choose the constants C_i so that $R(x, C_1, C_2, ..., C_m)$ differ from zero as little as possible (cf. [C 14]).

Each variant of the minimization of $R(x, C_1, C_2, ..., C_m)$ yields a 'method' of solving the boundary-value problem (5.37), (5.38).

Example 1 (the collocation method):

Choose m distinct abscissae $x_1, x_2, ..., x_m$ in the interval $a \leqslant x \leqslant b$. Requiring that R should vanish at these abscissae leads to a system of m algebraic equations

$$R(x_i, C_1, C_2, ..., C_m) = 0 \quad (i = 1, 2, ..., m). \tag{5.41}$$

Each independent set of roots $C_i = C_{i0}$ of (5.41) defines an approximate solution of form (5.39) of (5.37), (5.38).

Example 2 (mean-square error method):

Consider the integral

$$\mathscr{J} = \int_a^b R^2(x, C_1, C_2, ..., C_m)\, dx, \tag{5.42}$$

and suppose that C_i are to make $\mathscr{J}$ a local minimum. The system of algebraic equations fixing the values of the C_i becomes

$$\frac{\partial \mathscr{J}}{\partial C_i} = 0 \quad (i = 1, 2, ..., m). \tag{5.43}$$

Example 3 (discrete minimum error method):

Instead of the integral (5.42) consider the finite sum

$$\mathscr{J}_n = \sum_{i=1}^n N(0, R(x_i, C_1, C_2, ..., C_m)), \tag{5.44}$$

where the x_i, $i = 1, 2, ..., n$, are distinct abscissae in the interval $a \leqslant x \leqslant b$ and $N(u(x), v(x))$ is a given norm function. If $\mathscr{J}_n$ is a continuously differentiable function of the C_i, the C_i are roots of the system of equations

$$\frac{\partial \mathscr{J}_n}{\partial C_i} = 0 \quad (i = 1, 2, ..., m). \tag{5.45}$$

If the norm function N in (5.44) is such that $\mathscr{J}_n$ is not a continuously differentiable function of the C_i, the values of the C_i must be found directly from (5.44), by requiring that $\mathscr{J}_n$ take the least possible value.

Example 4 (Galerkin's method):

Suppose that (5.39) has the form

$$y_m(x) = \sum_{i=1}^m C_i \psi_i(x), \tag{5.46}$$

where the $\psi_i(x)$ is a set of m linearly independent functions chosen from some complete set Ψ. In order to fix the C_i it is sufficient to require (cf. [C 14]) that

$$\int_a^b \psi_i(x)\, R(x, C_1, C_2, ..., C_m)\, dx = 0 \quad (i = 1, 2, ..., m). \tag{5.47}$$

The preceding four examples do not exhaust all possibilities, and the 'guiding idea' of each example can be modified in various ways, yielding other approximate methods. The precision of the approximation can be tested by computing $y_m(x)$ for two different values of m, and then comparing the $y_m(x)$ and the corresponding values of $I(y_m)$, defined by (5.1).

Suppose that the result of the substitution of $y_m(x), y_{m+1}(x), ..., y_{m+k}(x)$ into $I(y)$ is already known. The comparison of the values of

$$I(y_m), \quad I(y_{m+1}), \quad ..., \quad I(y_{m+k})$$

may suggest a way of determining the form of $y_{k+m+1}(x)$, so that the sequences $y_n(x)$ and $I(y_n)$, $n > m+k$, converge at least asymptotically. If this is the case, $y_n(x)$ can be expressed as a recurrence equation

$$y_{n+1}(x) = \eta(n, x, y_n(x), ..., y_{n-k}(x)) \quad (n \geqslant m+k), \tag{5.48}$$

and this equation can be used till $|y_n(x) - y_{n-1}(x)|$ passes through a minimum or till the required precision is achieved. A considerable simplification occurs when (5.48) is linear and $k = 1$.

21.2   The method of finite differences

Consider the boundary-value problem (5.37), (5.38) and suppose that a sufficient accuracy will be obtained by subdividing the interval $a \leqslant x \leqslant b$ into m equal subintervals of length $h = (b-a)/m$ and by replacing the derivatives $\dot{y}(x)$ and $\ddot{y}(x)$ in (5.37) by the finite differences

$$\dot{y}_i = \frac{y_{i+1} - y_{i-1}}{2h}, \quad \ddot{y}_i = \frac{y_{i+1} - 2y_i + y_{i-1}}{h^2} \quad (i = 1, 2, ..., m-1), \tag{5.49}$$

where $y_i = y(x_i)$, $x_i = a + ih$. Let $y(a) = y_0$ and $y(b) = y_m$. For the endpoints $x = a$ and $x = a + mh = b$ suppose that

$$\dot{y}(a) = \frac{y_1 - y_0}{h}, \quad \dot{y}(b) = \frac{y_m - y_{m-1}}{h}. \tag{5.50}$$

Substituting (5.49) into (5.37) and (5.50) into (5.38) yields a system of $m+1$ algebraic equations

$$\left.\begin{aligned} G\left(x_i, y_i, \frac{y_{i+1} - y_{i-1}}{2h}, \frac{y_{i+1} - 2y_i + y_{i-1}}{h^2}\right) &= 0 \quad (i = 1, 2, ..., m-1), \\[2mm] g_1\left(y_0, y_m, \frac{y_1 - y_0}{h}, \frac{y_m - y_{m-1}}{h}\right) &= 0, \\[2mm] g_2\left(y_0, y_m, \frac{y_1 - y_0}{h}, \frac{y_m - y_{m-1}}{h}\right) &= 0, \end{aligned}\right\} \tag{5.51}$$

which are sufficient to fix the $m+1$ unknown constants y_i. The solution of (5.51) furnishes a table of the approximate solution $\bar{y}_m(x)$ at the abscissae x_i. The non-linear algebraic system can be solved by a variety of methods, the advantages and inconviences of which have been discussed, for example, by Durand[D 8].

The problem (5.37), (5.38) is easily solved when it is linear, i.e. when it can be written in the form

$$\ddot{y}(x) + A(x)\dot{y}(x) + B(x)y(x) = C(x), \tag{5.52}$$

$$\alpha_0 y(a) + \alpha_1 \dot{y}(a) = \alpha, \quad \beta_0 y(b) + \beta_1 \dot{y}(b) = \beta, \tag{5.53}$$

where $A(x)$, $B(x)$, $C(x)$ are continuous functions in $a \leqslant x \leqslant b$, and $\alpha, \alpha_0, \alpha_1, \beta, \beta_0, \beta_1$ are real constants. To avoid ambiguity assume

$$|\alpha_0| + |\alpha_1| \neq 0, \quad |\beta_0| + |\beta_1| \neq 0. \tag{5.54}$$

As before let $h = (b-a)/m$, $x_i = a+ih$ and $y_i = y(x_i)$, $i = 0, 1, ..., m$. Using (5.50), (5.51) and the abbreviations

$$p_i = \frac{2h^2 B(x_i) - 4}{2 + hA(x_i)}, \quad q_i = \frac{2 - hA(x_i)}{2 + hA(x_i)}, \quad r_i = \frac{2h^2 C(x_i)}{2 + hA(x_i)}, \tag{5.55}$$

the problem (5.52), (5.53) is approximated by

$$\left.\begin{aligned}
y_{i+2} + p_i y_{i+1} + q_i y_i = r_i, \quad (i = 0, 1, ..., m-2),& \\
\frac{\alpha_1}{h} y_1 + \left(\alpha_0 - \frac{\alpha_1}{h}\right) y_0 = \alpha,& \\
\frac{\beta_1}{2h} y_m + \beta_0 y_{m-1} - \frac{\beta_1}{2h} y_{m-2} = \beta.&
\end{aligned}\right\} \tag{5.56}$$

The system (5.56) is linear in y_i, $i = 0, 1, ..., m$, and it can be solved by any standard method.

A particularly effective method of solving (5.56) is the method of 'transfer' developed by Gelfand and Lokucievsky (cf. [B 7]). This method is based on the set of iterations

$$\left.\begin{aligned}
\gamma_i = \frac{1}{p_i - q_i \gamma_{i-1}}, \quad \gamma_0 = \frac{\alpha_1 - \alpha_0 h}{p_0(\alpha_1 - \alpha_0 h) + q_0 \alpha_1}& \\
\delta_i = r_i - q_i \gamma_{i-1} \delta_{i-1}, \quad \delta_0 = r_0 + \frac{q_0 \alpha h}{\alpha_1 - \alpha_0 h}&
\end{aligned}\right\} \begin{aligned}(i = 0, 1, ..., m-1),\end{aligned} \tag{5.57}$$

$$\left.\begin{aligned}
y_i = \gamma_i(\delta_i - y_{i+1}) \quad (i = m-1, m-2, ..., 1, 0),& \\
y_m = \frac{2h\beta - \beta_1(\delta_{m-1} - \gamma_{m-2}\delta_{m-2})}{2h\beta_0 + \beta_1(\gamma_{m-2} - \gamma_{m-1}^{-1})}.&
\end{aligned}\right\} \tag{5.58}$$

The recurrence equations (5.57) are first used to compute the coefficients γ_i, δ_i in an ascending order (forward transfer of (i)), thus permitting to calculate the value of y_m, and the y_i are then computed in a descending order (backward transfer of (i)) by means of the recurrence equation (5.58). The iterations defined by (5.57), (5.58) are computationally stable.

Consider now the more particular boundary-value problem

$$\ddot{y}(x) = f(x, y(x), \dot{y}(x)) \quad (a \leqslant x \leqslant b), \tag{5.59}$$

$$\alpha_0 y(a) - \alpha_1 \dot{y}(a) = \alpha, \quad \beta_0 y(b) + \beta_1 \dot{y}(b) = \beta, \tag{5.60}$$

where the constants $\alpha_0, \alpha_1, \beta_0, \beta_1$ are non-negative. The function f in (5.59) is continuous with respect to its arguments and satisfies a Lipschitz condition with respect to y and $z = \dot{y}$:

$$|f(x, y, z) - f(x, \bar{y}, \bar{z})| \leqslant L_1 |y - \bar{y}| + L_2 |z - \bar{z}|, \tag{5.61}$$

where L_1, L_2 are non-negative constants. Subdividing $a \leqslant x \leqslant b$ as usual into m subintervals of width $h = (b-a)/m$ and using finite differences to express $\dot{y}$ and $\ddot{y}$, yields

$$\left.\begin{aligned}
\frac{y_{i+1} - 2y_i + y_{i-1}}{h^2} &= f\left(x_i, y_i, \frac{y_{i+1} - y_{i-1}}{2h}\right) \quad (i = 1, 2, \ldots, m-1),\\[2mm]
R_0(y) &= \left(\alpha_0 + \frac{\alpha_1}{h}\right) y_0 - \frac{\alpha_1}{h} y_1 = \alpha,\\[2mm]
R_m(y) &= \left(\beta_0 + \frac{\beta_1}{h}\right) y_{m-1} - \frac{\beta_1}{h} y_m = \beta.
\end{aligned}\right\} \tag{5.62}$$

Instead of solving (5.62) directly it is possible to use successive iterations of the form

$$\left.\begin{aligned}
L(y_i^{(k+1)}) &= f(y_i^{(k)}),\\
R_0(y^{(k+1)}) = 0, \quad &R_m(y^{(k+1)}) = 0,
\end{aligned}\right\} \tag{5.63}$$

where for conciseness

$$y_i = y(x_i), \quad x_i = a + ih, \quad L(y_i) = \frac{1}{h^2}(y_{i+1} - 2y_i + y_{i-1}),$$

$$f(y_i) = f\left(x_i, y_i, \frac{y_{i+1} - y_{i-1}}{2h}\right).$$

The superscripts k, $k+1$ in (5.63) designate the order of the iteration i.e. the order of the approximation of y_i. Using the substitutions

$$\left.\begin{aligned}
y_i^{(k+1)} = z_i + t_i, \quad R_0(z) = R_m(z) &= 0,\\
z_{i+1} - 2z_i + z_{i-1} &= h^2 f(y_i^{(k)}),\\
t_{i+1} - 2t_i + t_{i-1} = 0, \quad R_0(t) = \alpha, \quad R_m(t) &= \beta,
\end{aligned}\right\} \tag{5.64}$$

it can be shown (cf. [B 7]) that (5.63) becomes

$$y_i^{(k+1)} = h\Delta[\alpha\beta_0(b-a) + \alpha\beta_1 + \alpha_1\beta] + \frac{i}{\Delta}(\alpha_0\beta - \alpha\beta_0) + h^2\sum_{j=1}^{m-1} g_{ij}f(y_j^{(k)}),$$
(5.65)

where
$$\Delta = \frac{1}{h}[\alpha_0\beta_0(b-a) + \alpha_0\beta_1 + \alpha_1\beta_0] \neq 0,$$

$$g_{ij} = g_{ji} = \left.\begin{cases} \frac{1}{\Delta}\left(j\alpha_0 + \frac{\alpha_1}{h}\right)\left(i\beta_0 - m\beta_0 - \frac{\beta_1}{h}\right) & (j \leqslant 1) \\ \frac{1}{\Delta}\left(i\alpha_0 + \frac{\alpha_1}{h}\right)\left(j\beta_0 - m\beta_0 - \frac{\beta_1}{h}\right) & (j \geqslant 1) \end{cases}\right\}.$$
(5.66)

The iterations (5.65) converge to the exact solution of (5.62), provided the initial approximation y_i^0 is not too far from the exact solution. A suitable initial approximation y_i^0 can be deduced from a coarse direct solution of (5.62), i.e. a solution of (5.62) with m small.

21.3   The method of close boundary-value problems

Whenever the direct solution of the boundary-value problem (5.37), (5.38) is inconvenient, it is possible to seek it by means of an imbedding into a parametric family of close problems

$$\left.\begin{aligned} G_\lambda(x, y(x), \dot{y}(x), \ddot{y}(x)) &= 0 \quad (a \leqslant x \leqslant b), \\ g_{1\lambda}(y(a), y(b), \dot{y}(a), \dot{y}(b)) &= 0, \\ g_{2\lambda}(y(a), y(b), \dot{y}(a), \dot{y}(b)) &= 0, \end{aligned}\right\}$$
(5.67)

provided (5.67) reduces to (5.37), (5.38) as the real parameter λ approaches a limit value λ_0. The parametric imbedding (5.57) is of course practically useless unless its solution is easier to obtain than that of (5.37), (5.38). Fortunately such a favourable situation does occur, and this is so in particular when (5.67) happens to be a linear system.

Consider therefore again the linear boundary-value problem (5.52), (5.53). Let us seek its solution in the form

$$y(x) = Mu(x) + v(x),$$
(5.68)

where M is an undetermined real constant and $u(x)$, $v(x)$ are particular solutions of the equations

$$\ddot{u}(x) + A(x)\dot{u}(x) + B(x)u(x) = 0$$
(5.69)

and
$$\ddot{v}(x) + A(x)\dot{v}(x) + B(x)v(x) = C(x),$$
(5.70)

respectively. Equation (5.68) is obviously a particular solution of (5.52) regardless of the value of M. In order to specify the initial conditions to be associated with (5.69) and (5.70) it is possible to require that a solution of (5.69) should satisfy the first boundary condition in (5.53) regardless of the value of M. Since from (5.68) it follows that

$$M[\alpha_0 u(a) + \alpha_1 \dot{u}(a)] + \alpha_0 v(a) + \alpha_1 \dot{v}(a) = \alpha,$$

the above requirement yields

$$\alpha_0 u(a) + \alpha_1 \dot{u}(a) = 0, \quad \alpha_0 v(a) + \alpha_1 \dot{v}(a) = \alpha. \tag{5.71}$$

Let δ be a non-vanishing real constant. A particular solution of (5.71) is then given by

$$u(a) = \delta \alpha_1, \quad \dot{u}(a) = -\delta \alpha_0, \tag{5.72}$$

$$\left.\begin{array}{ll} v(a) = \dfrac{\alpha}{\alpha_0}, & \dot{v}(a) = 0, \quad \text{if} \quad \alpha_0 \neq 0, \\[3mm] v(a) = 0, & \dot{v}(a) = \dfrac{\alpha}{\alpha_1}, \quad \text{if} \quad \alpha_1 \neq 0. \end{array}\right\} \tag{5.73}$$

Solving the two linear initial-value problems (5.69), (5.72) and (5.70), (5.73) in the interval $a \leqslant x \leqslant b$ by any standard method, and substituting the result into the second boundary condition in (5.53), yields

$$M[\beta_0 u(b) + \beta_1 \dot{u}(b)] + \beta_0 v(b) + \beta_1 \dot{v}(b) = \beta. \tag{5.74}$$

If $\beta_0 u(b) + \beta_1 \dot{u}(b) \neq 0$ the undetermined constant in (5.68) is given by

$$M = \frac{\beta - [\beta_0 v(b) + \beta_1 \dot{v}(b)]}{\beta_0 u(b) + \beta_1 \dot{u}(b)}. \tag{5.75}$$

The solution of the linear boundary-value problem (5.52), (5.53) is therefore equivalent to the solution of two initial-value problems (5.69), (5.72) and (5.70), (5.73), and the calculation of the constant M given by (5.75).

If in (5.52), (5.53) $C(x) = 0$ and $\alpha = 0$, then $v(a) = \dot{v}(a) = 0$, $v(x) \equiv 0$, and (5.68) is replaced by

$$y(x) = M u(x), \tag{5.76}$$

where

$$M = \frac{\beta}{\beta_0 u(b) + \beta_1 \dot{u}(b)}. \tag{5.77}$$

As an illustration of the parametric imbedding (5.67) consider the non-linear differential equation

$$\ddot{y}(x) = f(x, y(x), \dot{y}(x)) \quad (a \leqslant x \leqslant b). \tag{5.78}$$

Using the notation $\dot{y}(x) = z(x)$, (5.78) can be written in the matrix form

$$\dot{Y}(x) = F(x, Y), \tag{5.79}$$

where

$$Y(x) = \begin{pmatrix} y(x) \\ z(x) \end{pmatrix}, \quad F(x, Y) = \begin{pmatrix} z(x) \\ f(x, y(x), z(x)) \end{pmatrix}.$$

Let the boundary-value problem to be solved be given by (5.79), (5.53), and suppose that this problem admits a unique continuous solution

$$\overline{Y}(x) = \begin{pmatrix} \overline{y}(x) \\ \overline{z}(x) \end{pmatrix}.$$

As a first example of a parametric imbedding consider the enumerable set of linear differential systems

$$\dot{Y}_{n+1}(x) = \mathscr{J}(x, Y_n)\,[Y_{n+1}(x) - Y_n(x)] + F(x, Y_n)$$
$$(a \leqslant x \leqslant b, \quad n = 0, 1, \ldots), \quad (5.80)$$

where n takes the place of the parameter λ in (5.67), and $\mathscr{J}$ is the Jacobian matrix of $F(x, Y)$. Explicitly

$$Y_n = \begin{pmatrix} y_n(x) \\ z_n(x) \end{pmatrix}, \quad \mathscr{J}(x, Y_n) = \begin{pmatrix} 0 & (\partial f/\partial y)_n \\ 1 & (\partial f/\partial z)_n \end{pmatrix}.$$

Suppose that the boundary-value problem (5.80), (5.53) is 'close' to the boundary-value problem (5.79), (5.53), i.e. suppose that

$$\lim_{n \to \infty} Y_n(x) = \overline{Y}(x). \quad (5.81)$$

The solution of the set of boundary problems (5.80), (5.53) is straight-forward, except for the minor difficulty of finding a suitable initial approximation $Y_0(x)$. It was shown by Kantorovich[K 5] that the relation (5.81) does indeed hold, provided $Y_0(x)$ is sufficiently close to $\overline{Y}(x)$. The closeness between the exact and the approximate solution can be expressed by means of the norm function

$$N(\overline{Y}(x), Y_n(x)) = \max_{a \leqslant x \leqslant b} |\overline{y}(x) - y_n(x)| + \max_{a \leqslant x \leqslant b} |\overline{z}(x) - z(x)|. \quad (5.82)$$

Because of its formal resemblance to the Newton method of finding roots of algebraic functions, the approximation (5.80) is called the Newton method of solving the boundary-value problem (5.79), (5.53). This approximation is of course not limited to differential systems of order two. A modification of (5.80) has been recently proposed by Bakhalov[B 1].

As a second example of a parametric imbedding consider the somewhat special boundary-value problem

$$\frac{d}{dx}[p(x)\dot{y}(x)] = f(x, y) \quad (0 \leqslant x \leqslant 1, \quad y(0) = \alpha, \quad y(1) = \beta), \quad (5.83)$$

where $p(x)$ and $f(x, y)$ are continuous functions admitting continuous second-order derivatives. Using the results of Tonelli[T 7] and Bernstein[B8] it was shown by Klokov[K 16] that, provided

$$p(x) > 0, \quad (\partial/\partial y)f(x, y) > 0 \quad \text{for} \quad 0 \leqslant x \leqslant 1, \quad -\infty < y < +\infty$$

the non-linear boundary-value problem (5.83) is a limit of the sequence
of linear boundary-value problems

$$\left. \begin{array}{l} \dfrac{d}{dx}\,[p(x)\,\dot{y}_{n+1}(x)] = (1-\sigma)\dfrac{d}{dx}\,[p(x)\,\dot{y}_n(x)] + \sigma f(x, y_n(x)), \\[2mm] y_{n+1}(0) = \alpha, \quad y_{n+1}(1) = \beta, \quad n = 0, 1, \ldots, \end{array} \right\} \quad (5.84)$$

where $\sigma > 0$ is a suitably chosen constant. The solution

$$\bar{y}(x) = \lim_{n \to \infty} y_n(x) \qquad (5.85)$$

of (5.84) satisfies (5.83) independently of the value of σ. The value of σ
determines only the speed of convergence of the sequence $y_n(x)$,
$n = 0, 1, \ldots$.

As a third example let us recall the classical result of Picard[P 7] which
states that a solution of the boundary-value problem

$$\ddot{y}(x) = f(x, y(x), \dot{y}(x)), \quad y(a) = \alpha, \quad y(b) = \beta \quad (a \leqslant x \leqslant b), \quad (5.86)$$

can be obtained by means of the successive iterations

$$\ddot{y}_{n+1}(x) = f(x, y_n(x), \dot{y}_n(x)), \quad y_{n+1}(a) = \alpha, \quad y_{n+1}(b) = \beta$$

$$(n = 0, 1, \ldots). \quad (5.87)$$

The sufficient conditions, independent of the extremal problem (5.1), for
the solution (5.86) to exist and to be expressible in the form (5.87) are
(cf. [P 7])

$$\left. \begin{array}{l} |\dot{y}(x)| \leqslant L, \quad |f(x, y, \dot{y})| \leqslant M, \\[2mm] |f(x, y, \dot{y}) - f(x, \bar{y}, \ddot{y})| \leqslant L_1|y - \bar{y}| + L_2|\dot{y} - \ddot{y}|, \\[2mm] \dfrac{L_1}{8}\,(b-a)^2 + \dfrac{L_2}{2}\,(b-a) < 1, \quad \left|y(x) - \dfrac{\alpha+\beta}{2}\right| \leqslant K, \\[2mm] \dfrac{|\beta-\alpha|}{b-a} + \dfrac{M}{2}\,(b-\alpha) \leqslant L, \quad \tfrac{1}{2}(b-a) + \dfrac{M}{8}\,(b-a)^2 \leqslant K. \end{array} \right\} \quad (5.88)$$

The conditions (5.88) have been relaxed by various authors (see for
example [A 8],[A 9],[L 13],[P 2],[P 5]).

Let us note finally that, provided the solution $\bar{y}(x)$ of the boundary-
value problem (5.37), (5.38) is known to exist, at least in principle it is
possible to transform (5.37), (5.38) into an initial value problem, by a
trial-and-error method. Suppose for example that the initial conditions

$$y(a) = c_0, \quad \dot{y}(a) = c_0',$$

c_0, c_0' being real constants, are chosen to define a particular solution $y_0(x)$
of (5.37). The values of the constants c_0, c_0' are simply guessed. The
solutions $y_0(x)$ will exist in a certain interval $a \leqslant x \leqslant a + h_0$, $h_0 > 0$. If

$a + h_0 < b$ the guess of c_0, c_0' was wrong, because it is then impossible to determine the values $y_0(b) = d_0$, $\dot{y}_0(b) = d_0'$. If $a + h_0 \geqslant b$, the values d_0, d_0' can be computed by any standard method. Substituting c_0, c_0', d_0, d_0' into (5.38) yields an estimate of the error in the initial guess c_0, c_0'. Let the improved initial values be c_1, c_1'. Computing $y_1(b) = d_1$, $\dot{y}_1(b) = d_1'$ and substituting into (5.38) yields an estimate of the error in c_1, c_1'. The improvement of the initial values c_i, c_i' can be continued till the error in (5.38) is less than a given tolerance. The corresponding solution $y_i(x)$ of (5.37) is an approximation of the exact solution $\bar{y}(x)$.

The trial-and-error method just described is quite straightforward in theory, but quite unreliable in practice, because the sequence of initial values c_i, c_i' may converge very slowly. Furthermore, the differences

$$|c_{i+1} - c_i|, \quad |c_{i+1}' - c_i'|, \quad |d_{i+1} - d_i|, \quad |d_{i+1}' - d_i'|$$

carry very little information about the differences

$$|y_{i+1}(x) - y_i(x)|, \quad |\bar{y}(x) - y_i(x)| \quad (a \leqslant x \leqslant b). \tag{5.89}$$

The uncertainty about (5.89) can be partially removed by computing simultaneously with each $y_i(x)$ the sensitivity coefficients (cf. [R 1])

$$\xi_i(x) = \frac{\partial y_i(x)}{\partial c_i} \quad \text{and} \quad \eta_i(x) = \frac{\partial y_i(x)}{\partial c_i'} \quad (a \leqslant x \leqslant b),$$

defined by the initial value problems

$$\left.\begin{array}{l} \left(\dfrac{\partial G}{\partial \ddot{y}}\right)_i \ddot{\xi}_i(x) + \left(\dfrac{\partial G}{\partial \dot{y}}\right)_i \dot{\xi}_i(x) + \left(\dfrac{\partial G}{\partial y}\right)_i \xi_i(x) = 0, \\[2mm] \xi_i(a) = 1, \quad \dot{\xi}_i(a) = 0, \end{array}\right\} \tag{5.90}$$

and

$$\left.\begin{array}{l} \left(\dfrac{\partial G}{\partial \ddot{y}}\right)_i \ddot{\eta}_i(x) + \left(\dfrac{\partial G}{\partial \dot{y}}\right)_i \dot{\eta}_i(x) + \left(\dfrac{\partial G}{\partial y}\right)_i \eta_i(x) = 0, \\[2mm] \eta_i(a) = 0, \quad \dot{\eta}_i(a) = 1, \end{array}\right\} \tag{5.91}$$

respectively. The subscript i in the parentheses denotes a derivative evaluated for $y = y_i(x)$.

A more systematic method of finding the unknown initial conditions c_i, c_i' has been recently proposed [G 1]. This method is based on a geometrical interpretation of the domain of attainability.

21.4   The method of admissible variations

Let the extremal problem to be solved be a modified version of the problem (2.41):

$$I(a, b) = \min_{y \in Y} I(y) = \min_{y \in Y} \int_a^b F(x, y(x), \dot{y}(x))\, dx, \tag{5.92}$$

$$g_1(y(a), y(b), \dot{y}(a), \dot{y}(b)) = 0, \quad g_2(y(a), y(b), \dot{y}(a), \dot{y}(b)) = 0, \tag{5.93}$$

where as usual Y is the set of continuously differentiable functions and $I(y)$ is a convergent Riemann integral. Let $y_n(x)$ be an admissible function satisfying the boundary conditions (5.93). The corresponding value of $I(y)$ is

$$I(y_n) = \int_a^b F(x, y_n(x), \dot{y}_n(x))\, dx. \qquad (5.94)$$

In order to construct an iteration method to solve the boundary-value problem (5.92), (5.93), it is necessary to deduce from $y_n(x)$, $I(y_n)$ and the general properties of $I(y)$ a modification $\Delta y_n(x)$, $a \leqslant x \leqslant b$, so that

$$y_{n+1}(x) = y_n(x) + \Delta y_n(x) \in Y \qquad (5.95)$$

satisfies (5.93) and leads to a smaller value of $I(y)$ than $y_n(x)$, i.e. which results in

$$I(y_{n+1}) - I(y_n) = \int_a^b [F(x, y_n + \Delta y_n, \dot{y}_n + \Delta \dot{y}_n) - F(x, y_n, \dot{y}_n)]\, dx < 0. \qquad (5.96)$$

To every way of choosing $\Delta y_n(x)$ which yields (5.96) corresponds a 'method' of solving the boundary-value problem (5.92), (5.93).

If in addition to (5.93) the admissible functions $y(x)$ are subject to some supplementary constraint, say

$$f(x, y(x), \dot{y}(x)) \leqslant 0, \qquad (5.97)$$

then the modification $\Delta y_n(x)$ of $y_n(x)$ must simply be chosen in conformity with this constraint. In this method there is thus no essential difference between extremal problems with and without constraints. Modifications $\Delta y_n(x)$ which lead to (5.96) are usually called 'admissible variations'. The admissible variations must lead to a minimal sequence, i.e. given a norm-function N, one must have simultaneously

$$\lim_{n \to \infty} N(\Delta y_n(x), 0) = 0 \qquad (5.98)$$

and

$$\lim_{n \to \infty} [I(y_{n+1}) - I(y_n)] = 0. \qquad (5.99)$$

If the sequence $y_n(x)$, $n = 0, 1, ...$, does not converge, an approximate solution of (5.92), (5.93) is given by the method of admissible variations provided

$$N(y_{n+1}(x), y_n(x)) \quad \text{and} \quad |I(y_{n+1}) - I(y_n)| \quad (n = 0, 1, ...) \qquad (5.100)$$

pass simultaneously through a minimum. An efficient choice of $\Delta y_n(x)$ depends of course on the specific features of the extremal problem. In many cases the use of the very general inequality (5.96) may turn out to be inconvenient, and if this occurs (5.96) may be replaced by the more restrictive inequality

$$F(x, y_n + \Delta y_n, \dot{y}_n + \Delta \dot{y}_n) - F(x, y_n, \dot{y}_n) < 0 \quad (a \leqslant x \leqslant b). \qquad (5.101)$$

As an illustration of the use of (5.96) suppose that the functional $I(y)$, $y \in Y$, is continuous with the given norm function N. If $I(y)$ admits a functional derivative, then the inequality (5.96) simplifies into

$$I(y_{n+1}) - I(y_n) = \int_a^b \left(\frac{\delta I}{\delta y}\right)_n \Delta y_n \, dx + \dots < 0, \qquad (5.102)$$

where the subscript n inside the integral signifies that $\delta I/\delta y$ has been evaluated for $y = y_n$. In the simplest case when [D 7]

$$N(y_{n+1}, y_n) = N(\Delta y_n, 0) = |\Delta y_n|, \qquad (5.103)$$

the functional derivative $\delta I/\delta y$ is identical with the Euler 'derivative'

$$\frac{\partial F}{\partial y} - \frac{d}{dx}\left(\frac{\partial F}{\partial \dot{y}}\right). \qquad (5.104)$$

A sufficient first-order condition for (5.102) to be satisfied is

$$\left(\frac{\delta I}{\delta y}\right)_n \Delta y_n < 0 \quad (a \leqslant x \leqslant b). \qquad (5.105)$$

The inequality (5.105) does not determine a unique form of $\Delta y_n(x)$. Since $\delta I/\delta y$ can be interpreted as a gradient in the functional space Y, and (5.105) implies a certain relation between $(\delta I/\delta y)_n$ and Δy_n, all iteration methods based on (5.105) can be considered as particular versions of the 'functional' gradient method. The simplest choice of Δy_n is probably

$$\Delta y_n(x) = -k_n(x) \operatorname{sgn}(\delta I/\delta y)_n \quad (k_n(x) \geqslant 0, \quad a \leqslant x \leqslant b), \qquad (5.106)$$

where $k_n(x)$ is an arbitrary function. When $(\delta I/\delta y)_n$ does not change sign in the interval $a \leqslant x \leqslant b$, (5.106) is inefficient and it can be replaced by

$$\Delta y_n(x) = -k_n(x)\left(\frac{\delta I}{\delta y}\right)_n, \qquad (5.107)$$

or more generally by

$$\Delta y_n(x) = -k_n(x) G_n[(\delta I/\delta y)_n] \quad (G_n(x) \geqslant 0, \quad G_n(0) = 0), \qquad (5.108)$$

where $G_n(x)$ is also an arbitrary function.

As a concrete example of the use of (5.102) consider the Mayer problem

$$
\left.
\begin{aligned}
&\dot{y}(x) = u(x), \quad \dot{z}(x) = F(x, y(x), u(x)) \quad (a \leqslant x \leqslant b), \\
&y(a) = \alpha, \quad y(b) = \beta, \quad z(a) = 0, \\
&I(a, b) = \min_{y \in Y} z(b),
\end{aligned}
\right\} \qquad (5.109)
$$

where Y is the set of continuously differentiable functions. If $F(x, y, u)$ is a continuously differentiable function with respect to y and u, and the admissible variations of the initial approximation $u_0(x)$, $y_0(x)$ are Δu, Δy, then

$$\left.\begin{array}{l} \Delta \dot{y} = \Delta u, \quad \Delta \dot{z} = \left(\dfrac{\partial F}{\partial y}\right)_0 \Delta y + \left(\dfrac{\partial F}{\partial u}\right)_0 \Delta u + \ldots, \\[2mm] \Delta y(a) = 0, \quad \Delta y(b) = 0. \end{array}\right\} \quad (5.110)$$

The subscript $_0$ signifies that the partial derivatives in (5.110) have been evaluated for $y = y_0(x)$, $u = u_0(x)$. Consider the two auxiliary systems

$$\frac{dz}{dz} = 1, \quad \frac{dz}{dy} = \frac{F(x, y(x), u(x))}{u(x)} \quad (5.111)$$

and

$$\frac{d\lambda_1(x)}{dx} = -\left(\frac{\partial F}{\partial y}\right)_0 \lambda_2(x), \quad \frac{d\lambda_2(x)}{dx} = 0. \quad (5.112)$$

The general solution of (5.112) is

$$\lambda_2(x) = C_2, \quad \lambda_1(x) = C_1 - C_2 \int \left(\frac{\partial F}{\partial y}\right)_0 dx,$$

where C_1, C_2 are constants of integration. To determine these constants suppose that

$$\lambda_2(b) = -1, \quad \lambda_1(b) = \left(\frac{dz}{dy}\right)_{x=b} = \frac{F(b, y_0(b), u_0(b))}{u_0(b)}. \quad (5.113)$$

Because

$$\frac{d}{dx}(\lambda_1(x)\Delta y + \lambda_2(x)\Delta z) = \left(\lambda_1(x) + \lambda_2(x)\frac{\partial F}{\partial u}\right)\Delta u \quad (a \leqslant x \leqslant b) \quad (5.114)$$

from (5.109) to (5.112), it follows that in the linear approximation

$$\Delta z(b) = -\int_a^b \left[\lambda_1(x) + \lambda_2(x)\left(\frac{\partial F}{\partial u}\right)_0\right] \Delta u(x)\, dx. \quad (5.115)$$

Hence, as an admissible variation of $u_0(x)$, $y_0(x)$ it is possible to choose

$$\left.\begin{array}{l} \Delta u(x) = -k\left[\lambda_1(x) + \lambda_2(x)\left(\dfrac{\partial F}{\partial u}\right)_0\right] = -k\int\left[\left(\dfrac{\partial F}{\partial y}\right)_0 - \dfrac{d}{dx}\left(\dfrac{\partial F}{\partial u}\right)_0\right]dx, \\[4mm] \Delta y(x) = -k\int\int\left[\left(\dfrac{\partial F}{\partial y}\right)_0 - \dfrac{d}{dx}\left(\dfrac{\partial F}{\partial u}\right)_0\right]dx\, dx \quad (k = \mathrm{ct} > 0). \end{array}\right\} \quad (5.116)$$

The argument leading from (5.109) to (5.116) has been used by many authors (see for example [K 8],[K 26],[B 18],[D 7],[R 3]). It differs only in notation from the classical argument leading from (5.92), (5.93) to (5.108).

It is also possible to look for admissible variation Δu, Δy by starting from the assumption that $u = u(x)$ in (5.109) is only indirectly a function of x, in other words that $u(x)$ is a function of the dependent variable $z = z(x)$ (see for example [M 20],[V 2]). Nothing is changed in the basic argument, but the assumption $u = u(z)$ has the practical advantage that it solves automatically the so-called feedback controller synthesis problem (cf. §§ 15.6 and 5[P 11]).

§ 22. THE METHOD OF AN ASSOCIATED PARTIAL BOUNDARY-VALUE PROBLEM

It was shown in chapter 3 that solutions of extremal problems can be obtained by means of the Carathéodory formulation. Consider, for example, the problem (2.41) written in a slightly different form

$$I(a,b) = \min_{y \in Y} I(y) = \min_{y \in Y} \int_a^b F(x, y(x), \dot{y}(x)) \, dx$$

$$(y(a) = \alpha, \quad y(b) = \beta), \quad (5.117)$$

where Y is the set of piecewise differentiable functions and $I(y)$ is a convergent Riemann integral. Letting

$$I(y) = S(x, y), \tag{5.118}$$

the solution of (5.117) can be written in the form (cf. §16.1)

$$I(a, b) = \min_{g \in G} S(b, \beta), \tag{5.119}$$

$$\left. \begin{array}{l} \Phi\left(x, y, \dfrac{\partial S}{\partial x}, \dfrac{\partial S}{\partial y}\right) = 0, \\[2mm] S(x, y) = 0 \quad \text{on} \quad g(x, y) = 0, \quad g(a, \alpha) = 0, \end{array} \right\} \tag{5.120}$$

where G is the set of admissible initial functions $g(x, y)$. When the values a, α and b, β in (5.117) are given explicitly, the boundary-value problem (5.119), (5.120) is degenerated in the sense that the initial curve $g(x, y) = 0$, need only be known in an infinitesimal neighbourhood of the point (a, α). In other words the knowledge of $\partial \bar{g}/\partial x$ and $\partial \bar{g}/\partial y$, where $\bar{g}(x, y)$ is the minimizing initial curve, at the point (a, α) is sufficient to compute $I(a, b)$ (cf. §16.2). In such a case it is usually advantageous to solve (5.119), (5.120) numerically by the method of characteristics (cf. §16.2), which for all practical purposes amounts to solving a boundary-value problem associated with a system of ordinary differential equations.

The situation is entirely different when at least one pair of values a, α or b, β, say a, α, is not given explicitly, but is instead specified by the condition that the point (a, α) is located on a given curve $g(x, y) = 0$. This statement implies that the set G in (5.119), (5.120) consists of a single element. It might be thus worthwhile to compute $S(x, y)$ numerically in an appropriate part of the strip $a \leqslant x \leqslant b$, $c \leqslant y \leqslant d$ to the right of $g(x, y) = 0$ where the constants c, d are fixed by the condition that the solution $S(x, y)$ of (5.120) must reach the point (b, β). Once $S(b, \beta)$ is known, it is straightforward to compute a curve $y = \bar{y}(x)$ which is transversal to the family of curves $S(x, y) = \delta$, $0 \leqslant \delta \leqslant S(b, \beta)$ or $S(b, \beta) \leqslant \delta \leqslant 0$, and which passes through the point (b, β). The intersection of $y = \bar{y}(x)$ and $g(x, y) = 0$ defines the unknown coordinates of the point (a, α) (cf. §16.4).

At present only Bellman's iterations in the criterion-functional space (cf. §18.1), and the method of differences appear to be appropriate for the computation of $S(x, y)$. The implementation of the method of finite differences is essentially the same as that used for the solution of the second-order parabolic boundary-value problem

$$\left.\begin{aligned}
\epsilon \frac{\partial^2 S}{\partial x^2} + \Phi\left(x, y, \frac{\partial S}{\partial x}, \frac{\partial S}{\partial y}\right) &= 0, \\
S(x, y) = 0 \quad \text{on} \quad g(x, y) = 0, \quad g(a, \alpha) &= 0,
\end{aligned}\right\} \tag{5.121}$$

where ϵ is a real constant. Since the theory of parabolic boundary-value problem is relatively well explored (see for example [E 1],[B 7],[F 5]), Hadamard's method of descent (cf. §7) can be used to deduce the solution of (5.120) from that of (5.121). Let $S(x, y, \epsilon)$ be a solution of (5.121). Formally the solution of (5.120) can be written

$$S(x, y) = \lim_{\epsilon \to 0} S(x, y, \epsilon), \tag{5.122}$$

but this relation should be used with care because it implies the presence of stability with respect to perturbations of dimensionality (cf. §4.3).

It is probably simplest to compute the solution $S(x, y)$ of (5.120) directly. As an illustration of the procedure to follow consider the particularly simple case when the curve $g(x, y) = 0$ in (5.120) is a straight line segment $x = a$, $c \leqslant y \leqslant d$. This assumption involves often no loss of generality, because when $g(x, y) = 0$ can be written in the form $y = f(x)$ with $f'(x) \neq \infty$, it is sufficient to replace the variables x, y by the variables $\eta = x$, $\xi = y - f(x) + a$.

Having fixed the constants c, d suppose that the rectangular region $a \leqslant x \leqslant b, c \leqslant y \leqslant d$ has been covered by a uniform rectangular grid with the spacings h and l, respectively (Fig. 20). The continuous variables x, y are thus replaced by the discrete variables

$$\left.\begin{aligned}
x = a + ih \quad (i = 0, 1, \ldots, m), \quad y_j = \frac{d-c}{2} + jl \quad (j = 0, \pm 1, \ldots, \pm n), \\
x_0 = a, \quad x_m = b, \quad y_{-n} = c, \quad y_n = d.
\end{aligned}\right\} \tag{5.123}$$

The partial derivatives in (5.120) can be replaced by the approximations

$$\left(\frac{\partial S}{\partial x}\right)_{i,j} = \frac{S_{i+1,j} - S_{i,j}}{h}, \quad \left(\frac{\partial S}{\partial y}\right)_{i,j} = \frac{S_{i,j+1} - S_{i,j-1}}{2l}, \tag{5.124}$$

where the subscripts indicate the node in the grid at which $S(x, y)$, $\partial S/\partial x$ and $\partial S/\partial y$ are evaluated (Fig. 20). Substituting (5.124) into (5.120) amounts to transforming the continuous boundary-value problem (5.120) into the discrete boundary-value problem

$$\left.\begin{array}{l} \Phi_{i,j} = \Phi\left(x_i, y_j, \dfrac{S_{i+1,j} - S_{i,j}}{h}, \dfrac{S_{i,j+1} - S_{i,j-1}}{2l}\right) = 0, \\[2mm] S_{0,j} = S(x_0, y_j) = 0 \quad (i = 0, 1, ..., m, \quad j = 0, \pm 1, ..., \pm (n-1)). \end{array}\right\} \quad (5.125)$$

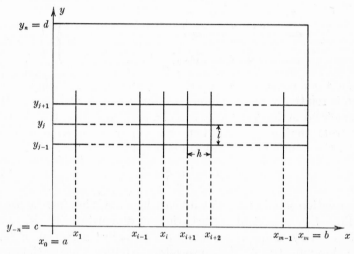

Fig. 20

Letting $i = 0$ in $\Phi_{i,j} = 0$ and taking account of the boundary condition $S_{0,j} = 0$ yields a system of algebraic equations

$$\Phi\left(x_0, y_j, \frac{S_{1,j}}{h}, 0\right) = 0, \qquad (5.126)$$

which defines the unknowns $S_{1,j}, j = 0, \pm 1, ..., \pm (n-1)$. Since the equations in (5.126) are not coupled with each other, each unknown $S_{1,j}$ can be computed independently. The missing values $S_{1,-n}$ and $S_{1,n}$ can be determined by means of the difference equations

$$\left(\frac{\partial S}{\partial x}\right)_{i,j} = \frac{S_{i,j} - S_{i-1,j}}{h}, \quad \left(\frac{\partial S}{\partial y}\right)_{i,j} = \frac{S_{i,j+1} - S_{i,j-1}}{2l}, \qquad (5.127)$$

and the corresponding discrete boundary-value problem

$$\left.\begin{array}{l} \Phi_{i,j} = \Phi\left(x_i, y_j, \dfrac{S_{i,j} - S_{i-1,j}}{h}, \dfrac{S_{i,j+1} - S_{i,j-1}}{2l}\right) = 0, \\[2mm] S_{0,j} = S(x_0, y_j) = 0 \quad (i = 1, 2, ..., m, \quad j = 0, \pm 1, ..., \pm (n-1)). \end{array}\right\} \quad (5.128)$$

Letting $i = 1$ in (5.128) yields

$$\Phi\left(x_1, y_j, \frac{S_{1,j}}{h}, \frac{S_{1,i+1} - S_{1,j-1}}{2l}\right) = 0. \tag{5.129}$$

Substituting the values $S_{1,j}$ determined from (5.126) and letting $j = \pm (n-1)$ in (5.129) gives two independent equations, the roots of which are $S_{1,-n}$ and $S_{1,n}$. The process just described can be continued step by step till the abscissa $x = b$ has been reached. At the end of the computation, when x is close to x_m, it is unnecessary to have recourse to (5.129) unless the point (b, β) happens to be near a corner of the region $a \leqslant x \leqslant b, c \leqslant x \leqslant d$. A minor difficulty occurs when the point (b, β) does not coincide with a node of the grid. Suppose for example that

$$y_k < \beta < y_{k+1}.$$

Knowing the values

$$\ldots S(b, y_{k-1}), \quad S(b, y_k), \quad S(b, y_{k+1}), \quad S(b, y_{k+2}) \ldots,$$

the value of $S(b, \beta)$ can be computed by a suitable interpolation formula. By interpolation it is also possible to compute the curves $S(x, y) = \text{ct.}$

It is well known that the step-by-step computation of $S(x_i, y_j)$ will have a satisfactory accuracy only if the computation process is stable with respect to rounding errors and errors due to the approximation of the boundary data. From the stability theory of second-order parabolic boundary-value problems (cf. [E 1],[B 7],[F 5]) it can be inferred that the computation processes (5.123), (5.124) are stable at least if

$$l \leqslant 2h. \tag{5.130}$$

The accuracy of the computation increases as the grid size diminishes.

It is of course quite easy to devise more complex difference schemes than (5.125) and (5.128), but, in view of the rather limited amount of significant numerical data available at present, the systematic study of such schemes appears to be premature.

APPENDIX

Some elementary definitions often used in functional analysis

Enumerable and non-enumerable sets

A set containing no elements is called an empty set. A set containing a finite number of elements (for example the members of a certain club, the grains of sand of a certain beach) is called a finite set. Sets which are not finite are called infinite.

Two finite or infinite sets A and B are said to be equivalent when to each element a of A, in short $a \in A$, there corresponds one element b of B, in short $b \in B$, and vice versa. When two sets are equivalent they are said to have the same cardinal number, i.e. the same 'number' of elements. This definition applies regardless of whether the sets are finite or infinite, and it permits thus to classify infinite sets according to their 'size'.

Example. The set of positive integers $1, 2, 3, \ldots$ is equivalent to the set of even integers $2, 4, 6, \ldots$, or to the set of odd integers $1, 3, 5, \ldots$. A set which is equivalent to the set of positive integers is called enumerable.

Example. The set of rational numbers

$$m/n \quad (m = 0, \pm 1, \pm 2, \ldots, \quad n = \pm 1, \pm 2, \ldots),$$

is enumerable.

The set of points in the interval $0 \leqslant x \leqslant 1$ is not equivalent to the set of positive integers, and because of this property it is called non-enumerable. A continuous interval, no matter how small, has therefore 'more' elements than the set of positive integers.

The set of points in the interval $a \leqslant x \leqslant b$ is equivalent to the set of points in the interval $0 \leqslant x \leqslant 1$. The 'number' of points in a continuous interval of any size is therefore the same.

A non-enumerable set can be discrete. For example, the set of irrational numbers in the interval $0 \leqslant x \leqslant 1$ is non-enumerable. Non-enumerable sets are said to have the 'power of continuum'. There exist infinite sets which have a larger power than that of a continuum. For example the set of single-valued functions $y(x)$ defined in the interval $0 \leqslant x \leqslant 1$ has 'more' elements than the set of points of this interval, i.e. the set of $y(x)$ has a larger cardinal number than that of a continuum.

The set of cardinal numbers is unbounded or, in other words, there does not exist a set having the 'largest number' of elements.

The set of all limits of real numbers has the power of continuum. The set of continuous functions $y(x)$, defined in a closed interval $a \leqslant x \leqslant b$ has also the power of continuum.

Zermelo's axiom of choice

The statement of this axiom is: If the elements of a set A are sets B, which are non-empty and disjoint (i.e. the sets B have no common elements), then there exists at least one set C which contains one and only one element of each set B. As a consequence of this axiom the elements of any finite or infinite set can be ordered, but in the latter case this ordering will require an infinite number of operations.

Metric space

Consider a set Y of real-valued functions $y(x)$, defined in a closed interval $a \leqslant x \leqslant b$. If there exists at least one function $N(y_i(x), y_j(x))$ such that for each pair of elements $y_i, y_j \in Y$ the following properties are satisfied

$$(1) \quad N(y_i(x), y_j(x)) = N(y_j(x), y_i(x)) \geqslant 0,$$

$$(2) \quad N(\alpha y_i(x), 0) = |\alpha| N(y_i(x), 0) \quad (\alpha = \text{ct}),$$

$$(3) \quad N(y_i(x), y_j(x)) \leqslant N(y_i(x), 0) + N(0, y_j(x)),$$

then the set Y is said to form a metric space. The value $N(y(x), 0)$ is called the measure of the element $y(x) \in Y$. A metric space has a power which is not larger than that of continuum.

Cauchy sequences

A sequence of elements $y_i(x)$, $i = 1, 2, \ldots$, of a metric space Y is said to be a Cauchy sequence if for any $k > 0$

$$\lim_{i \to \infty} N(y_i(x), y_{i+k}(x)) = 0.$$

Complete space

A metric space the elements of which admit only convergent Cauchy sequences is said to be complete.

Uniform convergence of a sequence of functions

A sequence $y_i(x)$, $i = 1, 2, \ldots$, is said to converge uniformly to a unique and finite function $\bar{y}(x)$, defined in a closed interval $a \leqslant x \leqslant b$, if, and only if, for every $\epsilon > 0$ no matter how small, there exists an integer N_0, such that

$$N(\bar{y}(x), y_n(x)) < \epsilon, \quad a \leqslant x \leqslant b, \quad n > N_0.$$

REFERENCES

A

1 P. Althammer, 'Eine Erweiterung des Orthogonalitätsbegriffes bei Polynomen und deren Anwendung auf die beste Approximation', Dissertation, Freie Universität, Berlin, 1961.

2 A. A. Andronov and A. A. Witt, 'Discontinuous periodic solutions and the theory of the Abraham–Bloch multivibrator', *Dokl. Akad. Nauk SSSR*, no. 8 (1930), p. 189; and *Complete Works of Andronov*, Acad. Sci. U.S.S.R. (1956), p. 65.

3 A. A. Andronov, A. A. Witt and L. S. Pontryagin, 'Statistical examination of dynamical systems', *J. exp. theor. Phys. U.S.S.R.*, 3, no. 3 (1933), 165; and *Complete Works of Andronov*, Acad. Sci. U.S.S.R. (1956), p. 142.

4 A. A. Andronov and L. S. Pontryagin, 'Inert systems', *Dokl. Akad. Nauk SSSR*, **14** (1937), 247; and *Complete Works of Andronov*, Acad. Sci. U.S.S.R. (1956), p. 181.

5 A. A. Andronov, 'L. I. Mandelshtam and the theory of non-linear oscillations', *Complete Works of Andronov*, Acad. Sci. U.S.S.R. (1956), pp. 441–72.

6 A. A. Andronov, A. A. Witt and S. Ie. Khaikin, *Theory of Vibrations*, Fitzmatgiz, Moscow, 1959.

7 E. A. Andronova-Leontovich and L. N. Beliustina, 'Theory of bifurcation of second-order dynamical systems and its application to the investigation of nonlinear problems of the theory of vibrations', *International Symposium on Non-linear Vibrations, Kiev, September* 1961 (*Proc.*), **2** (1963), 7–28.

8 V. G. Avakumović, 'Ueber die Randwertaufgabe zweiter Ordnung', *Publs Inst. math. Acad. Serbe Sci.* 4 (1952), 1–8.

9 V. G. Avakumović, 'Sur le problème aux limites des équations différentielles du second ordre non linéaires', *Bull. Acad. Serbe Sci.* 5 (1952), 183–7.

B

1 N. S. Bakhalov, 'On the solution of boundary-value problems for systems of ordinary differential equations', *Computing Methods and Programming*, University of Moscow (1966), pp. 9–16.

2 J. F. Barret, 'Application of Kolmogorov's equations to randomly distributed automatic control systems', *Proc. IFAC Congress, Moscow* (1960), vol. II, pp. 724–33.

3 N. N. Bautin, *The Behaviour of Dynamical Systems Near their Stability Boundary*, Gostekhizdat, Moscow, 1949.

4 R. Bellman, *Dynamic Programming*, Princeton University Press, 1957.

5 R. Bellman, *Adaptive Control Processes: A Guided Tour*, Princeton University Press, 1961.

6 R. Bellman and S. E. Dreyfus, *Applied Dynamic Programming*, Princeton University Press, 1962.

7 I. S. Berezin and N. P. Zhidkov, *Methods of Computation*, vol. II, Fizmatgiz, Moscow, 1962.

8 S. N. Bernstein, 'On the equations of the calculus of variations', *Usp. mat. Nauk*, 8 (1940), 32–74.

9 L. D. Berkovitz, 'Variational methods in problems of control and programming', *J. math. Analysis Applic.* **3** (1961), 145–69.

10 L. D. Berkovitz, 'On control problems with bounded state variables', *J. math Analysis Applic.* **4** (1962), 488–98.

11 G. A. Bliss, 'Sufficient conditions for a minimum with respect to one-sided variations', *Trans. Am. Math. Soc.* **5** (1904), 477–92.

12 G. A. Bliss and A. L. Underhill, 'The minimum of a definite integral for unilateral variations in space', *Trans. Am. Math. Soc.* **15** (1914), 291–310.

13 G. A. Bliss, 'The problem of Lagrange in the calculus of variations', *Am. J. Math.* **52** (1930), 673–744.

14 G. A. Bliss, *Lectures on the Calculus of Variations*, University of Chicago Press, 1961. (§31, pp. 77–80, Carathéodory equidistants.)

15 N. N. Bogoliubov and Iu. A. Mitropolsky, *Asymptotic Methods in the Theory of Non-linear Oscillations*, Fizmatgiz, Moscow, 1958.

16 O. Bolza, *Vorlesungen über Varationsrechnung*, Teubner, 1909.

17 E. Borel, *Leçons sur les séries divergentes*, Gauthier-Villars, Paris, 1901.

18 A. E. Bryson and W. E. Denham, 'A steepest-ascent method for solving optimum programming problems', *J. appl. Mech.* June 1962.

19 J. Burt and I. Gumowski, 'Sur quelques propriétés d'un circuit à réaction permettant l'extraction de la racine carrée', *C. r. hebd. Séanc. Acad. Sci., Paris,* **253** (1961), 2207–9.

20 A. G. Butkovsky, 'The generalized maximum principle for problems of optimal control', *Avtomatika Telemekh.,* **24**, no. 3 (1963), 314–27.

C

1 C. Carathéodory, 'Ueber das allgemeine Problem der Variationsrechnung', *Nachr. Ges. Wiss. Göttingen (Math.-Phys. Klasse),* **1** (1905), 83–90.

2 C. Carathéodory, 'Ueber die starken Maxima und Minima bei einfachen Integralen', *Math. Annln,* **62** (1906), no. 12, 449–503.

3 C. Carathéodory, 'Sur une méthode directe du calcul des variations', *Rc. Circ. mat. Palermo,* **25** (1908), 36–49.

4 C. Carathéodory, 'Die Methode der geodätischen Aequidistanten und das Problem von Lagrange', *Acta Math.* **47** (1926), 199–236.

5 C. Carathéodory, 'Ueber die Einteilung der Variationsprobleme von Lagrange nach Klassen', *Comment. math. helvet.* **5** (1933), 1–19.

6 C. Carathéodory, *Variationsrechnung und partielle Differentialgleichungen erster Ordnung*, Teubner, Leipzig–Berlin, 1935.

7 C. Carathéodory, 'Exemples particuliers et théorie générale dans le calcul des variations', *Enseign. math.* **34** (1935), 255–61.

8 W. Cauer, 'Die Verwircklichung von Wechselstromwiderständen vorgeschriebener Frequentzabhängigkeit', Dissertation, Technische Hochschule Berlin, published in *Arch. Elektrotechnik,* **15** (1926), 355–88.

9 W. Cauer, *Theorie der linearen Wechselstromschaltungen*, Berlin, 1940; translated into English under the title: *Synthesis of Linear Communication Networks*, McGraw-Hill, 1958.

10 L. Cesari, 'An existence theorem in the problem of optimal control', *J. Siam Control,* A, **3**, no. 1 (1965), 7–22.

11 L. Cesari, 'Existence theorems for optimal solutions in Pontryagin and Lagrange problems', *J. Siam control,* A, **3**, no. 3 (1966), 475.

12 S. S. L. Chang, 'General theory of optimal processes', *J. Siam Control,* **4**, no. 1 (1966), 46–55.

13   N. Ch. Chetaiev, *Stability of Motion*, Gostekhizdat, Moscow, 1955.
14   L. Collatz, *Numerische Behandlung von Differentialgleichungen*, Springer Verlag, 1955. English edition: *The Numerical Treatment of Differential Equations*, Springer Verlag, 1960.
15   R. Courant and D. Hilbert, *Methoden der mathematischen Physik*, vols. I and II, Berlin, 1937. English translation: *Methods of Mathematical Physics*, J. Wiley (Interscience), vol. I, 1955, vol. II, 1962. Chapter 7 of vol. II of the German edition is missing in the English translation. Its content has been published separately in R. Courant, *Dirichlet's Principle*, Interscience, 1950.

D

1   W. Damköhler and E. Hopf, 'Ueber einige Eigenschaften von Kurvenintegralen und über die Aequivalenz von indefiniten mit definiten Variationsproblemen', *Math. Annln*, **120** (1947), 12–20.
2   W. Damköhler, 'Ueber die Aequivalenz indefiniter mit definiten isoperimetrischen Variationsproblemen', *Math. Annln*, **120**, no. 3 (1948), 297–306.
3   G. Darboux, *Leçons sur la théorie générale des surfaces*, Gauthier-Villars, Paris, 1st ed. 1889, 2nd ed. 1914, 1915.
4   P. Dedecker, 'Sur un problème inverse du calcul des variations', *Bull. Sci. Acad. R. Belge* (5 ser.), **36** (1950), 63–70.
5   C. A. Desoer, 'The bang-bang servo problem treated by variational techniques', *Information and Control*, **2** (1959), 333–48.
6   A. Douglis, 'The continuous dependence of generalized solutions of nonlinear partial differential equations upon initial data', *Communs pure appl. Math.* **14** (1961), 267–84.
7   A. La. Dubovycky and A. A. Miliutin, 'Extremal problems in the presence of constraints', *Dokl. Akad. Nauk SSSR*, **149**, no. 4 (1963), 759–62.
8   E. Durand, *Solutions numériques des équations algébriques*, Masson, Paris, 1961.

E

1   S. D. Eidelman, *Parabolic Systems*, Nauka, Moscow, 1964.
2   G. von Escherich, 'Die zweite Variation der einfachen Integrale', *Sber. Akad. Wiss. Wien (Math. naturw. Kl.)*, **107**, Abt. II A (1898), 1191–250, 1267–326, 1381–428, **108**, Abt. II A (1899), 1269–340 and **110**, Abt. II A (1901), 1355–421.
3   G. Ewing and M. Morse, 'The variational theory in the large including the non-regular case', *Ann. Math.* **44**, no. 3 (1943), 339–74.

F

1   A. A. Feldbaum, *Foundations of the Theory of Optimal Automatic Systems*, Fizmatgiz, Moscow, 1963.
2   A. F. Filipov, 'On some problems of optimal control theory', *Vestnik Mosk. Univ.* no. 2 (1959), 25–32.
3   B. Flodin, 'Ueber diskontinuierliche Lösungen bei Variationsproblemen mit Gefällbeschränkung', *Acta Soc. Sci. fenn.* **3**, no. 10 (1945), 3–32.
4   O. Föllinger, 'Diskontinuierliche Lösungen von Variationsproblemen mit Gefällbeschränkung', *Math. Annln*, **126** (1953), 466–80.
5   G. E. Forsythe and W. R. Wasow, *Finite Difference Methods for Partial Differential Equations*, Wiley, 1960.

6   L. Fox, *Numerical Solutions of Ordinary and Partial Differential Equations*, Pergamon Press, 1962.

7   K. Friedrichs, 'Ein Verfahren der Variationsrechnung das Minimum eines Integrals als das Maximum eines anderen Ausdruckes darzustellen', *Nachr. Ges. Wiss. Göttingen (Math. Phys. Kl.)* (1929), pp. 13–20.

8   K. Friedrichs and H. Lewy, 'Ueber fortsetzbare Anfangsbedingungen bei hyperbolischen Differentialgleichungen in drei Veränderlichen', *Göttinger Nachrichten* (1932), pp. 135–43.

9   A. T. Fuller, 'Study of an optimum nonlinear control system', *J. Electron. Control*, **15** (1963), no. 1, 63–71.

10 A. T. Fuller, 'Directions of research in control', *Automatica*, **1** (1963), 289–96.

11 A. T. Fuller, 'Further study of an optimum nonlinear control system', *J. Electron. Control*, **17** (1964), no. 3, 283–300.

12 A. T. Fuller, 'Optimization of some nonlinear control systems by means of Bellman's equation and dimensional analysis', *Int. Jl Control*, **3**, no. 4 (1966), 359–94.

G

1   R. Gabasov and F. M. Kirillova, 'The construction of successive approximations for some problems of optimal control', *Avtomatika Telemekh.*, no. 2 (1966), 5–17.

2   R. A. Gambill, 'Generalized curves and the existence of optimal controls', *J. Siam Control*, A, **1**, no. 3 (1963), 246–60.

3   R. V. Gamkrelidze, 'On the general theory of optimal processes', *Dokl. Akad. Nauk SSSR*, **123**, no. 2 (1958), 223–6.

4   R. V. Gamkrelidze, 'Minimal time processes with phase-coordinate constraints', *Dokl. Akad. Nauk SSSR*, **125**, no. 3 (1959), 475–8.

5   R. V. Gamkrelidze, 'On optimal sliding regimes', *Dokl. Akad. Nauk SSSR*, **143**, no. 6 (1962), 1243–5.

6   R. V. Gamkrelidze, 'On some extremal problems in the theory of differential equations with applications to the theory of optimal control', *J. Siam Control*, A, **3**, no. 1 (1965), 106–28.

7   E. Goursat, *Cours d'analyse*, vol. III, Gauthier-Villars, Paris, 1923.

8   G. Grateloup, 'Sur l'indentification de la dynamique des processus en automatique', Dissertation no. 253, Université de Toulouse, 1965.

9   L. M. Graves, 'On the problem of Lagrange', *Am. J. Math.* **53** (1931), 547–54.

10 I. Gumowski, 'Some relations between frequency and time domain errors in network synthesis problems', *IRE Trans. Circuit Theory*, CT-5, no. 1 (1958), 66–9.

11 I. Gumowski, J. Lagasse and Y. Sevely, 'Mise en équation d'un amplificateur à transistor non linéaire', *C. r. hebd. Séanc. Acad. Sci., Paris*, **250** (1960), 1995.

12 I. Gumowski, 'Sur l'existence des solutions périodiques d'une équation différentielle-fonctionelle d'ordre un', *C. r. hebd. Séanc. Acad. Sci., Paris*, **256** (1963), 4828.

13 I. Gumowski, 'Sur le calcul des solutions périodiques d'une équation différentielle-fonctionelle d'ordre un', *C. r. hebd. Séanc. Acad. Sci., Paris*, **257** (1963), 2010.

14 I. Gumowski, 'Sensitivity analysis and Liapunov stability', *IFAC Symposium on Sensitivity Analysis*, Dubrovink, 1964 (see also [R 1]).

15 I. Gumowski, 'Sur les solutions périodiques de l'équation de Cherwell-Wright', *C. r. hebd. Séanc. Acad. Sci., Paris*, **258** (1964), 2738.

16 I. Gumowski, 'Sur l'interprétation physique des solutions extrémales obtenues au moyen de la programmation dynamique', *C. r. hebd. Séanc. Acad. Sci., Paris*, **260** (1965), 1096–9.

17 I. Gumowski, 'Sur une généralisation du calcul des variations', *C. r. hebd. Séanc. Acad. Sci., Paris*, **260** (1965), 1858–61.

18 I. Gumowski, 'Sur une principe du minimum dans le calcul des variations appliqué', *C. r. hebd. Séanc. Acad. Sci., Paris*, **260** (1965), 3279–82.

19 I. Gumowski, 'Sur une relation entre le calcul des variations et la programmation dynamique', *C. r. hebd. Séanc. Acad. Sci., Paris*, **260** (1965), 4912–15.

20 I. Gumowski and C. Mira, 'Sur un algorithme de détermination du domaine de stabilité d'un point double d'une récurrence non linéaire du deuxième ordre á variables réèlles', *C. r. hebd. Séanc. Acad. Sci., Paris*, **260** (1965), 6524.

21 I. Gumowski and C. Mira, *Optimization by means of the Hamilton–Jacobi theory*, Third IFAC Congress, London, June 1966.

22 I. Gumowski, 'Sur les conditions de validité du critère de Weierstrass', *C. r. hebd. Séanc. Acad. Sci. Paris*, **265** (1967), 344.

23 V. I. Gurman, 'On optical processes involving singular controls', *Avtomatika Telemekh.* **24**, no. 5 (1965), 782–91.

H

1 J. Hadamard, *Calcul des variations*, Hermann, Paris, 1910.

2 J. Hadamard, 'Principe de Huyghens et prolongement analytique', *Bull. Soc. Math. Fr.* **52** (1924), 241–78.

3 J. Hadamard, *Lectures on Cauchy's Problem*, Yale University Press, 1923; re-edited in French under the title: *Le problème de Cauchy et les équations aux dérivées partielles hyperboliques*, Hermann, Paris, 1932.

4 W. Hahn, *Theory and Application of Liapunov's Direct Method*, Prentice-Hall, 1963.

5 A. Haar, 'Sur l'unicité des solutions des équations aux dérivées partielles', *C. r. hebd. Séanc. Acad. Sci., Paris*, **187** (1928), 23–5.

6 M. G. Hestenes, 'Variational theory and optimal control theory', article in *Computing Methods in Optimization Problems*, Academic Press, 1964.

7 M. G. Hestenes, 'On variational theory and optimal control theory', *J. Siam Control*, A, **3**, no. 1 (1965), 23–48.

8 D. Hilbert, 'Ueber das Dirichletsche Prinzip', *Jber. dt. Mat Verein*, **8** (1900), 184.

9 D. Hilbert, 'Zur Variationsrechnung', *Göttinger Nachr.* (1905), pp. 159–80.

10 D. Hilbert, 'Zur Variationsrechnung', *Math. Annln*, **62** (1906), 351–70.

11 D. Hilbert, 'Wesen und Ziele einer Analysis der unendlich vielen unabhängigen Variablen', *Rc. Circ. mat. Palermo*, **27** (1909), 59–73.

12 D. Hilbert, 'Mathematische Probleme', *Gesamm. Abh.* **3** (1935), 290–329.

13 A. Hirsch, 'Ueber eine charakteristische Eigenschaft der Differentialgleichungen der Variationsrechnung', *Math. Annln*, **49** (1897), 49–72.

14 L. van Hove, 'Sur les champs de Carathéodory et leur construction par la méthode des caractéristiques', *Bull. Acad. R. Belge (Classe Sci., V série)*, **31** (1945), 625–738.

K

1 E. Kamke, *Differentialgleichungen reeller Funktionen*, Leipzig, 1930.

2 E. Kamke, 'Bemerkungen zur Theorie der partiellen Differentialgleichungen erster Ordnung', *Math. Z.* **49**, no. 2 (1943), 256–84.

3   E. Kamke, *Differentialgleichungen, Lösungsmethoden und Lösungen*, Leipzig, 1943.

4   E. Kamke, *Differentialgleichungen Lösungsmethoden und Lösungen, Partielle Differentialgleichungen erster Ordnung für eine gesuchte Funktion*, Leipzig, 1944.

5   L. V. Kantorovich, 'Functional analysis and applied mathematics', *Usp. mat. Nauk*, **3**, no. 6 (1948), 89–185.

6   L. V. Kantorovich and V. I. Krylov, *Approximate Methods of Higher Analysis*, 5th ed., Fizmatgiz 1962 (1st edition, 1936), English translation, Interscience, 1958.

7   E. Kasner, 'Systems of extremals in the calculus of variations', *Bull. Am. Math. Soc.* **13** (1907), 289–92.

8   H. J. Kelley, 'Gradient theory of optimal flight path', *J. Am. Rocket Soc.* **3**, no. 10 (1960).

9   M. K. Kerimov, 'On the necessary conditions of an extremum in discontinuous variational problems with mobile end-points'. *Dokl. Akad. Nauk SSSR*, **79** (1951), no. 4, 565–8.

10 M. K. Kerimov, 'On the Jacobi conditions for discontinuous variational problems with mobile end-points', *Dokl. Akad. Nauk SSSR*, **79** (1951), no. 5, 719–22.

11 M. K. Kerimov, 'On the sufficient conditions of an extremum in discontinuous variational problems with mobile end-conditions', *Dokl. Akad. Nauk SSSR*, **84** (1952), no. 2, 213–16.

12 M. K. Kerimov, 'On the theory of discontinuous variational problems with floating end-points', *Dokl. Akad. Nauk SSSR*, **136**, no. 3 (1961), 542–5.

13 M. K. Kerimov, 'On the theory of the second variation of discontinuous variational problems in space', *Dokl. Akad. Nauk SSSR*, **140**, no. 1 (1961), 41–4.

14 M. K. Kerimov, 'Sur la théorie des problèmes variationnels discontinus aux extrémités variables dans l'espace', *Czech. math. J.* **11** (86) (1961), no. 1, 1–23.

15 F. M. Kirillova, 'On the limit process in the solution of one problem of optimal regulation', *Prikl. Mat. Mekh.* **24**, no. 2 (1960), 277.

16 Iu. A. Klokov, 'On a boundary-value problem associated with a second order ordinary differential equation', *Sib. mat. Zh.* **4**, no. 1 (1963), 86–96.

17 A. Kneser, *Lehrbuch der Variationsrechnung*, Vieweg, Braunschweig, 1900.

18 N. N. Krasovsky, 'On the theory of optimal regulation', *Avtomatika Telemekh.*, **18**, no. 11 (1957), 960–70.

19 N. N. Krasovsky, *Some Problems of the Theory of Stability of Motion*, Fizmatgiz, Moscow, 1959.

20 V. F. Krotov, 'Discontinuous solutions of variational problems', *Izv. vyssh. ucheb. Zaved.* **18**, no. 5 (1960), 86–97.

21 V. F. Krotov, 'On discontinuous solutions of variational problems', *Izv. vyssh. ucheb. Zaved.* **2**, no. 2 (1961), 75–89.

22 V. F. Krotov, 'The fundamental problem of the calculus of variation for the simplest functional defined on the set of discontinuous functions', *Dokl. Akad. Nauk SSSR*, **137**, no. 1 (1961), 31–4.

23 V. F. Krotov, 'Methods of solving variational problems on the basis of a sufficient condition for an absolute minimum', *Avtomatika Telemekh.*, **23**, no. 12 (1962), 1571–83.

24 V. F. Krotov, 'Methods of solving variational problems. Sliding regimes', *Avtomatika Telemekh.*, **24**, no. 5 (1963), 581–98.

25 N. Krylov and N. Bogoliubov, *Introduction to Nonlinear Mechanics*, Kiev (1937), p. 106.

26 R. Kulikowski, 'Synthesis of optimal control systems with area bounded control signal', *Bull. Acad. Polon. Sci. (sér. Sci. Tech.)*, **8**, no. 4 (1960),

L

1 O. A. Ladyzhenskaia and N. N. Uralceva, *Linear and Quasilinear Equations of the Elliptic Type*, Nauka, Moscow, 1964.

2 J. H. Lanning and R. H. Battin, *Random Processes in Automatic Control*, McGraw-Hill, 1956.

3 M. Lavrentieff, 'Sur quelques problèmes du calcul des variations', *Annali Mat.* (ser. 4), **4** (1926), 7–28.

4 M. Lavrentiev and L. Liusternik, *Foundations of the Calculus of Variations*, vol. I, part II, Onti, 1935.

5 H. Lebesgue, 'Intégrale, longuer, aire', *Annali Mat.* **7** (1902), 231–359.

6 H. Lebesgue, *En marge du calcul des variations*, Institut de Mathématiques de l'Université de Genève, 1963.

7 E. Leimainis and N. Minorsky, *Dynamics of Nonlinear Mechanics*, Wiley, 1958.

8 G. Leitman, *Optimization Techniques with Applications to Aerospace Systems*, Academic Press, 1962.

9 Iu. P. Leonov, 'Statistical description of a system', *Avtomatika Telemekh.* **23**, no. 7 (1962), 901.

10 A. M. Letov, 'The state of the stability problem in automatic control theory', *Trans. of the 2nd Soviet Symposium on Automatic Control Theory*, Moscow, vol. I (1955), pp. 79–104.

11 A. M. Letov, *Stability of Nonlinear Regulating Systems*, Gostekhizdat, Moscow, 1955.

12 A. M. Letov, 'On the gap between theory and practice', *Avtomatika Telemekh.* **27**, no. 2 (1966), 152–57.

13 F. Lettenmeyer, 'Ueber die von einem Punkt ausgehenden Integralkurven einer Differentialgleichung zweiter Ordnung', *Dt. Math.* **7** (1944), 56–74.

14 D. C. Lewis, 'Polynomial least square approximations', *Am. J. Math.* **69** (1947), 273–8.

15 A. Liapunov, 'Problème général de la stabilité du mouvement', *Annales de la Faculté des Sciences de Toulouse*, 2, vol. 9 (1907), pp. 203–474. Reprinted by Princeton University Press, *Ann. Math. Studies*, no. 17, 1948.

16 A. I. Lure, *Some Nonlinear Problems of Automatic Control Theory*, Gostekhizdat, Moscow, 1951.

17 A. I. Lure, 'Thrust Programming in a central gravitational field', *IFAC Symposium on the Automatic Control in the Peaceful Uses of Space*, Stavanger, Norway, June 1965.

18 N. N. Luzin, *Theory of Functions of a Real Variable*, Uchpedgiz, Moscow, 1940.

M

1 I. G. Malkin, 'Stability in the presence of constantly acting perturbations', *Prikl. Mat. Mekh.* **8**, no. 3 (1944).

2 I. G. Malkin, *Theory of Stability of Motion*, Gostekhizdat, Moscow, 1952. English translation by the U.S. AEC: tr-3352. German translation: *Theorie der Stabilität einer Bewegung*, Oldenbourg, München, 1959.

3 J. D. Mancil, 'On the Carathéodory condition for unilateral variations', *J. Am. Math. Soc.* **46** (1940), 363–6.

4 A. Marchand, 'Sur le champ de demi-droites et les équations différentielles du premier ordre', *Bull. Soc. Math. Fr.* **62** (1934), 1–38.

5 L. Markus, 'Controllability of nonlinear processes', *J. Siam Control*, A, **3** (1965), 78–90.

6 P. Massé, 'Sur les principes de la régulation d'un débit aléatoire par un réservoir', *C. r. hebd. Séanc. Acad. Sci., Paris*, **219** (1944), 19–21.

7 P. Massé, 'Sur les effets de la régulation d'un débit aléatoire par un réservoir', *C. r. hebd. Séanc. Acad. Sci., Paris*, **219** (1944), 150–1.

8 P. Massé, 'Sur un cas particulier remarquable de la régulation d'un débit aléatoire par un réservoir', *C. r. hebd. Séanc. Acad Sci., Paris*, **219** (1944), 173–5.

9 P. Massé, *Les réserves et la régulation de l'avenir dans la vie économique*, vols. I and II, Hermann, Paris, 1946.

10 I. L. Massera, 'On Liapunov's condition of stability', *Ann. Math.* **50**, no. 3 (1943).

11 M. Mason and G. Bliss, 'A problem of the calculus of variations in which the integrand is discontinuous', *Trans. Am. Math. Soc.* **7** (1906), 325–36.

12 E. J. McShane, 'On multipliers for Lagrange problems', *Am. J. Math.* **61** (1939), 809–19.

13 E. J. McShane, 'Necessary conditions in generalized-curve problems of the calculus of variations', *Duke Math. J.* **7** (1940), 1–27.

14 E. J. McShane, 'Existence theorems for Bolza problems in the calculus of variations', *Duke Math. J.* **7** (1940), 28–61.

15 E. J. McShane, 'Generalized curves', *Duke Math. J.* **6** (1940), 513–36.

16 K. J. Merklinger, 'Numerical analysis of nonlinear control systems using the Fokker–Planck–Kolmogorov equation', *Proc. IFAC Congress, Basel, 1963*, vol. I, pp. 89–90.

17 D. Middleton, *An Introduction to Statistical Communication Theory*, McGraw-Hill, 1960.

18 A. Miele, *Theory of Optimum Aerodynamic Shapes*, Academic Press, 1965.

19 S. G. Mikhlin, *Numerical Realization of Variational Methods*, Nauka, Moscow, 1966.

20 G. N. Milstein, 'Application of successive iterations to the solution of one optimal problem', *Avtomatika Telemekh.* **25**, no. 3 (1963), 321–9.

21 N. Minorsky, 'Self-excited oscillations in dynamical systems possessing retarded actions', *J. appl. Mech.* **9**, no. 1, March 1942, A 65–A 71.

22 N. Minorsky, 'Sur une classe d'oscillations auto-entretenues', *C. r. hebd. Séanc. Acad. Sci., Paris*, **226** (1948), 1122–4.

23 N. Minorsky, *Non-linear Oscillations*, D. Van Nostrand, 1962.

24 A. Mishkis, 'Uniqueness of solutions of Cauchy problems', *Usp. mat. Nauk*, **3**, no. 2 (1948), 3–46.

25 M. Monge, 'Supplément', *Histoire de l'Académie Royale des Sciences, Paris* (1784), pp. 502–76.

26 C. Mira, 'Détermination pratique du domaine de stabilité d'un point d'equilibre d'une récurrence non linéaire du deuxième ordre á variables réelles', *C. r. hebd. Séanc. Acad. Sci., Paris*, **261** (1965), 5314.

27 C. Mira, 'Méthode de détermination du domaine de stabilité asymptotique d'un point double d'une récurrence non-linéaire', *Congrès d'Automatique Théorique, Paris*, May 1965.

O

1 O. A. Oleinik, 'On the Cauchy problem for nonlinear equations in the class of discontinuous functions', *Dokl. Acad. Nauk SSSR*, **95**, no. 3 (1954), 451–4.

2 O. A. Oleinik, 'On discontinuous solutions of non-linear differential equations', *Dokl. Akad. Nauk SSSR*, **109**, no. 6 (1956), 1098–101.

3 O. A. Oleinik, 'The Cauchy problem for nonlinear differential equations of first order with discontinuous initial conditions', *Trudy Mosk. mat. Obshch.* **5** (1956), 433–54.

4 O. A. Oleinik, 'Discontinuous solutions of nonlinear differential equations', *Usp. mat. Nauk*, **12**, no. 3 (1957), 3–73.

5 O. A. Oleinik, 'On uniqueness and stability of a generalized solution of a problem for a quasi-linear equation', *Usp. mat. Nauk*, **14**, no. 2 (86), (1959), 165–70.

6 O. A. Oleinik, 'On the construction of a generalized solution of a Cauchy problem for a quasi-linear first order equation by means of a vanishing viscosity', *Usp. mat. Nauk*, **14**, no. 2 (86), (1959), 159–64.

7 G. M. Ostrovski, Iu. M. Volin and I. I. Malkiss, 'On one method of solving optimal problems with boundary conditions', *Izvyestya Akad. Nauk* (ser. Tech. Cybern.), no. 6 (1965), 146–51.

P

1 R. Pallu de la Barrière, 'Extension du principe de Pontryagin au cas des liaisons instantanées entre l'état et la commande', *C. r. hebd. Séanc. Acad. Sci., Paris*, **258** (1964), 3961–3.

2 A. I. Perov and A. V. Kibenko, 'On one general method of studying boundary-value problems', *Izv. Akad. Nauk SSSR (Ser. mat.)*, **30** (1966), 249–64.

3 Peschon, *Disciplines and Techniques of System Control*, Blaisdell Publ. Co., 1965.

4 Iu. P. Petrov, *Variational Methods of the Theory of Optimal Controls*, Energia, Moscow, 1965.

5 W. Petry, 'Das Iterationsverfahren zum Lösen von Randwertproblemen gewöhnlicher Differentialgleichungen zweiter Ordnung', *Math. Z.* **87** (1965), 323–33.

6 A. Pfarr, 'Der Reguliervorgang bei Turbinen mit indirekt wirkenden Regulator', *Z. Ver. dt. Ing.* **43** (1899), 1553.

7 E. Picard, *Leçons sur quelques problèmes aux limites de la théorie des équations différentielles*, Paris, 1930.

8 L. S. Pontryagin, 'Some mathematical problems arising in connection with optimal automatic control systems', *Proceedings of the October 1956 Session of the USSR Acad. of Science on Automatic Regulations and Control*, Moscow, 1956, pp. 107–17.

9 L. S. Pontryagin, 'Asymptotic behaviour of solutions of differential equation systems when the higher derivatives contain a small parameter as a factor', *Izv. Akad. Nauk (Ser. mat.)*, **21** (1957), 605.

10 L. S. Pontryagin, 'Optimal regulation processes', *Usp. mat. Nauk*, **14**, no. 1 (85), (1959), 3–54.

11 L. S. Pontryagin, V. G. Boltiansky, R. V. Gamkrelidze and E. F. Mishchenko, *Mathematical Theory of Optimal Processes*, Fizmatgiz, Moscow, 1961. English translation: Interscience (Wiley), 1962.
12 V. S. Pugachev, *Theory of Random Functions*, Fizmatgiz, Moscow, 1960.

R

1 L. Radanović, *Sensitivity Methods in Control Theory*, Pergamon Press, 1966.
2 A. Razmadze, 'Sur les solutions discontinues dans le calcul des variations', *Math. Annln*, **94** (1905), 1–52.
3 R. Rosenbaum, 'Convergence technique for the steepest descent method of trajectory optimization', *A.I.A.A. J.* **1** (1963), 1703–05.
4 L. I. Rozonoer, 'Pontryagin's maximum principle in the theory of optimal systems', *Avtomatika Telemekh.*, **20** (1959), 1320–34, 1441–58, 1561–78.
5 H. Rund, *The Hamilton–Jacobi Theory in the Calculus of Variations*, D. van Nostrand Co., 1966.

S

1 L. Schwartz, *Théorie des distributions*, Hermann, Paris, vol. ɪ (1951), vol. ɪɪ, (1952).
2 G. Ye. Shilov, *Mathematical Analysis*, Pergamon Press, 1965; *Matematicheski Analiz, Specialnyi Kurs*, Fizmatgiz, Moscow, 1961.
3 V. I. Smirnow, *Course of Higher Mathematics*, Fizmatgiz, Moscow, 1958.
4 S. L. Sobolev, *Applications of Functional Analysis to Mathematical Physics*, Leningrad, 1960.
5 A. Sommerfeld, *Mechanics*, Academic Press, 1952.
6 G. Stampacchia, 'Il principo di minimo nel calcolo delle variazioni', *Atti del convegno Lagrangiano, Accad. Sci. Torino (Suppl. al)*, **97** (1963–1964), 152–171.
7 H. P. F. Swinnerton-Dyer, 'On an extremal problem', *Proc. Lond. Math. Soc.* **3** (1957), 568–83.

T

1 A. N. Tikhonov, 'Systems of differential equations containing a small parameter with the higher order derivatives', *Mat. Sbornik*, **31, 73** (1952), 575.
2 A. N. Tikhonov, 'On the regularization of incorrectly formulated problems', *Dokl. Akad. Nauk SSSR*, **153** (1963), no. 1, 49–52.
3 I. Todhunter, *Researches in the Calculus of Variations*, McMillan, 1871.
4 R. Tomović, *Sensitivity Analysis of Dynamic Systems*, McGraw-Hill, 1963.
5 R. Tomović, 'The role of sensitivity analysis in engineering problems', *IFAC Symposium on Sensitivity Analysis*, Dubrovnik, 1964.
6 L. Tonelli, *Fondamenti di calcolo delle variazioni*, Zanichelli, Bologna, vol. ɪ (1921), vol. ɪɪ (1923).
7 L. Tonelli, 'Sul equazione differenziale $\ddot{y} = f(x, y, \dot{y})$', *Ann. Sci. norm. Sup. Pisa*, **8** (1939), 75–88.
8 A. Turowicz, 'Remarques sur un travail de R. V. Gamkrelidze relatif aux régimes optimaux glissants', *Bull. Acad. Polon. Sci. (Ser. Math. Astr. Phys.)*, **10** (1962), 557–8.

U

1 A. Unterkircher, 'Verallgemeinerte orthogonale Polynomsysteme die gleich-
zeitig eine Funktion und deren erste Ableitung approximieren', Dissertation,
Leopold-Franzens Universität, Innsbruck (1966).

V

1 F. A. Valentine, 'The problem of Lagrange with differential inequalities as
side conditions', Dissertation, University of Chicago, 1937; and *Contributions
to the Calculus of Variations*, University of Chicago Press (1933–37), pp.
407–48.

2 I. B. Vapniarski, 'Some examples of solution of optimal regulation problems
by the method of variation in phase space', *Izv. Akad. Nauk SSSR (Technical
Cybernetics)*, no. 4 (1966), 39–44.

3 A. B. Vasileva, 'Differential equations containing a small parameter', *Mat.
Sb.* **31**, 73 (1952), 587.

4 V. Volterra, *Leçons sur les fonctions de lignes*, Gauthier-Villars, Paris, 1913.

5 V. Volterra, 'Sopra le funzioni che dipendono da altre funzioni', *Atti Accad.
naz. Lincei Rc.* **3**, 2 sem. (1887), 153–8.

W

1 J. Warga, 'Relaxed variational problems', *J. math. Anal. Appl.* **4** (1962),
111–28.

2 J. Warga, 'Minimax problems and unilateral curves in the calculus of variations', *J. Siam Control*, A, **3**, no. 1 (1965), 91–105.

3 J. Warga, 'Variational problems with unbounded controls', *J. Siam Control*,
A, **3**, no. 3 (1965), 424–38.

4 T. Ważewski, 'Sur l'appréciation du domaine d'existence des intégrales de
l'équation aux dérivés partielles du premier ordre', *Ann. Soc. Polon. Math.*
14 (1935), 149–77.

5 T. Ważewski, 'Ueber die Bedingungen der Existenz der Integrale partieller
Differentialgleichungen erster Ordnung', *Math. Z.* **43**, no. 4 (1938), 522–32.

6 T. Ważewski, 'Système de commande et équation au contingent', *Bull Acad.
Polon. Sci. (Sér. Sci. Math. Astr. Phys.)* **9**, no. 3 (1961), 151–5.

7 T. Ważewski, 'Sur une condition equivalente a l'équation au contingent',
Bull. Acad. Polon. Sci. (Sér. Sci. Math. Astr. Phys.) **9**, no. 12 (1961), 865–7.

8 T. Ważewski, 'Sur une généralisation de la notion des solutions d'une équation au contingent', *Bull. Acad. Polon. Sci. (Sér. Sci. Math. Astr. Phys.)* **10**,
(1962), 11–15.

9 T. Ważewski, 'Sur les systèmes de commande non linéaires dont le contre-
domaine de commande n'est pas forcément convexe', *Bull. Acad. Polon. Sci.
(Sér. Sci. Math. Astr. Phys.)*, **10** (1962), 17–21.

10 N. Wiener, 'Certain notions in potential theory', *J. Math. Phys.* **3** (1924),
24–51.

11 N. Wiener, 'The Dirichlet problem', *J. Math. Phys.* **3** (1924), 127–46.

12 M. M. V. Wilkinson, 'Review of Todhunter's researches on the calculus of
variations', *Messenger of Mathematics, London*, **3** (1874), 184–192.

13 M. W. Wonham, 'Note on a problem in optimal nonlinear control', *J.
Electron. Contr.* **15** (1963), 59–62.

X

1 'Theory and application of discrete systems', *Transaction of a* 1958 *Conference, Acad. Sci. USSR*, 1960, pp. 387–569.

Y

1 L. C. Young, 'On approximation by polygons in the calculus of variation', *Proc. Roy. Soc.* A, **141** (1933), 325–41.
2 L. C. Young, 'Generalized curves and the existence of attained absolute minimum in the calculus of variations', *C. r. Soc. Sci. et Lettres de Varsovie (Classe* 3), **30** (1937), 212–34.
3 L. C. Young, 'Necessary conditions in the calculus of variations', *Acta Math.* **69** (1938), 229–58.

Z

1 S. K. Zaremba, 'Sur une extension de la notion d'équation différentielle', *C. r. hebd. Séanc. Acad. Sci., Paris*, **199** (1934), 545–8.
2 S. K. Zaremba, 'Sur les équations au paratingent', *Bull. Sci. Math. (Paris)*, **60** (1937), 139–60.
3 R. I. Zarossky, 'On the problem of necessary and sufficient conditions for an absolute minimum', *Dokl. Akad. Nauk SSSR*, **163**, no. 1 (1965), 26–29.
4 E. Zermelo, 'Beweis dass jede Menge wohlgeordnet werden kann', *Math. Ann.* **59** (1904), 514–16.

INDEX